About Island Press

Island Press is the only nonprofit organization in the United States whose principal purpose is the publication of books on environmental issues and natural resource management. We provide solutions-oriented information to professionals, public officials, business and community leaders, and concerned citizens who are shaping responses to environmental problems.

Since 1984, Island Press has been the leading provider of timely and practical books that take a multidisciplinary approach to critical environmental concerns. Our growing list of titles reflects our commitment to bringing the best of an expanding body of literature to the environmental community throughout North America and the world.

Support for Island Press is provided by the Agua Fund, The Geraldine R. Dodge Foundation, Doris Duke Charitable Foundation, The Ford Foundation, The William and Flora Hewlett Foundation, The Joyce Foundation, Kendeda Sustainability Fund of the Tides Foundation, The Forrest & Frances Lattner Foundation, The Henry Luce Foundation, The John D. and Catherine T. MacArthur Foundation, The Marisla Foundation, The Andrew W. Mellon Foundation, Gordon and Betty Moore Foundation, The Curtis and Edith Munson Foundation, National Fish and Wildlife Foundation, Oak Foundation, The Overbrook Foundation, The David and Lucile Packard Foundation, Wallace Global Fund, The Winslow Foundation, and other generous donors.

The opinions expressed in this book are those of the author(s) and do not necessarily reflect the views of these foundations.

About the Pacific Institute for Studies in Development, Environment, and Security

The Pacific Institute for Studies in Development, Environment, and Security, in Oakland, California, is an independent, nonprofit organization created in 1987 to conduct research and policy analysis in the areas of environmental protection, sustainable development, and international security. Underlying all of the Institute's work is the recognition that the urgent problems of environmental degradation, regional and global poverty, and political tension and conflict are fundamentally interrelated, and that long-term solutions dictate an interdisciplinary approach. Since 1987, we have produced more than sixty research studies, organized roundtable discussions, and held widespread briefings for policymakers and the public. The Institute has formulated a new vision for long-term water planning in California and internationally, developed a new approach for valuing well-being in local communities, worked on transborder environment and trade issues in North America, analyzed ISO 14000's role in global environmental protection, clarified key concepts and criteria for sustainable water use in the lower Colorado basin, offered recommendations for reducing conflicts over water in the Middle East and elsewhere, assessed the impacts of global warming on freshwater resources, and created programs to address environmental justice concerns in poor communities and communities of color.

For detailed information about the Institute's activities, visit www.pacinst.org, www.worldwater.org, and www.globalchange.org.

THE WORLD'S WATER
2008-2009

THE WORLD'S WATER

2008-2009

The Biennial Report on Freshwater Resources

Peter H. Gleick

with Heather Cooley, Michael J. Cohen, Mari Morikawa,
Jason Morrison, and Meena Palaniappan

ISLANDPRESS

Washington • Covelo • London

ISLAND PRESS is a trademark of the Center for Resource Economics.

Library of Congress Card Catalog Number 98024877
ISBN 10: 1-59726-504-7 (cloth)
ISBN 13: 978-1-59726-504-1 (cloth)

ISBN 10: 1-59726-505-5 (paper)
ISBN 13: 978-1-59726-505-8 (paper)

ISSN 15287-7165

Printed on recycled, acid-free paper ✪

Manufactured in the United States of America.
10 9 8 7 6 5 4 3 2 1

Contents

DATA SECTION

WATER UNITS, DATA CONVERSIONS, AND CONSTANTS   343

COMPREHENSIVE TABLE OF CONTENTS   353

COMPREHENSIVE INDEX   373

Foreword

When it comes to freshwater— the bloodstream of both the biosphere and the social sphere—the world is on the threshold to a new era. Humanity's profound dependence on freshwater makes basic water security a necessary condition for improving living conditions, securing food production to eradicate hunger, and providing employment and income to eliminate poverty.

Looking at countries from the perspective of water security reveals some important differences. Industrial countries have already tackled their hydroclimatic vulnerability, and have come far in expanding economic development and improving the quality of life for their populations. Emerging economies, on the other hand, remain hampered by water-related challenges such as flooding, drought, and severe water pollution. And the poorest economies in semi-arid climates still remain hostage to water problems and suffer large-scale poverty, disease, and uncurbed population growth.

To make matters worse, we now see evidence signalling the end of the era of easy access to freshwater. In many places, the perception of water security is starting to dissolve: overpumped aquifers under breadbasket regions of the world constitute economic "bubbles" ready to burst; excessive water consumption by irrigation is depleting rivers and deltas; neglect of pollutants in return flows from water use is contaminating potential raw water sources; and some city water-supply systems have started to crumble because of neglect of critical maintenance. We even hear about drought-driven farmer suicides in places where farmers find themselves unable to repay loans they had to take to deepen failing irrigation wells.

A fundamental tool in mastering the variability that characterizes the hydroclimatic water delivery system has been water storage. Statistics shows that North Americans have access to over 6,000 cubic meters per person per year stored in reservoirs, while the poorest African countries have less than 700. Ethiopia, for example, has less than 50 cubic meters per person per year of water storage. This adds a trace of myopia to Northerners' anti-dam campaigns in the South. Instead of asking how to simply stop dams, a fairer question would be "how much per-capita water storage does a poor semiarid country need to eradicate poverty and hunger in line with the Millennium Development Goals, and how can it be built to protect both people and the environment?"

Furthermore, we are slowly starting to realize that water is the key vector of climate change. The models developed and summarized by the scientific community and the Nobel Prize-winning Intergovernmental Panel on Climate Change show how we can expect altered atmospheric circulation patterns, expanding aridity, changes in the patterns of rainfall and river flow seasonality, vanishing winter snow, and much more. We can no longer make the comfortable assumption that history will be repeating itself and that planning of new water infrastructure can be based on hydrologic averages from the past. In response to this new non-stationary situation, universities, water managers, and others will have to develop new methods and planning tools to allow safe water infrastructure in the future, and the water community will have to rewrite their textbooks.

A shift in thinking will be essential. First, we have to realize that the true water resource is not just what happens to go in the river today – it may not be there tomorrow – but includes the precipitation and the way it recharges both the soil moisture feeding plants and trees and the water flowing through rivers and aquifers. To this aim, the distinction between *green water* in the soil and *blue water* in rivers and aquifers is now spreading across the water community. We now know that most crop production is in fact supported by green water—rainfall—not by the blue water we apply through irrigation. Sustainable water management will thus require a broader approach than used over the past century.

A rethinking is also necessary to answer the question "*what happens to water after use?*" We must distinguish between water withdrawals based on taking water from rivers and aquifers and then returning it after use, often loaded with pollutants, and water consumption, where a single use prevents any later uses in a watershed because the water is consumed, evaporated, or lost. Indeed, the treatment and reuse of water can be repeated over and over again. But the unlimited consumption of water can lead to disasters like the Aral Sea, and we must not let new such disasters happen.

Freshwater is fundamental for the biosphere as well as the social spheres. When water gets increasingly scarce, more attention will have to be paid to the compatibility between upstream and downstream uses of water in a catchment in an integrated way. A key tool for this is catchment-based Integrated Land Water Resources Management. We must balance the "green water" component in the soil, influenced by land use patterns and other factors, and the "blue water" component in rivers and aquifers, influenced by societal factors, consumptive uses, and return flow. After all, humanity is living at the mercy of the water cycle shared with the equally water-dependent terrestrial and aquatic ecosystems.

All of the volumes of *The World's Water*—the biennial series initiated by my long-time friend and colleague Peter Gleick in 1998—highlight these problems and coming transitions. This volume is the sixth in the series and, among other things, addresses the new concept of "peak water" in the context of growing scarcity. It addresses water as a vector of climate change, as it has done regularly since the first volume. It addresses water pollution in response to blind and inappropriate water policies in advancing industrialized nations, such as China. And it addresses the particular problems of water security in growing cities and how improving efficiency is a critical element for moving along the "soft path" for water that Gleick has developed. Like the earlier volumes, *The World's Water* continues to provide a highly interesting and up-to-date place to go to explore the new water transitions that are coming. Once again, Peter Gleick and his colleagues are to be congratulated.

Finally, *The World's Water* continues to carry a fundamental message: humanity finds itself on the threshold to a new era related to its dependence on, and interaction with, the global water cycle, the bloodstream of the biosphere. This new era demands that we further develop our thinking and approaches so that we adequately prepare for a better future and lay the basis for successfully coping with the increasingly complex challenges that will face our children and grandchildren.

Malin Falkenmark
Professor
Stockholm International Water
Institute and Stockholm Resilience Center

Acknowledgments

Thanks up front to my dear friend and colleague, Malin Falkenmark, for her kind foreword to this volume. Malin is a Swedish water expert who has worked on water forever, and who continues to challenge the water community and policymakers to do the right thing, faster. I still fondly remember visiting her when my wife and I were living in Sweden in 1987 and being fed real Moose Mousse while we talked water.

Special thanks to my co-authors, but especially to Heather Cooley. Heather has continued to write and co-author several of the chapters, and she also did much of the hard work of preparing many of the data tables. She better watch out or I may end up turning over to her the whole show. Continued thanks to Todd Baldwin, my editor at Island Press, whose enthusiastic support of this project from the beginning has helped spur me on.

Finally, thanks to all my readers and colleagues who write, call, or email me with comments, suggestions, and ideas for The World's Water. You've all helped make it a bearable load and better product.

Peter Gleick

Introduction

Water has become a hot topic and I'd like to think that the five previous volumes of our biennial book, *The World's Water,* have helped improve global understanding of the water challenges and solutions that face us, at least a little. But it is also true that a series of growing crises in water have finally raised enough awareness that more policy makers, scientists, and members of the public are paying attention. The water crisis is a bad thing; but more attention and effort focused on solving it is good.

The need for *The World's Water* is more evident to me than ever. Good information and analysis on critical issues, new thinking about solutions, and better data on water are more valuable than they've ever been. While I believe that we know enough to take far more aggressive and extensive actions now to solve water problems, I also believe in the value of debate, discussion, and data.

This is the sixth volume of *The World's Water.* As with the first five, my colleagues and I have chosen a subset of water issues to explore based on timeliness, urgency, and our own experience. There is no shortage of important water topics to address, and as always, it is a challenge to try to choose among them for inclusion in the books.

Among the topics acquiring a growing urgency, as the deadline for meeting them closes in, is the Millennium Development Goals for water. The MDGs define the ongoing effort of nations, international organizations, and communities to seriously address unmet needs for water and sanitation by 2015, specifically the inexcusable failure to provide safe water and adequate sanitation for billions of the world's poorest. In this volume, Meena Palaniappan offers us both good news and bad news. Some remarkable progress is being reported in meeting needs for safe water. We are failing miserably, however, in meeting sanitation needs. And regional progress has been spotty, with Africa, once again, being neglected and falling further and further behind the rest of the world.

This volume also offers a comprehensive assessment of the water catastrophe rapidly unfolding in China. Previous volumes looked briefly at some of the major water projects underway there, such as the Three Gorges Dam project and the massive South-North water diversions. But here I devote a full chapter to the implications of China's extraordinary economic expansion for that nation's water resources, and to the country's gross failure to tackle pollution, overuse, and mismanagement. As a result, economic productivity, human health, and ecosystem survival are all directly threatened. It remains unclear whether China will be able to avoid a water nightmare.

Part of China's problem, like that in the U.S., stems from a belief that we can engineer our way out of water scarcity with "hard" solutions like dams. But we now know that there are limits to our water resources, and that we will have to learn to make do with less. This requires "soft path" solutions. As part of our ongoing effort to explore the soft path for water– the subject of several past chapters—this volume looks at the concept of "peak water" as a parallel to the ongoing debate about "peak oil." Meena Palaniappan and I find there are both strong similarities and strong differences between oil and water. Some regions of the world are truly approaching limits to sustainable extraction and use of water. Understanding these limits and the consequences for food production, economic well-being, and the environment may help us develop ways of managing and using water that will help us avoid those consequences.

Moving to a soft path for water is especially important in urban areas. Heather Cooley and I tackle the issue of wasteful and inefficient use of water in urban centers of the United States, with a chapter that offers lessons on both inefficient use and on ways to improve that efficiency and reduce demand for water. We focus on recent experience in a number of major western U.S. cities where water is increasingly scarce and valuable, and where efforts to understand and improve water-use efficiency are advancing. This chapter looks at a variety of technologies, pricing strategies, and conservation policies to highlight successful efforts and ways that urban centers everywhere can meet growing demands with the same, or even less, water than now being used.

In recent years, I and my colleagues Jason Morrison and Mari Morikawa, together with others, have begun to extensively explore the risks that water problems pose for corporations and industrial production, and the parallel risks that these organizations pose to water resources. The role of the private sector in using, managing, and even commercializing water is intensely controversial, and we've analyzed and written about these issues for many years. As part of that work (available at the Institute's website, www.pacinst.org), we recently completed an analysis of the kinds of information, water thinking, and reporting done by over 120 different leading companies, described and summarized in a chapter here. More and more companies are issuing sustainability or environmental reports of some kind – indeed, corporations that do not are now in the minority – and their use or management of water is one of the factors often discussed in them. But there is no consistency in how water use is reported or described, and this chapter offers suggestions for improving corporate reporting on water.

The complex and growing implications of climate change for water resources continues to be a critical issue. We take another look at the problem here, with a chapter by Heather Cooley updating the state-of-knowledge related to water and climate. We continue to believe that among the most severe consequences of unavoidable climate change will be changes in water availability, quality, and demand. National and local water managers and planners continue to be grossly underprepared. Read this chapter to learn the latest.

As in the previous volumes, the major chapters are supplemented with shorter "In Brief" reports on items of interest. Heather Cooley offers an update to the status of the Tampa Bay desalination plant, which is finally up and running after years of delay and many millions of dollars in overruns. The experience at Tampa Bay offers lessons to anyone interested in the future of desalination. Michael Cohen, who has previous written about the remarkable saga of the Salton Sea brings new information about ambitious, but not-yet implemented plans to try to salvage a healthy ecosystem and environment while stabilizing the human and economic benefits of the Sea as well. I also update the reader on the status of the Three Gorges Dam project in China – the largest dam and reservoir in the world. This topic was originally covered in the very first volume of The World's Water in 1998 and in the subsequent decade, the Chinese moved aggressively to complete the project. As noted in 1998, Three Gorges Dam offered both benefits and risks, and these benefits and risks are now both increasingly obvious.

And we bring to the readers, once again, our tremendously popular Water Conflict Chronology, with hundreds of updated historical examples of conflicts related to water going back millennia. This Chronology is also on line at www.worldwater.org.

Finally, *The World's Water* has always offered readers a variety of data tables in the final section. In this volume we again present the popular data on water availability and water use by country, and offer new data on access to safe water and sanitation. We have also compiled new data on dams in the United States, including a growing number of dams removed for safety, economic, or environmental reasons, a summary of dams in Africa, data on the mortality rate in children under five years of age from water-related diseases, a remarkable dataset on the prices for water in a range of OECD (Organization for Economic Cooperation and Development) and non-OECD countries and cities, and much more. Over the next few years, we will continue to explore new issues, new data, and new ways of reaching out to all of you interested in the world of water.

Peter H. Gleick
Oakland, California, 2008

CHAPTER 1

Peak Water

Meena Palaniappan and Peter H. Gleick

In the past few years, discussions about the possibility of resource crises around water, energy, and food have introduced new terms and concepts into the public debate. Energy experts predict that the world is approaching, or has even passed, the point of maximum production of oil, or "peak oil." The implications of reaching this point for energy policy are profound, for a range of economic, political, and environmental reasons. More recently, there has been a growing discussion of whether we are also approaching a comparable point of "peak water," at which we run up against natural limits to availability or human use of freshwater.

To judge from recent media attention, the finite supply of freshwater on Earth has been nearly tapped dry, leading to a natural resource calamity on par with, or even worse than, running out of accessible, affordable oil. In this chapter, we evaluate the similarities and differences between water and oil to understand whether and how the concept of "peak water" is analogous to the idea of peak oil; how relevant this idea is to actual hydrologic and water management conditions; and the implications of limits on freshwater availability for human and ecosystem well-being.

Regional water scarcity is a significant and growing problem although there are many different (and often inconsistent) measures and indicators of water scarcity (Gleick et al. 2002). In some regions, water use exceeds the amount of water that is naturally replenished every year. About one-third of the world's population lives in countries with moderate-to-high water stress, defined by the United Nations to be water consumption that exceeds 10 percent of renewable freshwater resources. By this measure, some 80 countries, constituting 40 percent of the world's population, were suffering from water shortages by the mid-1990s (CSD 1997, UN/WWAP 2003). By 2020, water use is expected to increase by 40 percent, and 17 percent more water will be required for food production to meet the needs of the growing population. According to another estimate from the United Nations, by 2025, 1.8 billion people will be living in regions with absolute water scarcity, and two out of three people in the world could be living under conditions of water stress (UNEP 2007). Are we reaching natural limits to growth, long predicted by some observers? Are there peaks in availability or use of certain resources? These questions have long been debated in the energy field, and they are now being raised for other vital resources, particularly water.

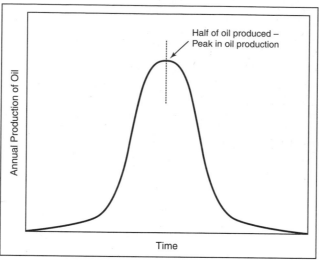

FIGURE 1.1 HUBBERT CURVE FOR AN OIL-PRODUCING REGION.

Concept of Peak Oil

The theory of peak oil originated in the 1950s with the work of geologist M. King Hubbert and colleagues who suggested that the rate of oil production would likely be characterized by several phases that follow a bell-shaped curve. First, discovery and the rate of exploitation rapidly increase as demand rises, production becomes more efficient, and costs fall. Second, as oil is consumed, the resource becomes increasingly scarce, costs increase, and production levels off and peaks. Finally, increasing scarcity leads to a decline in the rate of production more quickly than new supplies can be found. This last phase would also be typically accompanied by the substitution of alternatives (Ehrlich et al. 1977). The phrase "peak oil" refers to the point at which approximately half of the existing stock of petroleum has been depleted and the rate of production peaks (Fig. 1.1).

In 1956, Hubbert predicted that oil production in the United States would peak between 1965 and 1970. And, in fact in 1970, oil production in the U.S. reached its height and began to decline (Fig. 1.2). The concept of a bell-shaped oil production

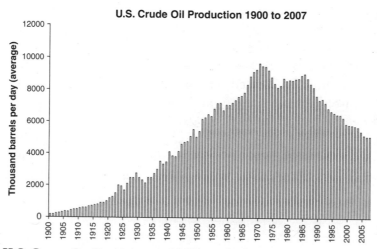

FIGURE 1.2 U.S. CRUDE OIL PRODUCTION 1900 TO 2007.
Source: USEIA 2007.

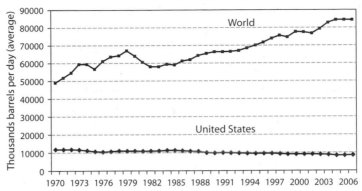

FIGURE 1.3 WORLD AND U.S. OIL PRODUCTION 1970 TO 2007.
Source: USEIA 2008.

curve has been proven for a well, an oil field, a region, and is thought to hold true worldwide. The theory of peak oil also envisions that once half of oil reserves have been produced, oil would become increasingly more difficult and expensive to extract because the most accessible sources of petroleum had already been tapped.

In recent years, the concept of peak oil has received renewed attention because of growing concern that the world as a whole is approaching the point of declining petroleum production. No one knows when global oil production will actually peak, and forecasts of the date range from early in the 21st century to after 2025. One of many recent estimates suggests that oil production may peak as early as 2012 at 100 million barrels of oil per day (Gold and Davis 2007). The actual peak of production depends on the demand and cost of oil, the economics of technologies for extracting oil, the rate of discovery of new reserves compared to the rate of extraction, the cost of alternative energy sources, and political factors. Figure 1.3 shows total U.S. and global oil production from 1970 to 2007.

There are many reasons for growing concern over reaching the point of maximum production of oil. In particular, the population of the planet continues to grow rapidly, driving rising demand for energy in the form of liquid fuels. This growing demand, together with the fact that alternatives or substitutes for oil remain economically expensive and technologically immature, raises the specter of energy shortages, constraints on industrial activity, and economic disruptions. And in summer 2008, when the price of oil shot to $140 per barrel, the concept of peak oil began to feel all too tangible.

Comparison of Water and Oil

Does production or use of water follow a similar bell-shaped curve? In the growing concern about global and local water shortages and scarcity, is the concept of "peak water" valid and useful to water planners, managers, and users?

In the following sections, we consider the differences and similarities between oil and water to evaluate whether a peak in the production of water is possible, and in what contexts it may be relevant. We assess existing limits to the amount of water and

TABLE 1.1   Summary Comparison of Oil and Water

Characteristic	Oil	Water
Quantity of resource	Finite	Literally finite, but practically unlimited at a cost
Renewable or non-renewable	Non-renewable resource	Renewable overall, but with locally non-renewable stocks
Flow	Only as withdrawals from fixed stocks	Water cycle renews natural flows
Transportability	Long-distance transport is economically viable.	Long-distance transport is not economically viable.
Consumptive versus non-consumptive use	Almost all use of petroleum is consumptive, converting high-quality fuel into lower quality heat.	Some uses of water are consumptive, but many are not. Overall, water is not "consumed" from the hydrologic cycle.
Substitutability	The energy provided by the combustion of oil can be provided by a wide range of alternatives.	Water has no substitute for a wide range of functions and purposes.
Prospects	Limited availability; substitution inevitable by a backstop renewable source	Locally limited, but globally unlimited after backstop source (e.g., desalination of oceans) is economically and environmentally developed.

oil available on earth. Oil and water are also compared in terms of the renewability of the resource, whether the substance is consumed or not during use, and whether its use is global or local in scale. We also look at whether substitutes for the resources are possible. Our major findings are summarized in Table 1.1. Based on this analysis, in the next section we evaluate the utility of the term "peak water."

First, we look at the question of limits on total water availability. While it is clear that we will at some point in the future run out of oil (or, to be more precise, economically and environmentally accessible oil), will we run out of water? Considering this question on a planet covered with water may seem odd, but as the following section illustrates, there are distinct differences in the amount of water that exists in stocks versus that which is available in flows of the hydrologic cycle.

Are We Running Out of Water?

The total quantity of both water and oil on Earth are literally limited, though the more important question is whether they are practically limited. The origins of petroleum rest with biological and chemical processes that turned decaying plant carbon into stocks of liquid and solid "fossil fuels" over the geologic time of millions of years. The origins of water on Earth are less certain, but most geologists agree that the water on the planet is of cosmic origins from around the time when the planet itself was formed (Box 1.1).

How much water is there on Earth and where is it? Table 1.2 shows the distribution of the main components of the world's water. The Earth has a stock of approximately

---

Box 1.1 The Origins of Water on Earth

A healthy academic debate continues over the origins of water on Earth. At present, the evidence suggests that at least a substantial amount of the world's water originated billions of years ago during the formation of the planet itself. Drake and Campins argue that the evidence is strong that the Earth had a proto-atmosphere and large bodies of water as far back as 4.45 billion years ago, but there is also evidence of later accretion of water from comets and meteors (Robert 2001, NASA 2001, Drake and Campins 2006). Among the ideas about the origins of the planet's water resources are:

- As the Earth cooled, the temperature reached a point where gases released from cooling terrestrial materials could be retained in an atmosphere whose pressure permitted the formation of liquid water.

- Water was delivered to the surface by large comets, trans-Neptunian objects, or water-rich asteroids. The presence of water in comets and outer solar system planetoids has long been confirmed, and the composition of some of this water is similar to the composition of water in the Earth's oceans. In particular, the distribution of the hydrogen isotopic ratio in carbonaceous meteorites is the most similar to the isotopic ratio found in water on Earth.

- Water was delivered to the surface by very small comets over a very long period of time. These comets continue to deliver water to the Earth.

- The release of water stored in hydrous minerals of the planet over time.

1.4 billion cubic kilometers of water, spread over a wide variety of forms and locations. Of this water, the vast majority (nearly 97%) is salt water in the oceans. The world's total freshwater reserves are estimated at around 35 million cubic kilometers. Most of this, however, is locked up in glaciers and permanent snow cover, or in deep groundwater, inaccessible to humans.

Considering the total volume of water on Earth, the concept of running out of water at the global scale is of little practical utility. There are huge volumes of water—many thousands of times the volumes that humans appropriate for all purposes. In the early 2000s, total global withdrawals of water were approximately 3,700 km^3 per year, a tiny fraction of the estimated stock of 35 million km^3 of water (Gleick 2006).

A more accurate, and sobering, way to evaluate human uses of water, however, would look at the total impact of human appropriations through the use of rainfall, surface and groundwater stocks, soil moisture, and so on. An early effort to evaluate these uses estimated that humans already appropriate over 50% of all renewable and

TABLE 1.2 Major Stocks of Water on Earth

	Distribution Area (10^3 km^2)	Volume (10^3 km^3)	Percent of Total Water (%)	Percent of Fresh Water (%)
Total water	510,000	1,386,000	100	
Total freshwater	149,000	35,000	2.53	100
World oceans	361,300	1,340,000	96.5	
Saline groundwater		13,000	1	
Fresh groundwater		10,500	.76	30
Antarctic glaciers	13,980	21,600	1.56	61.7
Greenland glaciers	1,800	2,340	.17	6.7
Arctic islands	226	84	.006	.24
Mountain glaciers	224	40.6	.003	.12
Ground ice/permafrost	21,000	300	.022	.86
Saline lakes	822	85.4	.006	
Freshwater lakes	1,240	91	.007	.26
Wetlands	2,680	11.5	.0008	.03
Rivers (as flows on average)		2.12	.0002	.006
In biological matter		1.12	.0001	.0003
In the atmosphere (on average)		12.9	.0001	.04

Source: Shiklomanov (1993).

"accessible" freshwater flows (Postel et al. 1996), including a fairly large fraction of water that is used instream for dilution of human wastes. It is important to note, however, that these uses are of the "renewable" flows of water, which we explain later. In theory, this use can continue indefinitely without any effect on future availability because of the renewability of the resource. Still, while water itself is renewable, many uses of water will degrade its quality to such an extent that this theoretically "available" water is practically useless. Improving the quality of this water for reuse will require the input of energy, technology, biological treatment, or dilution with more water.

Renewable vs. Nonrenewable Resources

In any comparison between oil and water, it is vital to distinguish between renewable and nonrenewable resources. The key difference between these is that renewable resources are flow (or rate) limited; nonrenewable resources are stock limited (Ehrlich et al. 1977). Stock-limited resources, especially fossil fuels, can be depleted without being replenished on a time-scale of practical interest. Stocks of oil, for example, accumulated over millions of years. How long oil lasts depends on our ability to find it, the rate we use it, and the cost of removing and using it; the volume of oil stocks is effectively independent of any natural rates of replenishment because such rates are so slow.

Flow-limited resources can be virtually inexhaustible over time, because their use does not diminish the production of the next unit. Such resources, such as solar energy, are, however, limited in the flow rate, i.e., the amount available per unit time. Our use of solar energy has no effect on the next amount produced by the sun, but our ability to capture solar energy is a function of the rate at which it is delivered.

Water is a unique renewable natural resource that demonstrates characteristics of both flow-limited and stock-limited resources, because of the wide range of forms and locations for freshwater. This dual characteristic of water has implications for the applicability of the term peak water. Overall, water is a renewable resource with rapid flows from one stock and form to another, and the production of water typically has no effect on natural recharge rates. But there are also fixed or isolated stocks of local water resources that can be consumed at rates far faster than natural rates of renewal, or for which the rate of recharge is extremely slow. Most of these are groundwater aquifers—often called "fossil" aquifers because of their slow recharge rates—but some surface water storage in the form of lakes or glaciers can also be used at rates exceeding natural renewal, a problem that may be worsened by climate change, as noted later and in Chapter 3.

Consumptive vs. Non-Consumptive Uses

Another key factor in evaluating the utility of the concept of a resource peak is whether water and oil are used in consumptive or non-consumptive ways. Practically every use of petroleum is consumptive; once the energy is extracted and used, it is degraded in quality.[1] Almost every year, the amount of oil consumed matches the amount of oil produced, and sometimes we consume more than is produced that year. Thus a production curve for oil is solely dependent on access to new oil.

Not all uses of water are consumptive and even water that has been "consumed" is not lost to the hydrologic cycle or to future use—it is simply recycled by natural systems. Consumptive uses of water only refer to uses of water that make that water unavailable for immediate or short-term reuse within the same watershed. Such consumptive uses include water that has evaporated, transpired, been incorporated into products or crops, heavily contaminated, or consumed by humans or animals. As discussed in the section on the renewability of water resources, some stocks of water can be effectively consumed locally. When withdrawals are not replaced on a timescale of interest to society, eventually that stock becomes depleted. The water itself remains in the hydrologic cycle, in another stock or flow, but it is no longer available for use in the region originally found. There are also many non-consumptive uses of water, including water used for cooling in industrial and energy production, and water used for washing, flushing, or other residential uses if that water can be collected, treated, and reused.

Transportability of Water

Because the Earth will never "run out" of freshwater, growing concerns about water scarcity must, therefore, be the result of something other than a concern about literally consuming a limited resource. And, of course, they are; water challenges are the result of the tremendously uneven distribution (due to both natural and human factors) of water on earth, the economic and physical constraints on tapping some of the largest volumes (such as deep groundwater and ice in Antarctica and Greenland) of freshwa-

---

1. Due to the law of conservation of energy, energy is never "consumed"—simply converted to another form. But in this case, the use of oil converts concentrated, high-quality energy into low-quality, unusable waste heat, effectively "consuming" the oil.

ter, human contamination of some readily available stocks, and the high costs of moving water from one place to another.

This last point—the "transportability" of water—is highly relevant to the concept of peaking. Oil is transported around the world because it has a high economic value compared to the cost of transportation. As a result, there is, effectively a single global stock of oil that can be depleted, and regional constraints can be overcome by moving oil from the point of production to any point of use. In contrast, water is expensive to move any large distance, compared to its value. As a result, there is no single, fungible global stock of water, and regional constraints become a legitimate and serious concern.

Media attention to the concept of "peak water" has focused on local water scarcity and challenges, for good reason. But, there has been little or no academic research or analysis on this concept. In regions where water is scarce, the apparent nature of water constraints—and hence, some of the real implications of a "peak" in availability—are already apparent. Because the costs of transporting bulk water from one place to another are so high, once a region's water use exceeds its renewable supply, it will begin tapping into non-renewable resources such as slow-recharge aquifers. Once extraction of water exceeds natural rates of replenishment, the only long-term options are to reduce demand to sustainable levels, move the demand to an area where water is available, or shift to increasingly expensive sources, such as desalination.

A few exceptions to the economic limits on transporting water exist. Bottled water, for example, is sometimes consumed vast distances from where it was produced because it commands a premium far above normal costs. Growth in bottled water consumption may expand in some markets, but overall, long-distance transfers of bulk water are not likely to become a significant export in commercial markets.

Substitutes for Oil and Water

An important characteristic of peak oil discussions is the inevitable substitution of alternative energy sources for oil as production declines and prices increase. Oil serves particular functions in industrial society that can be satisfied by other means or resources (e.g., solar, natural gas, biofuels). In this sense, any depletable resource, such as fossil fuels, must be considered a transition option, useful only as long as its availability falls within economic and environmental limits.

The basic amount of water needed for drinking and growing food should be considered irreplaceable. There are also many ways that we use water that are unnecessary or highly inefficient. For example, using water to transport human waste is a choice, but not a necessary use of water.

Like energy, water is used for a variety of purposes. And like energy, the efficiency of water use can be greatly improved by changes in technologies and processes. Unlike oil, however, fresh water is the *only* substance capable of meeting certain needs. Thus, while other energy sources can substitute for oil, water has no substitutes for many uses.

A relevant concept to both peak water and peak oil, therefore, is the "backstop" price of substitutes—i.e., the price of the substitute capable of replacing or expanding the original source of supply of a resource. As oil production peaks and then declines, the price of oil will increase in the classic "supply/demand" economic response. Prices will continue to increase until the point when a substitute for oil becomes economically competitive, at which point prices will stabilize at the new "backstop" price.

Similarly, for water, as cheaper sources of water are depleted or allocated, more expensive sources must be found and brought to the user. Ultimately, the "backstop" price for water will be reached. Unlike oil, however, which must be backstopped by a different, renewable energy source, the ultimate water backstop is still water, from an essentially unlimited source—for example, desalination of ocean water. The amount of water in the oceans is limited only by how much we are willing to pay for it and the environmental constraints of using it. In some regions, desalination is already an economically competitive alternative, particularly where water is truly scarce, such as certain islands in the Caribbean and parts of the Persian Gulf (see Cooley et al. 2006).

Climate Change

Oil and water are also intricately tied to climate change, which affects the production and cost curves of both substances. Petroleum, as a fossil fuel that produces carbon dioxide when burned, is one of the major culprits driving global warming. Among the most significant consequences of climate change will be impacts on the hydrologic cycle (see Chapter 3). Such changes are already being experienced (IPCC 2007). As the climate changes, among the hydrologic impacts will be changes in precipitation intensity and duration, loss of snowpack and an acceleration of snowmelt in mountainous areas, loss of glaciers due to accelerated melt, and a risk of both more floods and droughts. Many of these factors will increase both water demand and water scarcity, affecting human and ecosystem health.

In some places, climate change will affect the renewability of local water resources. Where local communities are currently dependent on river runoff from glacier melt, the loss of glaciers over the next century will lead to a "peak water" effect: the diminishment of water supply over time. Communities dependent on groundwater recharge that suffer a decrease in recharge rate will also experience an effect akin to "peak water." In this case, the concept of "peak water" is slightly different: it is not affected by the magnitude of human use, but by climatic factors that diminish the rate of replenishment. Similar to peak oil, however, when the stock is gone, alternative sources will have to be found.

Utility of the Term "Peak Water"

Given the physical and economic characteristics of oil and water reviewed earlier, how relevant or useful is the concept of a peak in the production of water? As described in the previous sections, the fact that the volume of extractable oil is limited, while water is essentially unlimited, means that if global water use followed a bell-shaped curve, we would never reach a "peak" in global water production. A true "peak" in resource production followed by a decline is only possible for resources that are non-renewable and consumed in their use. We cannot reach a point globally where half of water resources have been tapped because water is a renewable resource that is not consumed in its use. For these reasons, the idea of global "peak water" is inaccurate.

However, the concept can be applied in some interesting ways. In the following sections, we explore cases in which "peak water" is useful. We also introduce a new term that is useful when thinking about maximizing the multiple services that water provides: "peak ecological water." And, we explore the value of the "peak water" concept for driving important paradigm shifts in how water is used and managed.

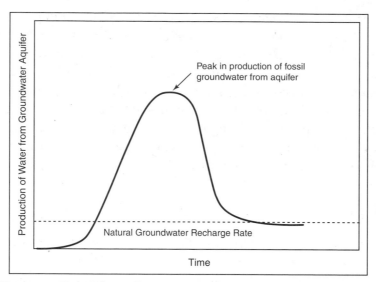

FIGURE 1.4 POTENTIAL PEAK WATER CURVE FOR PRODUCTION OF FOSSIL GROUNDWATER FROM AN AQUIFER. This theoretical curve shows the progression of water extraction from a groundwater aquifer, hypothesizing a peak oil type production curve for water after production rates surpass the natural groundwater recharge rate and production costs rise.

Fossil Groundwater

In most watersheds, there are renewable flows of water, such as rainfall, stream flows, and snow melt, and *effectively* non-renewable stocks of water, such as fossil groundwater. As defined previously, fossil groundwater is groundwater in an aquifer accumulated over many thousands of years, with a very slow recharge rate. When the use of water from a groundwater aquifer far exceeds the natural recharge rate, this stock of groundwater will be depleted quickly. In these particular situations, the groundwater aquifer is analogous to an oil field or oil-producing region. Continued production of water, beyond natural recharge rates, becomes increasingly difficult and expensive as groundwater levels drop, leading to a peak of production, followed by diminishing withdrawals and use. As shown in Figure 1.4, once withdrawals from the groundwater aquifer pass the natural recharge rate for the aquifer (shown as a dashed line), the production of water from the aquifer can continue to increase until a significant portion of the groundwater has been harvested. After this point, deeper boreholes and increased pumping will be required to harvest the remaining amount of water, potentially reducing the rate of production of water.

It is also possible that the production of water from the aquifer will continue to increase until all the economically affordable groundwater is harvested, after which the production of water drops quickly. In both these cases, the important point is that extraction will not fall to zero, but to the renewable recharge rate, where economically and physically sustainable pumping is possible.

"Peak Ecological Water"

For many watersheds, a more immediate and serious concern than "running out" of water is exceeding a point of water use that causes serious or irreversible ecological damage. Water provides many services: not only does it sustain human life and commercial and industrial activity, but it is also fundamental for the sustenance for animals, plants, habitats, and environmentally dependent livelihoods.

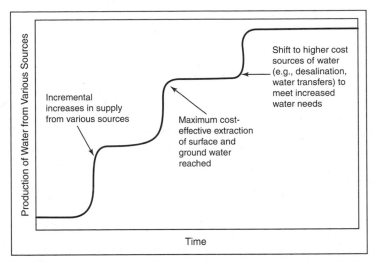

FIGURE 1.5 POTENTIAL WATER PRODUCTION SCENARIO IN A WATERSHED. This figure graphs a potential water production (supply) scenario in a watershed or region. As demand increases, incremental supply projects (dams, reservoirs, pumping) increase water availability. Once the maximum cost-effective extraction of surface and groundwater is reached, there is a final shift to a higher cost "backstop" supply of water such as desalination or water transfers.

Figure 1.5 graphs a potential water-production scenario in a watershed, where incremental supply increases through supply-side projects, e.g., groundwater harvesting, in-stream flow allocation, and reservoir construction are layered upon each other until the maximum cost-effective extraction of surface and ground water is reached. After this point, a final backstop supply of fresh water, such as desalination or water transfers, might be implemented.

Each new incremental supply project that captures water for human use and consumption decreases the availability of this water to support ecosystems and diminishes their capacity to provide services. The water that has been temporarily appropriated or moved was once sustaining habitats and terrestrial, avian, and aquatic plants and animals. As mentioned, by some estimates, humans already appropriate almost 50% of all renewable and accessible freshwater flows (Postel et al. 1996), leading to significant ecological disruptions. Since 1900, half of the world's wetlands have disappeared (Katz 2006). The number of freshwater species has decreased faster than the decline of species on land or in the sea. River deltas are increasingly deprived of flows due to upstream diversions, or receive water heavily contaminated with human and industrial wastes.

Figure 1.6 is a simplified graph of the value that humans obtain from water produced through incremental increases in supply (e.g., drinking water, irrigation), against the declining value of the ecological services (e.g., water for plants and animals) that were being satisfied with this water. The graph envisions that ecological services decrease as water is appropriated from watersheds. The pace or severity of ecological disruptions increases as increasing amounts of water are appropriated. Because ecological services are not well valued in dollar terms, the y-axis is labeled here simply as "value provided by water."

At a certain theoretical point, the value of ecological services provided by water is equivalent to the value of human services provided by water. After this point, increas-

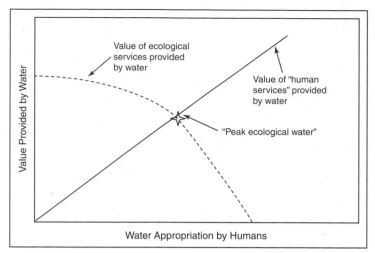

FIGURE 1.6 "PEAK ECOLOGICAL WATER": WHEN ECOLOGICAL DISRUPTIONS EXCEED THE BENEFITS OF NEW SUPPLY PROJECTS. Graph charts the value of water provided by increasing supply from various sources in a watershed against the loss in value of ecological services provided by this water. As water supply projects increase water production in a watershed (solid line), the ecological services provided by water are in decline (dashed line). At a certain subjectively determined point, the value of water provided through supply projects is equal to the value of the ecological services. Beyond this point ecological disruptions exceed the benefits of increased water extraction. We entitle this point "peak ecological water."

ing appropriation of water leads to ecological disruptions beyond the value that this increased water provides to humans (the slope of the decline in ecological services is greater than the slope of the increase in value to humans). At the point of "peak ecological water," society will maximize the ecological and human benefits provided by water. As shown in Figure 1.7, the overall value of water, combining ecological and social benefits, rises to a peak and then declines as human appropriation increases. Of

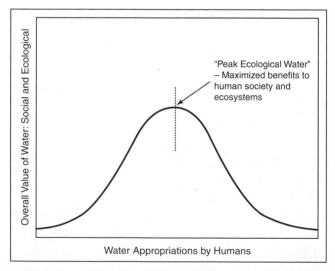

FIGURE 1.7 OVERALL VALUE OF WATER WITH INCREASING HUMAN APPROPRIATION OF WATER. Graph charts the overall value of water, a combination of social and ecological value, as water appropriation by humans increases. The value increases to a peak, where benefits to society and ecosystems is maximized, and then declines.

course, determining the point of "peak ecological water" is difficult to quantify and largely subjective based on different appraisals of the value of each unit of water in ecosystems and to humans.

Despite the difficulty in determining "peak ecological water," human societies make determinations as to what level of ecological disruption is acceptable to meet human needs (though they rarely do so with complete information about the true ecological consequences of their actions). The important point is that as human appropriations of water increase, there is a corresponding decrease in the ecological services this water can provide.

As human societies grapple with a water-constrained future, it is important to consider the many services that water provides. Whereas the use of the term "peak water" is flawed, the idea of maximizing both social and ecological benefits that water provides is more relevant. We propose the idea of "peak ecological water" as the point when maximum benefits to human society and the ecosystem can occur.

A New Water Paradigm: The Soft Path for Water

Real limits on oil production will, inevitably, stimulate efforts to identify and develop alternative energy sources capable of providing the same benefits as oil. And indeed, there are many substitutes for the different uses of oil for electricity, transportation fuels, lubricants, and the production of materials.

Real limits on water are far more worrisome, because water is fundamental for life, and for many uses, it has no substitutes. Absolute limits on affordable, accessible water will constrain the ability of regions to do certain things: in particular, limits to the availability of freshwater typically lead to the inability of a region to produce all the food required to meet domestic needs, and hence lead to a reliance on international markets for food. While this has been the subject of previous work in The World's Water (see, for example, "Water and Food" in the 2000–2001 volume, pp. 63–92), it is worth revisiting in the future. But limited water availability can also lead to more efficient use of water, better management of available resources, replacements with alternatives when possible, and increases in the resource productivity of water.

In the late 1970s, Amory Lovins coined the term the "soft path" for energy to denote an alternative approach for meeting human energy needs (Lovins 1977). The "soft path" recognizes that energy is a means to a certain end. People don't want energy itself, but transport, light, and warmth, as examples. The soft path for energy means reduction in wastage and inefficient use of energy, the deployment of renewable energy, and increased use of decentralized options, among other things.

Expanding this theme, Peter Gleick and others coined the concept of a "soft path for water" (Gleick 2002, 2003; Wolff and Gleick 2002; Brooks 2005). The "soft path" is a comprehensive approach to water management, planning, and use that uses water infrastructure, but combines it with improvements in the overall productivity of water use, the smart application of economics to encourage efficiency and equitable use, innovative new technologies, and the strong participation of communities and local water users in making decisions. Rather than seek endless sources of new supply, the soft path matches water services to the scale of the users' needs, and it takes environmental and social concerns into account to ensure that basic human needs and the needs of the natural world are both met.

A key insight behind the soft path for water is that people don't want to "use" water—they want to drink and bathe, produce goods and services, grow food, and meet human needs. Achieving this goal can be done the traditional "hard" way by building more dams, pipelines, and environmentally destructive infrastructure. Or, it can be done in a more integrated, sustainable, and effective way. The soft path can be distinguished from the traditional, hard path for water in six main ways:

1. **Focusing on ensuring water for human needs:** The soft path directs governments, companies, and individuals to meet the water needs of people and businesses, instead of just supplying water. People want clean clothes, or to be able to produce goods and services—they do not care how much water is used and may not care if water is used at all.

2. **Focusing on ensuring water for ecological needs:** The soft path recognizes that the health of our natural world and the activities that depend on it (like swimming, water purification, ecological habitat, and tourism) are important to water-users and people in general. The hard path, by not returning enough water to the natural world, ultimately harms human and other ecological users downstream.

3. **Matching the quality of water needed with the quality of water used:** The soft path leads to water systems that supply water of various qualities for different uses. For instance, storm runoff, gray water, and reclaimed wastewater are well suited to irrigate landscaping or for some industrial purposes that currently are supplied with more expensive potable water.

4. **Matching the scale of the infrastructure to the scale of the need:** The soft path for water recognizes that investing in decentralized infrastructure can be just as cost-effective as investing in large, centralized facilities. There is nothing inherently better about providing irrigation water from a massive reservoir instead of using decentralized rainwater capture and storage.

5. **Ensuring public participation in decisions over water:** The soft path requires water agencies, policy makers, or private entities to interact closely with water users and to engage community groups in water management. The hard path, governed by an engineering mentality, is accustomed to meeting generic needs with little transparency or public input.

6. **Using the power of smart economics:** The soft path recognizes the public and economic aspects of water, using the power of water economics to encourage equitable distribution and efficient use of water.

Conclusion

As the world anticipates a resource-constrained future, the specter of "peak oil"—a peaking in the production of oil—has been predicted. Similarly, many in the news media have begun referring to new limits on the availability of water, which some have termed "peak water." There are important differences between water resources and oil resources. Oil production will inevitably decline, while water uses within renewable limits can continue indefinitely. Oil is a finite, non-renewable resource that is consumed during its use; therefore, oil production will inevitably decline. Peak oil, thus, means the end of cheap, easy-to-access sources of petroleum. Any new sources of liquid fuel will be harder to reach and more expensive to extract. Water is a renewable resource and is not consumed in the global sense; therefore, water uses within renewable limits can continue indefinitely. Oil is routinely transported over long distances from extraction to use, making it a global resource. Conversely, water cannot be economically transported over long distances, making it primarily a local resource. These characteristics mean that there is a global limit to oil production; constraints on water are only manifested regionally. And while many water uses can be reduced or eliminated, a basic amount of water is necessary for life to exist and for which, unlike oil, there are no substitutes.

Despite the serious limitations in the concept of "peak water," as described in this chapter, there are some interesting and valid applications. Not all water use is renewable; indeed some water uses are non-renewable and unsustainable. Groundwater use beyond normal recharge rates follows a peak oil type curve with a peak and then precipitous decline in water production.

Considering the multiple roles that water provides as the fulcrum for ecosystems as well as human society, we suggest that the term "peak ecological water" better delineates an important crisis in the water sector. As human appropriation of water increases, the ecological services that water provides decrease. Once we begin appropriating more than "peak ecological water," ecological disruptions exceed the human benefit obtained. Defined this way, many regions of the world have already surpassed "peak ecological water"—humans use more water than the ecosystem can sustain without significant deterioration and degradation.

Another resonance in the concept of "peak water" is that similar to peak oil it signals the end of cheap and easy to access water. This recognition of the value of water can help drive towards an important and needed paradigm shift in the way water is managed and priced. In this way, the concept of "peak water" helps moves us towards using water in ways that improve the productivity, equity, and efficiency of water use.

What is exciting about the concept of "peak water" is that it may be an additional impetus for a new "soft path for water" paradigm to emerge. In places where peak water is a reality, managers are moving to recognize and manage water as a valuable and precious resource. True limits on regional water availability can also stimulate innovations and behaviors that can reduce water use and increase the productivity of water. Though the use of "peak water" is flawed in key ways, it shifts us in the direction of protecting and preserving precious water resources—a necessary step for a sustainable water future.

REFERENCES

Brooks, D.B. 2005. Beyond greater efficiency: the concept of water soft paths. *Canadian Water Resources Journal* 30(1):1–10.

Commission on Sustainable Development (CSD). 1997. *Comprehensive Assessment of the Freshwater Resources of the World.* Report of the Secretary-General. United Nations Economic and Social Council, New York.

Cooley, H. Gleick, P.H., and Wolff, G. 2006. *Desalination: With a Grain of Salt. A California Perspective.* A Report of the Pacific Institute for Studies in Development, Environment, and Security, Oakland, California.

Drake, M.J. and Campins, H. 2006. Origins of water on the terrestrial planets. In: *Asteroids, Comets, and Meteors.* Proceedings of the International Astronomical Union, August 7–12, 2005. pp. 381–394.

Ehrlich, P., Ehrlich, A., and Holdren, J.P. 1977. *Ecoscience: Population, Resources, Environment.* San Francisco: W.H. Freeman and Company.

Energy Information Administration, March 2008 International Petroleum Monthly, Posted: April 11, 2008, http://www.eia.doe.gov/emeu/international/oilproduction.html

Gleick, P.H. 2002. Soft water paths. *Nature* 418:373.

Gleick, P.H., Chalecki, E.L., and Wong, A. 2002. Measuring water well-being: Water indicators and indices. In: *The World's Water 2002–2003*. Gleick, P.H., editor. Washington, DC: Island Press, pp. 87–112.

Gleick, P.H. 2003. Global freshwater resources: Soft-path solutions for the 21st century. *Science* 302(28):1524–1528.

Gleick, P.H. 2006. Table 2: Freshwater withdrawal, by country and sector (2006 update). *The World's Water 2006–2007*. Washington, DC: Island Press.

Gold, R., and Davis, A. 2007. Oil officials see limit looming on production. *Wall Street Journal* 10:1.

IPCC. 2007. Summary for Policymakers. In: *Climate Change 2007: The Physical Science Basis. Contribution of Working Group I to the Fourth Assessment Report of the Intergovernmental Panel on Climate Change.* Solomon, S., Qin, D., Manning, M., Chen, Z., Marquis, M., Averyt, K.B, Tignor, M., and Miller, H.L., editors. United Kingdom and New York, NY: Cambridge University Press, Cambridge.

Katz, D. 2006. Going with the flow: Preserving and restoring instream water allocations. In: *The World's Water 2006–2007*. Gleick, P.H., editor. Washington, D.C.: Island Press, pp. 29–39.

Lovins, A.B. 1977. *Soft Energy Paths: Toward a Durable Peace.* San Francisco, California: Friends of the Earth International.

NASA Goddard Space Flight Center. 2001. A Dying Comet's Kin May Have Nourished Life on Earth. Viewed May 12, 2008. http://www.gsfc.nasa.gov/gsfc/spacesci/origins/linearwater/linearwater.htm.

Postel, S.L., Daily, G.C., and Ehrlich, P.R. 1996. Human appropriation of renewable fresh water. *Science* 271(2):785–788.

Power, M. 2008. Peak water: Aquifers and rivers are running dry. How three regions are coping. *Wired Magazine,* April 25, 2008. http://www.wired.com/science/planetearth/magazine/16-05/ff_peakwater

Robert, F. 2001. Isotope geochemistry: The origin of water on earth. *Science* 293(5532):1056–1058.

Shiklomanov, I.A. 1993. World fresh water resources. In: *Water in Crisis.* Gleick, P.H., editor. New York: Oxford University Press, pp. 13–24.

United States Energy Information Administration (USEIA). 2007. U.S. Crude Oil Production. Data available for download from: http://tonto.eia.doe.gov/dnav/pet/pet_crd_crpdn_adc_mbblpd_a.htm

United States Energy Information Administration (USEIA). 2008. International Petroleum Monthly, Data from April 2008. http://www.eia.doe.gov/emeu/international/oilproduc-tion.html

United Nations World Water Assessment Program (UN/WWAP). 2003. *World Water Development Report: Water for People, Water for Life.* http://www.unesco.org/water/wwap/wwdr/index.shtml

United Nations Environment Programme. 2007. *Global Environmental Outlook 4: Environment for Development.* Malta.

Wired Science. 2007. http://www.pbs.org/kcet/wiredscience/story/71-peak_water.html (viewed February 18, 2008). In America's southwest, more people plus less water equals trouble. Also, Water worries. 2007. http://www.celsias.com/2007/07/05/water-worries/ (viewed February 18, 2008).

Wolff, G., and Gleick, P.H. 2002. The soft path for water. In: *The World's Water 2002–2003*. Gleick, P.H., editor. Washington, D.C.: Island Press, pp. 1–32.

CHAPTER 2

Business Reporting on Water

Mari Morikawa, Jason Morrison, and Peter H. Gleick

Corporations have long published annual financial reports describing economic trends and opportunities in their sector, as well as the yearly performance of their own firms. Over the past two decades, a growing number of companies have begun publishing non-financial reports to describe their environmental and social performance to their stakeholders, with the understanding that these factors are increasingly tied to financial performance and company reputation. Unlike financial reporting, sustainability reports use diverse formats, and the information presented in them varies from company to company. This is partly because sustainability reporting is still in the process of being harmonized, but also because the information that is considered material to a company and its stakeholders varies greatly among companies and industry sectors. In other words, the contents of such reports reflect what companies (and sometimes their stakeholders) perceive as critical. We believe that new and consistent standards for reporting are needed.

In this chapter, we offer a comprehensive review of corporate sustainability or environmental reports, with a focus on evaluating how companies recognize, address, and report their water-related risks and practices.[1] The review of corporate reporting on water provides insights on the type and extent of water risks that businesses are recognizing as significant and how they are striving to manage those risks.

Water is a crucial resource for nearly all industry activities. Yet decreasing water availability, declining water quality, and growing water demands from non-industrial water users are creating new challenges to businesses that have traditionally taken clean and reliable water for granted. Around the world, corporations now face diverse water risks, including changing allotments, more stringent water-quality regulations, growing community interest and control over local resources, and increased public scrutiny of water-related activities. Confronted by these challenges, some businesses are starting to see the need to take more proactive and comprehensive strategic water management actions, and to report on those actions to shareholders and the public.

Corporate Reporting: A Brief History

Corporate reporting of non-financial information goes back to the 1970s, when companies started to put environmental information in their annual reports. These

---

1. This chapter summarizes a comprehensive assessment on how corporations report on water risks, water use, and water planning (see Morikawa et al. 2007).

early reports, however, mostly consisted of descriptive anecdotal information more consistent with corporate public relations and advertising, and contained little information on actual performance (Tepper Marlin and Tepper Marlin 2003). In the 1980s, regulatory requirements in the United States, such as the Superfund Amendments and Reauthorization Act and the Toxic Releases Inventory, began to mandate the disclosure of more performance data, in particular the volume of pollutants released into air and water (Baue 2004).

The trend toward broader reporting began in the 1990s as the concept of "sustainable development"—introduced by the Brundtland Commission in 1987—rapidly gained recognition among both the public and business sectors (Brundtland 1987). In addition to providing environmental information, corporate reports during this time increasingly included information on social issues, such as labor, human rights, and stakeholder engagement practices.

The creation of the Global Reporting Initiative (GRI) in 1997 further solidified the practice of publishing comprehensive corporate reports, today often called sustainability or CSR reports. The GRI established the first international guidelines for reporting on the "triple bottom line"—the economic, environmental, and social performance of companies. GRI has continued to revise its sustainability reporting guidelines, releasing its third version in October 2006. In the public policy sphere, the emergence of formal environmental management regulations such as Europe's Eco-Management and Audit Scheme[2] requires new measurement and reporting in the form of comprehensive environmental reports.

Today, the number of companies publishing non-financial information is continuing to increase rapidly as corporations and their stakeholders recognize the significance of social and environmental performance and how both can affect their financial performance. According to the database of CorporateRegister.com, the number of non-financial reports has grown from fewer than 50 in 1992 to over 1,900 in 2005 and 2,470 by the end of 2007. Companies that do not publish such information now are in the minority and are considered to be failing to meet a basic requirement of transparency and accountability.

Review of Corporate Non-Financial Reports

In 2007, we reviewed the non-financial reports of more than one hundred of the largest companies from 11 water-intensive industry sectors (Morikawa et al. 2007). Based on our research on industrial water use and water-related risks, we selected industry sectors that 1) are highly dependent on water resources or vulnerable to water risks; and 2) have different types of business models and water-use characteristics. For instance, some sectors require high-quality water as a key input to production, whereas others use water mainly for cooling or in-plant processes. Some businesses produce high volumes of wastewater. Some are more concerned about the quality of wastewater discharge. Some industry sectors have long and complex water-intensive supply chains, and others conduct most of their manufacturing in company-owned facilities.

---

2. The scheme has been available for participation by companies since 1995 (Council Regulation (EEC) No 1836/93 of 29 June 1993) and was originally restricted to companies in industrial sectors. Since 2001 EMAS has been open to all economic sectors including public and private services (Regulation (EC) No 761/2001 of the European Parliament and of the Council of 19 March 2001).

TABLE 2.1 Corporate Non-Financial Reports Reviewed by Sector

Apparel: Target, AEON, Federated, GAP, Nike, Adidas Group.

Automobile: GM, Ford, DaimlerChrysler, Toyota, Volkswagen, Honda, Nissan, Peugeot, Fiat, BMW, Renault, Hyndai, Volvo, Mazda, Suzuki.

Beverage: Pepsico, TCCC, Anheuser Busch, Heineken, Diageo, InBev, Cadbury Schweppes, Kirin, SABMiller, FEMSA.

Biotech/Pharmaceutical: Pfizer, Johnson & Johnson, Bayer, GlaxoSmithKline, Sanofi-Aventis, Novartis, Roche Group, AstraZeneca, Abbott, Merck, Wyeth, Bristol-Myers Squibb, Akzo Nobel Group.

Chemical: P & G, BASF, Dow, Dupont, Mitsubishi Chemical, Lyondell Chemical, Degussa, Huntsman, Asahi Kasei.

Food: Nestlé, Unilever, Kraft, Tyson Foods, McDonalds, DANONE, ConAgra.

Forest Product: International Paper, Weyerhaeuser, Georgia-Pacific, Kimberly-Clark, Stora Enso, SCA-Svenska Cellulose, Oji Paper, UPM-Kymmene, Nippon Paper, Smurfit Stone.

High-Tech/Electronic: IBM, HP, Panasonic, Samsung, Sony, Dell, Toshiba, NEC, LG, Fujitsu, Nokia, Intel, Motorola, Canon, Cisco.

Metal/Mining: Arcelor, Nippon Steel, BHP Billiton, Anglo American, Alcoa, JFE Holdings, POSCO, Alcan, Rio Tinto, Corus Group.

Refining: ExxonMobil, BP, Chevron, Shell, ConocoPhillips, Total, ENI, Petrobras-Betroleo Brasil, Statoil Group, Repsol-YPF, PetroChina, SK Corp

Utility: E.ON, Electricite de France, RWE Group, Suez Group, ENEL, Tokyo Electric Power, Veolia Environment, UES of Russia, Gaz de France, Korea Electric Power, Kansai Electric Power, Ecntrica, Endesa Group, Chubu Electric Power.

Corporate non-financial reports of the following companies were analyzed. They were accessed online in September/October 2006 (by sector, in order of size by 2005 sales).

We focused our analysis on the largest companies in each sector because they are likely to have larger water-related impacts and risks. Companies were selected based on the following criteria: 1) they had to be publicly traded[3]; and 2) they were among the ten largest companies in each sector (globally), with annual sales greater than 15 billion (US$) in 2005.[4] Altogether 139 companies were selected for the analysis.[5]

We reviewed both water-management information and water-related performance data presented in annual or biannual corporate non-financial reports.[6] Among the 139 companies selected for this study, 121 (87%) publish annual or biannual corporate non-financial reports. Table 2.1 lists the companies reviewed. The other 18 companies

---

3. An exception to this criterion is Levi Strauss, which is privately owned but a significant player in the apparel sector.

4. For sectors having more than 10 companies with annual sales of more than $US 15 billion, up to 15 companies were included in the analysis. The only exception is the apparel industry for which we selected equal numbers of apparel manufacturers and apparel retailers (seven companies each) for a total of 14 companies.

5. With the exception of two companies based in China and Brazil, the sample consists entirely of multinational corporations from developed countries.

6. Our research focused on the contents of published reports themselves. Additional information presented on companies' corporate website was *not* reviewed, unless a company specifically mentions in its published report that more detailed information on related topics should be accessed on its website.

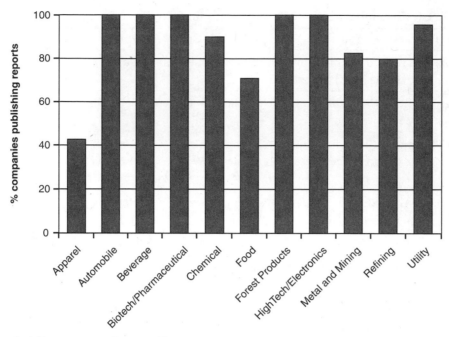

FIGURE 2.1 PUBLICATION RATE OF NON-FINANCIAL REPORTS, BY SECTOR.

(13%) either did not publish non-financial information at the time, or only had a one-time description (as opposed to regular/periodic reporting) of their environmental/social programs on their website. Figure 2.1 illustrates the percentage of the companies selected for this study that publish environmental or sustainability reports, by industry sector. As this figure shows, the worst reporting record is in the apparel sector.

In evaluating water reporting, we use a framework developed by two of the authors, focusing on 10 key practices for managing water-related business risks (Morrison and Gleick 2004). These activities are discussed in Box 2.1. The 121 reports were reviewed to see the degree to which they addressed the 10 recommended management practices, and if so, what kinds of examples are given.

What Companies Report on Water

Among the 121 companies that publish non-financial reports, 115 have sustainability/CSR reports that include information on their social or economic performance in addition to environmental performance. Only six companies opted for a more narrow environmental report. One company published an independent water report in 2006 in addition to its CSR report.

Ninety-seven percent of the reports provide information on water management. This information is presented in two broad forms: 1) descriptive information on water management policies, strategies, or activities; and 2) quantitative information on water-related performance, such as total water use and wastewater quality. About half of the reports contain both types of information, and 63% of the reports have a designated water section or chapter (Fig. 2.2). The following section examines the contents of both types of water reporting, using the 10 recommendations for water-risk management listed in Box 2.1 as a framework.

BOX 2.1 Ten Key Actions for Sustainable Corporate Water Management

Measure Current Water Use
Companies need to measure water use and wastewater discharges associated with their own operations and production. Although harder to track and quantify, they should also assess water use and discharges associated with key suppliers and inputs. This information will provide the baseline data for assessing risks, prioritizing management efforts, and measuring progress.

Assess Water Conditions and Water Risks
For key areas of operation and sourcing, companies should assess local water conditions, including hydrological, social, economical, and political factors. This assessment should flag risk areas of current shortage, rapidly growing demand, insufficient institutional and political water governance capacity, and large disparities in water access or prices between large commercial users and local communities. In high-risk regions, businesses should also have in place contingency plans to respond to water-supply and related risks, such as decreasing water quality, higher water prices, disruptions due to extreme hydrologic events, and local concerns about the scope and pace of economic development.

Consult and Engage Stakeholders
By engaging key stakeholders concerned with water resources, companies can better anticipate and respond to emerging issues, such as competing water demands by local communities or industry, or local concern over wastewater discharge. Open discussions with water providers and local communities are a key factor in preventing or reducing the risk of future water-related disputes. In addition, proactive efforts by the company to improve water quality or water availability can help build positive relations with regional stakeholders.

Engage the Supply Chain
Many companies' most significant water impacts and risks may be embedded in their supply chain. Companies should assess and evaluate water use in their supply chain and work collaboratively with key suppliers to reduce water use and minimize risks of water-related disruptions.

Establish a Water Policy and Set Corollary Goals and Targets
Top management should clearly articulate the organization's policies regarding water-resource issues. In addition, companies should establish supporting quantifiable goals and targets for water-use efficiency, conservation, and minimizing water impacts and associated water-related risks.

continues

Box 2.1 *continued*

Implement Best Available Technology

Companies should assess best available technology for reducing water use and wastewater discharges and commit to using such technology in new facilities and retrofitting existing facilities. There are numerous technologies that can reduce water use and improve water quality, including reclaiming and reusing process water, sophisticated filtration systems, replacing water cooling towers with air cooling, and more. Companies have often found that such technology investments can have short payback periods and generate high returns on investment.

Factor Water Risk into Relevant Business Decisions

Given its growing importance, companies should consider water scarcity and water-related risks as important criteria when making a range of strategic business decisions, from factory siting to new product development.

Report Performance

Companies should publicly report key metrics on their water use and impacts and track how their performance changes over time. This information improves transparency and feedback for investors, customers, local communities, and other key stakeholders, and is often a useful tool for engaging employees across the enterprise in supporting water programs.

Form Strategic Partnerships

Because many water-related issues can best be addressed on a regional scale involving multiple sectors and stakeholders, companies should consider building strategic partnerships through industry associations or multi-stakeholder programs formed to promote watershed protection and improve access to water for impoverished communities.

*Modified from Morrison and Gleick 2004.

Qualitative Information: Water Management Policies, Strategies, and Activities

Measure Current Water Use

One-hundred three companies (85%) say they measure water use. Most of these companies also publicly report the results of their measurements. How they report freshwater use performance is discussed in the next section.

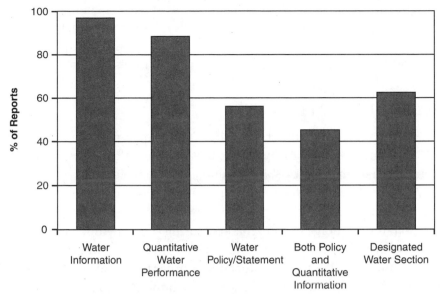

FIGURE 2.2 WATER INFORMATION IN NON-FINANCIAL REPORTS.

Assess Water Conditions and Water Risks

Only twenty-six companies (21%) mention water risk-assessment programs or describe water-related risks they are facing. Most risk assessments focus on water supply or availability, and to a lesser extent on water quality. Broadly, companies recognize two types of water risks: risks posed by local hydrologic or socioeconomic conditions, and risks business activities may impose on local and global water resources.

Nestlé reports that they conduct hydrogeological assessments of their bottled water sites and monitors source water quality and other environmental conditions and parameters, including water levels, spring flow, and rainfall data. Others pay more attention to risks caused by their business activities. Statoil, for example, assesses the environmental risk resulting from the discharge of "produced" water,[7] and Conoco-Phillips has environmental risk management systems to analyze the environmental impact of the offshore discharge of waste from drilling and production. Several companies have more comprehensive analyses, looking both at the types of risks caused by natural conditions and the business impacts on the environment. For example, AkzoNobel (a multi-national manufacturing corporation, active in healthcare products, coatings and chemicals, headquartered in the Netherlands) developed a sustainable water-management model that describes and quantifies indicators to ensure that the users of certain water sources consume less water than is replenished by natural process and that the source water quality is not affected. BHP Billiton (which specializes in metals and fossil fuels) recognizes that its business often competes with agriculture and other human activities for access to water resources and established water-management plans that require sites to identify water sources, water consumption, and opportunities to reduce water usage. Some companies conduct water risk assessment as a part of site selection process for new facilities; others do it as a part of ongoing water management.

---

7. Produced water is groundwater produced during oil or gas extraction.

Although most companies include some mention of climate change and report their effort to reduce greenhouse gas emissions, only four companies (3% of reports reviewed) discuss water risks associated with climate change. We expect that more companies will include climate-related water risks in future reports.

Consult and Engage Stakeholders

Most of the companies (87%) describe policies and programs to engage stakeholders, but only 10 of them (8%) have explicit examples of stakeholder consultations as a part of their water management efforts. This number is notably low, considering that many companies' water statements seem to recognize water as a local issue.

Most of the companies that provide water-related stakeholder consultation examples are in the beverage or food manufacturing sectors. Anheuser-Busch's "Water Council" works with key stakeholders to develop initiatives and pursue research in water management. It also has an educational program to raise awareness of water conservation and watershed protection issues. Groupe Danone (which specializes in water and fresh dairy products) reports that it cooperates with farmers, communities, and other local stakeholders to draw up guidelines for sustainable water management. It also reports appointing local managers to oversee operation of each spring in partnership with local communities and participants from the local economy.

Engage the Supply Chain

A majority of companies (88%) report supply-chain management policies or programs, such as supplier codes of conduct. But as is the case with stakeholder engagement, very few (9%) explicitly describe their efforts to factor water risks into supply-chain management. Some focus on efforts such as information sharing and educational programs for water conservation, whereas others have a more prescriptive approach, such as implementing water standards for suppliers and measuring and monitoring their water performance. The failure to adequately evaluate supply-chain water connections seems to us to be a serious potential risk for certain industries where vulnerabilities to water disruptions may move up the supply chain and affect production.

Anheuser-Busch collaborates with its suppliers to understand their water management and potential risks associated with their supply chain. The company surveys the water use of its suppliers and meets with the highest volume users and those located in areas that could potentially face limited water availability. Unilever's Sustainable Agriculture Programme measures water impacts at the farm level. Nike developed a company-wide water-quality standard called Nike Apparel's Global Water Quality Guidelines and encourages suppliers to meet their local legal requirement or the Nike standard, whichever is more stringent. In the area of water conservation, Nike also describes its work with textile suppliers to minimize the use of water resources and promote better water management in process operations.

Even though several companies measure or monitor their suppliers' water performance, only two companies reported actual data. Nike reports the percentage of suppliers that have water-efficiency programs and are in compliance with its Global Water Quality Guidelines. McDonald's reports changes in water use (from 2003 to 2004, in percentage values) of suppliers by five commodity areas: potatoes, poultry, pork, beef, and buns.

Establish a Water Policy and Set Corollary Goals and Targets

Just over half of the reports include some form of water-policy statement. These statements provide a background or an overview of the companies' water management and include a combination of the following statements:

- **General importance of water resources.** It is common for water-policy statements to include routine comments about how important and crucial water resources are for life and the environment. Some mention global water problems such as growing populations without access to safe drinking water or declining water supply in some parts of the world. For example, Veolia notes, "Water resources are becoming fragile and scarce in many parts of the world, as humans make increasing demands on them."

- **Importance of water resources for their business.** Some statements describe why water is important and how water is used in their business. "Fresh water is crucial at every stage of our product life-cycle, from production and processing of raw materials to consumers using and disposing of our products," says the Unilever report. "We use fresh water from many sources, including lakes, rivers, wells and municipal supplies, for cooling, steam generation and industrial processing. Water is a critical natural resource for BP and an aspect of the natural environment we want to protect," says British Petroleum (BP).

- **Impact of their business on water resources and challenges they face in water management.** Some companies describe how their use of water resources or wastewater discharge affects the environment and the community where they operate. "According to United Nations data, around 52% of our current production volume originates from plants operating in countries that have some nature of water vulnerability and this is expected to increase," according to SAB Miller.

- **High-level commitments related to water management.** Some companies describe their efforts to use water efficiently and protect water resources. "Our central focus is to continue to meet the water supply and wastewater needs of our customers, without harming the environment, despite the pressures of population growth and increased climate variability" (RWE). "To help foster long-term sustainable water resources, water conservation and management are integral functions at our mills" (International Paper).

Figure 2.3 summarizes reporting information on specific targets and goals. As this graph shows, reporting on specific water management goals and targets is weak and inconsistent. Seventy companies (58%) published explicit goals or targets in water management, such as reduction in water use, improvement in water-use efficiency (i.e., reduced water use per unit of production or sales), improvement in the rate of water recycling, reduction in wastewater discharge volumes, and improvement in wastewater quality. Among the companies that set management goals, about half have set quantitative targets (e.g., 5% reduction of water use per liter of product over the next 5 years), whereas the rest do not provide specific targets. Most quantitative targets are short-term (typically annual or less than 5 years) though some targets are long-

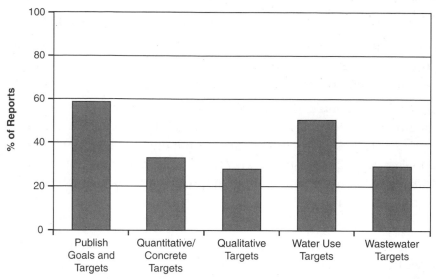

FIGURE 2.3 WATER MANAGEMENT GOALS AND TARGETS, AS PUBLISHED IN NON-FINANCIAL REPORTS.

term (5–10 years). Fifty-nine companies (49%) set goals or targets for water use; 33 companies (27%) set targets for wastewater discharge volume or wastewater quality.

Most of the published quantitative targets are at the overall corporate level, but some companies set separate targets for water-intensive business units, or facilities located in water-stressed areas. A few companies set quantifiable goals related to water-management practices, such as the percentage of facilities that meet the company's water standards or that implement management plans. Figure 2.4 shows the range and prevalence of water management policies reported by the companies surveyed.

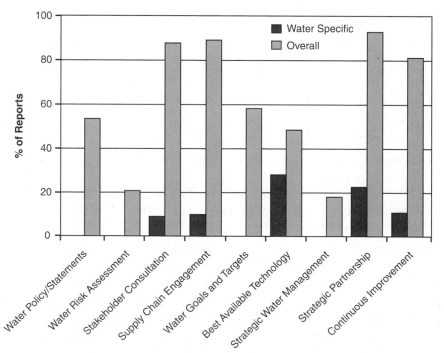

FIGURE 2.4 WATER POLICY AND MANAGEMENT STRATEGY REPORTING. For five parameters, dark gray bars indicate the percentages of reports that provide descriptions and examples specifically related to water management.

Implement Best Available Technology

Fewer than half the companies reviewed (56 companies or 46%) mention policies to use best available technology (BAT) or describe their use of innovative and cutting-edge technologies. Among those, 36 companies offer specific examples. Most describe their efforts in the form of case studies of sites and facilities where new technologies are used to treat water or to improve water-use efficiency. The case studies often include information on financial savings realized through the projects. About 20 companies report the amount of expense or investment for water conservation or wastewater treatment.

A majority of the technologies mentioned are wastewater treatment and recycling; a smaller number report on water-use efficiency technologies. A few companies mention the use of information technology, such as computerized data collection and analysis systems.

Factor Water Risk into Relevant Business Decisions

Only 23 companies (19%) explain how they incorporate water information into strategic business decisions. Some have policies and programs to integrate water issues into their overall management practices. For instance, Nike has a life-cycle matrix to identify water-conservation and water-quality measures that are or can be implemented for each life-cycle stage including: product creation; materials; manufacturing process; delivery, packaging and logistics; consumer end of life; and corporate operations. Alcan created a company-wide management system that takes a systematic approach to managing water resources, with a focus on resource efficiency, recycling, and reuse.

Others describe using water information for specific strategic decisions, including new site selection/evaluation, financial decisions, and new product/process development. As a part of its site development policy, Target Corporation specifies that the developer or contractor identify up to 10 LEED points (which are awarded for sustainability measures implemented in the context of ranking "green" building design and construction) from Sustainable Site and Water Efficiency categories as defined by the United States Green Building Council. SCA uses a risk assessment program to look at hydrologic considerations and water emissions when it evaluates acquisitions.

BMW conducted a Sustainable Value project, which calculated the savings from water conservation measured by profit per cubic meter of water use and compared it against 16 other automobile manufacturers. PepsiCo describes a Capital Expenditure Filter that ensures that sustainability issues, including water, are formally considered in all major capital expenditure proposals. Kirin, a Japanese beer brewer, uses an environmental accounting program to measure the impact of water use in business and the financial benefits derived from water-conservation projects. The pharmaceuticals manufacturer Merck is developing a methodology to assign a cost structure to water that reflects its true value. Merck and several others also mention the connection between water use and energy and plan to factor water information in energy decisions.

Several companies consider water risks when developing new products. Abbott, which produces pharmaceutical and nutritional products, and Unilever describe programs to develop products that require less water not only during production, but during the consumer use phase. Proctor & Gamble (P&G) recognizes the global problem with safe drinking water supply and has developed point-of-use water purification products, though they have run into problems marketing them commercially.

Bayer selected global water demand and climate change as issues to consider when developing products, and they describe their efforts to attain specific sustainability advantages by doing so.

Report Water Performance

One hundred and five companies (87%) publish quantitative water performance data (either water use or wastewater-related performance) in their reports.[8] While most companies (88, 73%) report water use over time, the performance reports are far from standardized. Among the 103 companies that measure water use, 91 (88%) report specific results. However, the companies use a variety of different terms and measures. For example, the following terms were used to refer to the use of water: water use, water consumption, water intake, freshwater withdrawal, specific water consumption, or total water withdrawal. Some only report a certain portion of their water use: process water use, industrial water use, manufacturing water use, process water input, industrial water input, or water used for primary activities. Others include the use of seawater.

Some companies provide specific definition of the terms or measurements they use, but the majority does not. According to the Global Reporting Initiative (GRI) Water Protocol,[9] which provides measurement guidelines for its water indicators, total water use is represented by three figures: total water withdrawal, storage, and consumption.[10] Total water consumption is the total amount of water consumed (i.e., not stored or discharged) by the organization. The Water Protocol recommends reporting all three of these figures for total water use, but most companies do not follow these recommendations.

Indicators Used for Reporting

Major indicators used to report water performance are: water use (91 companies, 75%); wastewater discharge quality (54 companies, 45%); wastewater discharge volume (40 companies, 33%); and water recycling (13 companies, 11%) (Fig. 2.5). Most companies reporting on wastewater discharge used only two of the five indicators recommended in the GRI Water Protocol: biological oxygen demand (BOD) and chemical oxygen demand (COD). Occasionally, companies report the three additional GRI-recommended indicators: total suspended solid (TSS), nitrate/nitrogen, and phosphorous concentrations or discharges.[11]

The popularity of the GRI Sustainability Reporting Guidelines may have contributed to the higher number of water use and wastewater reports. Ninety-one (75%) companies mention that they used GRI's Guidelines as a reference to develop their report. Among those, 78 include the GRI Index or reference table that shows which indicators are found on which pages of their report. The 2002 GRI Guidelines have six

---

8. Companies that state they improved water efficiency, but do not provide any supporting data, are not included in this number.

9. Global Reporting Initiative. Water Protocol—for use with the GRI 2002 Sustainability Reporting Guidelines. 2003.

10. Total water withdrawal is the sum of all water drawn into the facility from all sources for any uses over the course of the reporting period. Storage is the amount of water withdrawn and put into storage, minus any water taken out of storage for use by the facility.

11. Companies sometimes report discharge of other constituents, including metals (such as lead, copper, zinc, etc.), absorbable organic halogens (AOX), and hydrocarbons.

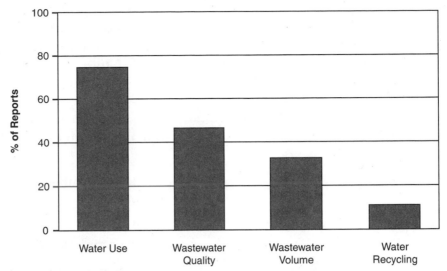

FIGURE 2.5 USE OF WATER PERFORMANCE INDICATORS.

water indicators (Box 2.2)[12]—two core indicators and four additional indicators. Core indicators are those identified to be relevant to most stakeholders, and companies are required to report core indicators to be "in accordance" with the GRI Guideline.[13] The two core indicators ask companies to report total water use and significant discharges to water by type.

Additional indicators represent emerging practices, or address topics that may be material to some organizations but not generally for a majority, and are considered optional. Some companies also use other indicators to report water performance, including rainwater use, water conservation contribution (i.e., estimated amount of groundwater replenished through the company's water cultivation projects), seawater use, water loss (percentage or amount of water lost during the distribution process reported by drinking water service providers), percentage of suppliers in compliance with the local or company's standard, and percentage of suppliers that have water-efficiency programs.

How the Data Are Reported

In addition to reporting a range of indicators, companies often use different units of measure and data formats relating to water. For example, most European companies use metric units, whereas many U.S. companies are still using non-metric units. Over 70% of the reporters present their water performance in absolute values (e.g., 20 billion gallons of water used, 8,000 tons of COD released into water in 2005). Forty percent present the data in normalized form, such as water use per unit of production or economic value produced. Because the companies choose a normalization method that can provide the best insight to their businesses, the data vary widely from

---

12. The most current version of the GRI Guideline released in October 2006 (G3) has five water indicators. One additional indicator in the 2002 version—EN 32 Water sources and related ecosystems/habitats significantly affected by discharges of water and runoff—was deleted in the latest version. However, all the reports reviewed in this study were published before the release of the G3, and it is assumed that 2002 version was used to develop the reports.

13. The "In accordance" system was developed for the 2002 version of the GRI Guideline. The G3 no longer uses "in accordance," and instead rates the level of GRI application from A to C.

BOX 2.2 Water-Related Performance Indicators in Global Reporting Initiative's Sustainability Reporting Guidelines

Performance Indicators in 2002 Version	Performance Indicators in 2006 Version (G3)
EN5. Total water use (Core)	**EN8 Total water withdrawal by source (Core)**
EN12. Significant discharges to water by type (Core). See GRI Water Protocol.	**EN21 Total water discharge by quality and destination (Core).**
EN20. Water sources and related ecosystems/habitats significantly affected by use of water (Additional). Include Ramsar-listed wetlands and the overall contribution to resulting environmental trends.	**EN9 Water sources significantly affected by withdrawal of water (Additional).**
EN21. Annual withdrawals of ground and surface water as a percent of annual renewable quantity of water available from the sources (Additional). Breakdown by region.	**This indicator was deleted in G3**
EN22. Total recycling and reuse of water (Additional). Include wastewater and other used water (e.g., cooling water).	**EN10 Percentage and total volume of water recycled and reused (Additional).**
EN32. Water sources and related ecosystems/habitats significantly affected by discharges of water and runoff (Additional). Include Ramsar-listed wetlands and the overall contribution to resulting environmental trends. See GRI Water Protocol.	**EN25 Identity, size, protected status, and biodiversity value of water bodies and related habitats significantly affected by the reporting organization's discharges of water and runoff (Additional).**

company to company, and from sector to sector. For instance, companies and sectors whose output can be well measured by weight or volume of product (such as beverages, metal/mining, and forest products) typically report water consumed (and other indicators) per liter or ton or cubic meter of products. Industries that have a distinctive unit of product typically use that unit, such as water consumed per pair of shoes or automobiles produced. Others report water use per unit sales (such as gallons per dollar). Just over 25% of companies publish water-related data using both absolute and normalized values.

A majority of the companies studied (88, or 84% of companies that report performance) include water data for more than 1 year, typically in the range of 3 to 5 years, to present trends. They often use charts to show the trend and performance against targets.

Twenty-one companies (17%) provide data on the amount or percentage of freshwater withdrawn from various sources (e.g., surface water, groundwater, or municipal water supply), a recommended practice according to GRI's most recent reporting guidelines. Twenty-seven companies (22%) publish water performance data by region or at the facility level. The majority of these companies are either from the forest products or metal/mining sectors, which tend to have a relatively small number of large facilities.

Other examples of data reported include wastewater discharge volume by destination (surface, ocean, ground, municipal treatment); water consumption by purpose (agriculture, cooling, mining, potable, process, steam, etc.); wastewater volume by source (process, cooling, production, group discharge, drilling discharge); water consumption by products (metal, aluminum, carbon, steel, stainless steel, coal, petroleum, diamonds). A few companies report water use by product lifecycle. For instance, Fujitsu (a multinational computer manufacturer headquartered in Japan) reports water consumption per personal computer (notebook and desktop) for four major lifecycle stages: manufacturing, distribution/sales, usage, collection/recycling. Nike reports breakdown of water use by three lifecycle stages: corporate level, manufacturing, and textile production stage.

Form Strategic Partnerships

Nearly all of the companies reviewed (112, 93%) report partnership programs with various stakeholder groups such as local communities, governments, NGOs, industry associations, and universities. Among those, 30 companies mention partnerships specifically in the area of water management. These programs fit into one of three types:

- Funding to groups or NGOs that are working on water issues, such as access to safe drinking water and watershed protection;

- Collaborating and sharing information with other companies through industry associations or working groups created by organizations such as the United Nations, World Business Council for Sustainable Development (WBCSD), and Business for Social Responsibility (BSR); or

- Partnering with local communities and municipal governments to work on specific problems or solutions, such as co-development of water-related infrastructure with a municipal authority.

Only nine companies—four beverage companies and others from the food, biotech, chemical, and metal sectors—also report programs or projects to improve access to safe drinking water in local communities and worldwide. For example, The Coca Cola Company, which has come under pressure in some countries to improve community efforts and reduce the impacts of their operations, described efforts to establish approximately 20 community and watershed partnerships with local and international bodies to help provide access to potable water and sanitation in communities where the company operates.

Most of the companies report water partnerships in case-study format, but few clearly state a commitment to use partnerships in water management. Abbott's water statement specifically mentions a policy to "engage with other water users and providers to promote appropriate water management principles and address challenges." Intel Corporation adopted a new water conservation strategy that focuses not only on its internal efforts but also on how it can share its expertise and learning with other businesses; promote water conservation education and awareness in its local communities; and collaborate with universities, water suppliers, governments, and water users to solve pressing regional water challenges.

Commit to Continuous Improvement

Ninety-eight companies (81%) mention their commitment to continuous improvement or have an environmental management system (EMS) in place. However, only 13 of them have policies or case studies that demonstrate such a commitment specifically in the area of water management. This is another area where we strongly recommended improvement—formal environmental management systems for critical resources such as energy and water will have to be adopted in coming years by all major companies (Morrison and Gleick 2004, Morikawa et al. 2007). Some of the reports reviewed mention the use of their ISO 14001 EMS for addressing water issues, whereas others have a stated corporate policy or goal to strive for continuous improvement in the management of water resources. GM states it uses its ISO 14001-certified EMS for its water management efforts, and Ford has a three percent year-over-year water-use reduction goal. Abbott describes a policy to continuously improve its water-use efficiency and water-discharge quality within its worldwide manufacturing operations. ExxonMobil says it "continually seeks ways to reduce water use and preserve water quality, through the design and operation of our facilities, recycling and reuse and aggressive measures to prevent water pollution."

Water Reporting Trends by Sector

Our analysis shows that there are distinctive sector-specific themes and characteristics in water reporting. Some sectors provide very detailed information on water use, whereas others pay more attention to the environmental impacts of their wastewater discharge. Some sectors tend to report water use in absolute values, and others provide the data in normalized values. Some have detailed water policies, while others just present performance data. These sector-specific variations are due to:

- Differences in how water resources are used (e.g., some consume water as main ingredient while others use water mainly for cooling purposes);

- Size and location of individual operations (e.g., some industries have a large number of small facilities while others operate a few very large facilities; some are located in industrial nations while others operate in developing countries); and

- Social and political environment (e.g., history of water-related activism or legal reporting requirements for certain substances).

A summary of the main water-related parameters reported by industry sector is presented in Table 2.2.[14] Below are the main findings from the sector-based analysis.

Water reporting is inconsistent across industrial sectors. While each industry sector tends to focus on different aspects of water reporting, the metal/mining industry provides the most comprehensive water reports (69% of the water-related information we tracked). By contrast, the apparel manufacturing sector provides the least comprehensive reports (18% of the water-related information we tracked). On average, the sectors studied provide 54% of the water-related information we tracked.

Industry sectors that use water as a main ingredient of their product or require high-quality water for production tend to undertake water reporting more comprehensively than other water-intensive industries. Our analysis found that a relatively high percentage of the beverage, biotech/pharmaceutical, and food sector businesses provide information on water policy/management and water-related performance metrics.

The high-technology/electronics, metal/mining, and utilities sectors have 100% reporting of the three parameters we use to measure interactions with external stakeholders and interested parties. All of the companies we reviewed in these three sectors provide at least some information on such activities: engage supply chain, consult with stakeholders, and form strategic partnerships. The beverage and food industries have, by far, the highest percentage of companies that report stakeholder interactions specifically related to water.

The metal/mining and forest products industries have the highest percentage of companies that include facility-level/regional water reporting. This may be explained by the fact that these sectors have relatively fewer but larger facilities, as well as a relatively high focus on wastewater issues.

Water reporting methods and report contents are inconsistent from company to company, even within the same industry sector. Although there are several industry-specific water reporting themes and characteristics, companies in the same industry often use different indicators and definitions to report their water performance, making water performance comparison and benchmarking difficult, even within sectors. Some of the sectors examined in this study have either developed their own reporting guidelines[15] or have GRI Reporting Guidelines sector supplements.[16] However, only a fraction of the large companies in these industries consistently apply these standards/guidelines.

---

14. Morikawa et al. (2007) selected and reviewed the 10–15 largest companies for each of the 11 industry sectors. In some sectors, not all the companies reviewed publish environmental/CSR reports. For instance, the study reviewed companies from the apparel/textile sector, but only six of them publish reports. When calculating the reporting percentage, we used for a denominator the number of companies publishing reports, as opposed to the number of companies reviewed for each sector.

15. Such as Oil and Gas Industry Guidance on Voluntary Sustainability Reporting and the chemical industry's Responsible Care Reporting.

16. The automobile and metal and mining industries have GRI sector supplements. GRI is also developing sector supplements for apparel/footwear and energy utilities.

TABLE 2.2 Reporting Rate (as Percent of all Reports in Each Sector) for Main Water-Related Measures

	Apparel	Automobile	Beverage	Biotech/ Pharma	Chemical	Food	Forest Products	HighTech/ Electronics	Metal and Mining	Refining	Utility	Average*
Measure Water Use	17	87	100	92	67	100	90	87	100	67	100	85
Assess Water Risks	0	7	50	15	0	43	30	13	40	33	14	21
Consult Stakeholders	33	87	90	85	89	71	70	100	100	92	100	87
Engage Supply Chain	50	87	90	77	78	86	80	100	100	92	100	88
Water Statement/ Policy	33	40	70	69	11	86	70	40	80	58	36	53
Water Goals and Targets	17	47	80	69	44	71	60	60	60	67	50	58
Quantitative Target	0	33	40	46	33	29	30	47	40	8	21	31
Target Water Use	0	47	80	38	33	71	30	47	60	67	50	49
Target Wastewater	17	20	10	23	22	43	50	20	30	58	14	27
Best Available Technology	0	33	50	46	11	57	70	33	70	58	64	46
Water Risk in Decision Making	33	27	30	31	11	14	10	7	30	17	7	19
Measure and Report Performance	17	87	90	92	89	100	100	87	100	75	93	87
Report Freshwater Use	0	87	80	92	67	100	70	80	80	58	79	75

Parameter												
Report Wastewater Quality	0	40	30	62	67	29	90	33	70	42	21	45
Report Wastewater Volume	17	33	30	23	22	29	50	47	50	33	21	33
Report Water Recycling	0	13	0	15	0	0	10	13	40	8	7	11
Report in Absolute Value	0	87	50	92	89	86	70	60	100	67	79	74
Report in Normalized Value	17	47	80	69	33	43	40	33	50	8	14	40
Report Both Normalized and Absolute	0	40	50	69	33	29	20	13	10	8	7	26
Trends Reporting	0	67	80	69	89	100	90	60	100	58	79	73
Regional/facility-based reporting	0	20	20	8	11	14	50	20	60	8	29	22
Use GRI	50	73	80	77	56	57	70	80	100	75	86	75
Strategic Partnership	83	93	80	85	89	86	100	100	100	92	100	93
Continuous Improvement	50	93	80	77	89	71	80	80	80	83	86	81
Average	18	54	60	59	47	59	60	53	69	51	52	54

Key: Italic: *Reporting rate = 0–25%*; Bold: **Reporting rate = 75–100%**

*This average was calculated as a percentage of companies among 121 reporting companies that publish information on each parameter.

**This number is the average of the reporting rate of 24 parameters listed in this table.

Conclusions and Recommendations

Global and regional water issues are increasingly affecting corporate risk, production, and decisions. For many companies and industrial sectors, the availability of reliable and clean water is vital for operations; for others, regional water problems are causing local communities to assess—and sometimes protest—local industrial water use and discharges. As a result, corporations have a growing need to understand and evaluate water issues, and stakeholders are calling on corporations to report on water use consistently and transparently. To meet these challenges, an increasing number of companies are expanding their annual or periodic reports to include information on water.

These reports, however, vary enormously in content, quality, detail, and format. The assessment presented summarizes our effort to understand the strengths and weaknesses of corporate water reporting and to call attention to how these reports can be made more effective and valuable to corporations and their stakeholders affected by industrial water management and use.

The review of corporate non-financial reports found that most of the reports reviewed (97%) provide some form of information on water performance or water-management practices and policies. However, the examination of the reports revealed several important gaps and inconsistencies in corporate water reporting.

Lack of context in water reporting is a problem. Most of the reports provide some water data, but they do not provide context to these numbers. Only about half of the reports offer information on the company's water policies or a description of its water-management objectives. Water performance—such as water use or wastewater discharge volume and load—and its impact on local environment and communities vary greatly according to the type of business and the water landscape where their facilities operate. Without the information on company- or industry-specific water challenges and associated water policies and management objectives, the readers of these reports can not fully understand or interpret the performance data presented.

Measurement methods and definitions are often inconsistent. The scope and range of water information and the level of detail presented in the reports differ greatly from company to company, even within the same sector. Moreover, different kinds of methods, scopes, boundaries, and units are used to report the same parameter (such as "water use"), often without an explicit definition of what they are reporting. Even for companies using the GRI Guidelines, there is no guarantee or confirmation that the measurement and reporting methods described in the GRI Water Protocol are being applied. Although the contents of the reports must reflect the needs of their audiences and stakeholders, further harmonization of core information is needed to improve the comparability and usability of the information, as well as the credibility and transparency of the corporate sector itself.

Information on the water landscape and associated risks is rarely reported. Only about 20% of the reporting companies mention their water risks and challenges or describe their programs to assess water risks. Businesses that are heavily dependent on water and have actually experienced problems such as conflict with local communities and disruptions of water supply are starting to see the importance of assessing and managing water-related risks. However, a review of their reports shows that few companies are managing water issues strategically. In addition, the water risk information that is available is mostly high level and qualitative and contains limited data on the type and level of the risks.

Quantitative targets are not widely published. Only about half of the companies that have water performance goals publish quantitative targets, and these often do not cover all the indicators they report. Quantitative targets are especially helpful for companies and stakeholders to evaluate water performance and improvements over time.

Gaps in water performance reporting make comparisons difficult. While total water use is published in most of the reports, important water-performance indicators are often not reported. Only 22% of the reports include regional or facility-level water performance information. Only 10% of the reports reviewed here mention supply-chain considerations in relation to water management, and no company reports on the actual water use or wastewater quality/quantity of their suppliers. Companies also rarely report on water recycling and reuse, though this is an increasingly important component of sustainable water management and use in the industrial sector.

Recommendations

In summary, our review of corporate reporting provides a snapshot of the current status of water reporting and offers a foundation for more detailed analyses of industry sector-specific water risks and water-management practices. We also present specific recommendations for corporate reporting and behavior in the drive toward more sustainable water management and use. Water information presented in corporate non-financial reports does not and cannot present a comprehensive and detailed picture of a company's water performance. Rather, it is a reflection of what a company finds material and critical for itself and for its stakeholders. Nevertheless, given stakeholders' increasing demands for more transparency and disclosure, growing water risks, and the introduction of the new GRI Guidelines,[17] the contents of corporate water reporting are likely to keep evolving. We offer the following suggestions to move this process forward more consistently and rapidly.

- **Go beyond water use reporting.**

    While an increasing number of companies are publishing their water performance information, it is often limited to annual total water use volume at a corporate level. The complex and multi-dimensional nature of water issues that affect companies and surrounding communities in different ways and levels, depending on the locations and type of business activities, requires more comprehensive water reporting, which includes the following:

    - Information on corporate water policy and objectives that explain how water issues affect business and how the company manages risks associated with it;

    - Clear and quantitative goals and targets on water performance, as well as performance data over time, which will enable senior management, employees, and stakeholders to evaluate progress and challenges;

---

17. The two core water indicators are slightly modified in the G3, providing more clarity on the definition of the indicators and how they should be reported. For instance, "EN5 Total water use" is revised to "Total water withdrawal by source (EN8)" and "EN12 Significant discharges to water by type" is modified to "Total water discharge by quality and destination (EN21)."

- Regional, local, and facility-level data that helps companies and local communities assess potential risks associated with water use and wastewater discharge;

- Explanation/breakdown of total numbers, such as by business unit or by purpose (e.g., heating, cooling, cleaning etc.) by water withdrawal source (ground, surface, municipal water etc.), discharge destination, percentage of recycled water;

- Information on water use and wastewater discharge "beyond company's fence line," looking at the water data for product/service's entire life cycle, such as water performance of suppliers, water use, and wastewater discharge during product use and disposal; and

- Assessment of the water situation facing any business sector, including overall risks and impacts on business, local environment/ecosystems and communities that will provide context for the quantitative water use and wastewater information, and help companies and stakeholders better understand and manage water-related challenges.

- **Harmonize definitions and reporting formats**

  While the contents and format of water reporting should reflect what is "material" to the company and its stakeholders, harmonization on how to report core information is needed to enable benchmarking and evaluation by companies and stakeholders. Companies should consider using accepted reporting standards such as GRI Sustainability Reporting Guideline, industry-specific reporting standard or facility-level reporting standard.[18] Companies should also clearly indicate the definition of indicators used, and methods and scope of measurement to allow more accurate and objective comparison of data published in corporate reports.

  Companies should commit to continuous improvement in assessing and managing water risks and lessening impacts of the company's water use on local communities and the environment. Doing so can help protect operations from unexpected water-related business disruptions. Such a commitment should be in written form and can be a stand-alone statement or part of an organization's overall environmental policy.

REFERENCES

Baue, W. 2004. A brief history of sustainability reporting: Trends in sustainability reporting. Article written for SocialFunds.com. July 2004.

Bruntland, G. (ed.). 1987. *Our Common Future: The World Commission on Environment and Development*. New York: Oxford University Press.

Morikawa, M., Morrison, J., Gleick, P.H. 2007. *Corporate Reporting on Water: A Review of Eleven Global Industries*. Oakland, California: Pacific Institute.

Morrison, J., and Gleick, P.H. 2004. *Freshwater Resources: Managing the Risks Facing the Private Sector*. Oakland, California: Pacific Institute.

Tepper Marlin, A., and Tepper Marlin, J. 2003. A brief history of social reporting. *Business Respect* 51.

---

18. The Facility Reporting Project Sustainability Reporting Guidance (2005). The guidance is aligned with the GRI Sustainability Reporting Guidelines.

CHAPTER 3

Water Management in a Changing Climate

Heather Cooley

Rising greenhouse gas concentrations from human activities are causing large-scale changes to the Earth's climate system. Because water is a fundamental element of our climate system, these changes will have important implications for the hydrologic cycle, including impacts on water availability, timing, quality, or demand. The water sector, however, is already faced with a variety of challenges, from meeting basic needs to providing for population and economic growth. Climatic change will make meeting these challenges even more difficult.

Early research on climate change and water was used to justify the implementation of various mitigation strategies. Given that some degree of climate change is unavoidable (and mitigation activities have been largely unsuccessful thus far), the discussion about climate change has shifted to include adaptation – actions or policies that reduce vulnerability or increase resilience to inevitable climate change impacts. As noted by Frederick and Gleick (1999), "The socioeconomic impacts of floods, droughts, and climate and non-climate factors affecting the supply and demand for water will depend in large part on how society adapts." Adaptation can not only serve to minimize our vulnerability to climate change impacts, it can also reduce our vulnerability to current climate variability and promote sustainable development.

This chapter will examine the projected impacts of climate change on water resources. It will discuss various adaptation practices, as well as the process of evaluating and implementing those practices, including issues surrounding financing and equity.

The Climate Is Already Changing

The Intergovernmental Panel on Climate Change (IPCC), established in 1988 by the World Meteorological Organization and the United Nations Environment Programme, draws upon scientific and socio-economic literature to produce an assessment of the potential impacts of human-induced climate change, as well as ways to adapt to and mitigate these impacts. In 2007, the IPCC released its fourth, and most strongly worded, assessment, citing increasing evidence for large-scale changes in the Earth's climate system: "Warming of the climate system is *unequivocal*, as is now evident from observations of increases in global average air and ocean temperatures, widespread melting of snow and ice, and rising global average sea level" (emphasis added) (Bernstein et al. 2007). According to the IPCC Fourth Assessment Report, the global mean surface temperatures have increased by 0.74°C (± 0.18°C) over the past 100 years. Additionally, warming rates are accelerating: the warming rate over the past 50 years (about 0.13°C per decade) was nearly twice as fast as over the past 100 years (about 0.07°C per decade). And, 11 of the 12 warmest years since 1850 occurred between 1995 and 2006 (Trenberth et al. 2007), providing further evidence that the rate of warming is increasing.

Precipitation patterns are also changing. According to the IPCC, observations show that there have been pronounced changes in the amount, intensity, frequency, and type of precipitation over the past 100 years. Between 1900 and 2005, precipitation increased over land north of 30°N, with the notable exception of northwest Mexico and southwest USA. In the tropics, precipitation has declined over the past 30 years. Observed changes in the southern hemisphere are highly variable and are likely associated with changes in monsoon features. Changes in precipitation, where significant, are consistent with observed changes in streamflow (Trenberth et al. 2007).

Extreme hydrologic events, such as droughts, hurricanes, and floods also appear to be becoming more common. While the majority of studies on climate change have emphasized changes in average conditions, scientists are increasingly investigating the impacts of climate change on hydrologic extremes. Observations indicate that there has been a likely increase in the number of heavy precipitation events, even in areas where total precipitation has declined. There has also been an observed increase over the past 30 years in the intensity, duration, and spatial extent of droughts, associated with higher temperatures, warmer sea surface temperatures, less precipitation, and a diminishing snowpack (Trenberth et al. 2007).

Projected Impacts of Rising Greenhouse Gas Concentrations

Scientists evaluate various greenhouse gas emissions scenarios using a suite of climate models (see Box 3.1). Current climate models project that rising greenhouse gas concentrations will "likely" increase global mean surface air temperature between 1.1 and 6.4°C by 2090–2099 relative to a 1980–1999 baseline (Table 3.1).[1] The models project greater warming over land and at high northern latitudes and lesser warming over the southern oceans and North Atlantic. Heat waves are "very likely" to increase in intensity, frequency, and spatial extent, whereas cold waves are projected to decrease significantly. Diurnal temperature ranges are expected to decrease as daily minimum temperatures are expected to increase faster than daily maximum temperatures (Meehl et al. 2007).

---

1. Terms such as "likely" and "very likely" have specific meaning associated with the expected probability of occurrence, given current knowledge. A "likely" outcome has more than a 66% probability of occurrence. A "very likely" outcome has more than a 90% probability of occurrence.

Box 3.1 Future Scenarios of Climate Change; Future Scenarios of World Demographics

Scientists use general circulation models (GCMs) to understand the potential impacts of various greenhouse gas emission scenarios on future climate conditions. As described by the IPCC, "Climate models are mathematical representations of the climate system, expressed as computer codes and run on powerful computers... model fundamentals are based on established physical laws, such as conservation of mass, energy and momentum, along with a wealth of observations" (Randall et al. 2007). Over time, these models have evolved to include better and more detailed representations of the land surface, ocean, atmosphere, and their connections. The GCMs are tested via comparisons of various simulations with features of the recent climate and the historical climate record. The Fourth Assessment Report includes results from 23 GCMs that output hourly temperature, precipitation, and other climatic data at a spatial grid of 125–400 km. In some regions, downscaling has been applied to GCM output to obtain estimates at a smaller spatial scale.

The impacts of climate change will ultimately depend on future greenhouse gas (GHG) concentrations. Future GHG emissions remain uncertain and are influenced by a variety of demographic, socio-economic, and technological factors. Scenarios can be a useful tool for examining how changes in these driving factors affect greenhouse gas concentrations. These scenarios can be useful for evaluating impacts associated with climate change as well as assessing adaptation and mitigation activities. The Special Report on Emissions Scenarios (SRES) outlines four storylines that differ according to demographics, social, economic, environmental, and technological factors and lead to different levels of greenhouse gas emissions. Each storyline has a number of different scenarios, referred to as a family. A total of 40 scenarios have been developed.

The four storylines are described below:

- The **A1** storyline is characterized by "a future world of very rapid economic growth, global population that peaks in mid-century and declines thereafter, and the rapid introduction of new and more efficient technologies. Major underlying themes are convergence among regions, capacity building, and increased cultural and social interactions, with a substantial reduction in regional differences in per capita income" (IPCC 2000) The A1 family is further divided into three subgroups that are differentiated according to energy source: fossil intensive (**A1FI**), non-fossil sources (**A1T**), and a mix of fossil and non-fossil sources (**A1B**).

continues

Box 3.1 *continued*

- The **A2** storyline is characterized by "self-reliance and preservation of local identities" (IPCC 2000). Population is expected to continuously increase, but economic growth and technological development are expected to be slow.

- The **B1** storyline has the same population projections as the A1 storyline but "rapid changes in economic structures toward a service and information economy, with reductions in material intensity, and the introduction of clean and resource-efficient technologies" (IPCC 2000).

- The **B2** storyline is characterized by "a world with continuously increasing global population at a rate lower than A2, intermediate levels of economic development, and less rapid and more diverse technological change than in the B1 and A1 storylines" (IPCC 2000).

The AIB scenario has high greenhouse gas emissions associated with high population growth, income convergence, a market-oriented economy, and a balanced mix of energy sources (both fossil and non-fossil fuels). By contrast, the B1 scenario has lower greenhouse gas emissions due to slower economic growth, an emphasis on economic, social, and environmental sustainability, and clean, resource efficient technologies.

TABLE 3.1   Projected Global Averaged Surface Warming for Selected IPCC Scenarios at the End of the 21st Century

	Approximate CO_2-eq Concentration (ppm)	Temperature Change (°C at 2090–2099 relative to 1980–1999)	
		Best Estimate	"Likely" Change
Constant year 2000 Concentration		0.6	0.3–0.9
B1 Scenario	600	1.8	1.1–2.9
A1T Scenario	700	2.4	1.4–3.8
B2 Scenario	800	2.4	1.4–3.8
A1B Scenario	850	2.8	1.7–4.4
A2 Scenario	1250	3.4	2.0–5.4
A1F1 Scenario	1550	4.0	2.4–6.4

See Box 3.1 for a general description of the IPCC scenarios.
Source: IPCC 2007a.

Warmer temperatures are expected to intensify the hydrologic cycle. In general, current climate models project that global precipitation will increase, although this is subject to significant spatial and temporal variability, with an increase in the tropics and high latitude and decrease in the subtropics and mid-latitude (Meehl et al. 2007). Furthermore, hydrologic studies generally agree that warming will lead to changes in the timing of runoff, with earlier peak flows, greater winter flows, and lower summer flows.

Climate Change and Water Resources

Changing climate conditions will affect the supply of and demand for water resources. Indeed, all of the IPCC reports conclude that freshwater systems are among the most vulnerable sectors. The Fourth Assessment Report notes that climate change will lead to "changes in all components of the freshwater system" (Kundzewicz et al. 2007). This section examines the impacts of climate change for the major components of our freshwater systems, i.e., surface water, groundwater, water quality, hydrologic extreme events, and water demand.

Surface Water

Climate change is expected to alter the quantity and timing of runoff. There is general agreement among models for greater runoff in the tropics and high latitudes and reduced runoff in some dry regions at mid-latitudes (the Mediterranean, southern Africa, and western USA/northern Mexico) and in the dry tropics. In snow-dominated basins, hydrologic studies have long agreed that warming will lead to changes in the timing of runoff, with earlier peak flows, greater winter flows, and lower summer flows. These effects are more pronounced in basins at or near the current snowline. In rain-dominated basins, studies suggest that changes in precipitation will have a greater effect on flows than warming temperatures and lead to greater winter flows and lower summer flows, but no change in timing.

Groundwater

Because our understanding of groundwater and its uses is limited, the potential impacts associated with climate change are less well understood than for surface hydrology. According to the IPCC, demand for groundwater may increase as a means of offsetting reduced surface water flows in some regions. Climate change will affect groundwater recharge rates, but these effects are site-specific, with some areas experiencing an increase and others experiencing a decrease. Higher evaporation rates will likely lead to salinization of groundwater. Sea level rise will likely lead to greater saltwater intrusion in coastal aquifers (Kundzewicz et al. 2007).

Hydrologic Extremes

Climate models suggest that warmer temperatures will very likely lead to greater climate variability and an increase in the risk of hydrologic extremes. The frequency and intensity of floods and droughts are expected to increase, e.g., a 1-in-100 year event could become a 1-in-10 year event. Mid-continental regions are expected to dry during the summer, thereby increasing the risk of drought. All areas are expected to see an increase in the intensity of precipitation events. In areas that are projected to dry, the increase in intensity of rainfall may still not produce more overall water because of an expected reduction in the frequency of precipitation events (Meehl et al. 2007). Our vulnerability to these events, however, depends on a range of socio-economic and political factors, such as development patterns and poverty. The World Meteorological Organization (2004) maintains that "hydrological uncertainty is perhaps subordinate to social, economic, and political uncertainties. For example, the biggest and unpredictable changes are expected to result from population growth and economic activity" (WMO 2004).

Water Quality

The connections between climate change and water quality are less well understood than impacts on quantity, although the literature on these connections is growing. Climate change is expected to increase water temperatures in lakes, reservoirs, and rivers, leading to more algal and bacterial blooms, which in turn lead to lower dissolved oxygen concentrations. More intense precipitation events could increase erosions rates and wash more pollutants and toxins into waterways. And in coastal systems, rising sea levels could push salt water further into rivers, deltas, and coastal aquifers, threatening the quality and reliability of these systems.

Water Demand

The effects of climate change on water demands are far less studied than impacts on hydrology. Yet, we know that water demands in some sectors are sensitive to climate, particularly agriculture and urban landscapes. Because agriculture accounts for 70–80% of global water use, demand changes in this sector may have broad implications for overall water availability. In some urban areas, lawns have become an increasingly important consumer of water, accounting for up to 70% of total residential water use in some hot, dry areas. These uses typically require more water as temperatures rise, although higher atmospheric CO_2 concentrations can reduce water requirements under some conditions. Warmer temperatures will also increase cooling water requirements, although this demand could be mitigated by efficiency improvements. Because of the uncertainty associated with water demand, net effects must be evaluated on a regional level.

Vulnerability to Climate Change

Many regions and sectors are vulnerable to current climate conditions. Among the most serious global problems is the failure to meet basic human needs for safe water and sanitation, addressed many times in The World's Water series, and in Chapter 4 of this volume. In 2002, an estimated 1.1 billion people lacked access to an improved water supply, and more than 2.6 billion people lacked access to improved sanitation (WHO/UNICEF 2000, 2004). These inadequacies have direct implications for human health, with large numbers of water-related diseases and deaths from diarrheal diseases, cholera, typhoid fever, and more (see Chapter 4 on the Millennium Development Goals for water). Because water is a key input for most economic and social activities, water scarcity also threatens economic development and political stability. Climate change, along with continued population growth, urbanization, and increasing competition among users, will make meeting basic water needs even more challenging over the next century.

Vulnerability to climate change will vary across regions, sectors, and socio-economic groups (Olmos 2001). The degree of vulnerability is a function of the magnitude of the impact, the sensitivity of the system to that impact, and the system's ability to adapt. In general, developing countries are most vulnerable to climate change impacts because:

> "they are more exposed by virtue of being at lower latitudes, where impacts such as increased disease and extreme heat and drought will be more pronounced, and because they derive a larger proportion of their economic output from climate-sensitive sectors such as agriculture, fishing, and tourism. In addition, developing countries generally have lower per capita incomes, weaker institutions, and less access to technology, credit and international markets—hence, lower adaptive capacity" (Burton et al. 2006)

Vulnerability to climate change, however, is not limited to developing countries. In many parts of the developed world, water managers have built vast networks of reservoirs, canals, and pipes to deliver water where and when it is needed. These systems are designed and operated based on historic climate conditions – a concept referred to as *stationarity*. As greenhouse gas concentrations continue to rise and the effects of climate change become more pronounced, however, the assumption that future water supply and availability will look much like the past is no longer valid. Relying on historic, stationary climatic conditions may result in poorly sited, inappropriately designed, and improperly operated water systems. The IPCC notes that "The current procedures for designing water-related infrastructures therefore have to be revised. Otherwise, systems would be over- or under-designed, resulting in either excessive costs or poor performance" (Kundzewicz et al. 2007).

Adaptation

Because some degree of climate change is unavoidable and all regions and sectors are vulnerable to climate change impacts, adaptation must be a central element of climate change policy. The IPCC defines adaptation as "initiatives and measures to reduce the vulnerability of natural and human systems against actual and expected climate change effects."

There are a variety of adaptation options for the various water-use sectors (Table 3.2). These options range from building or expanding reservoir capacity to improving water-use efficiency. Adaptation can take many forms, e.g., proactive or reactive, autonomous or planned, structural or nonstructural, supply-side or demand-side, and more.

Rigid, expensive, and irreversible actions in climate-sensitive areas can increase vulnerability and long-term costs. Given the uncertainty associated with climate change, some planners support policies that provide social, economic, and environmental benefits regardless of climate change impacts – referred to as "no regret" policies.

TABLE 3.2   Examples of Supply-side and Demand-side Adaptation Options for
Various Water-Use Sectors

Water-Use Sector	Supply-Side Measure	Demand-Side Measure
Municipal water supply	Increase reservoir capacity	Incentives to use less (e.g., through pricing or rebates)
	Extract more water from rivers or groundwater	Legally enforceable water use standards (e.g., for appliances)
	Alter system operating rules	Increase use of grey water
	Inter-basin water transfer	Reduce leakage
	Capture more rain water	Increase use of recycled water
	Desalination	Development of non-water-based sanitation systems
	Seasonal forecasting	
Irrigation	Increase irrigation source capacity	Increase irrigation-use efficiency
		Increase use of drought-tolerant plants
		Alter cropping patterns
Industrial and power station cooling	Increase source capacity	Increase water-use efficiency and water recycling
	Use of low-grade water	
Hydropower generation	Increase reservoir capacity	Increase efficiency of turbines; encourage energy efficiency
Navigation	Build weirs and locks	Alter ship size and frequency
Pollution control	Enhance treatment works	Reduce volume of effluents to treat (e.g., by charging for discharges)
		Catchment management to reduce polluting runoff
Flood management	Increase flood protection (levees, reservoirs)	Improve flood warning and dissemination
	Catchment source control to reduce peak discharges	Curb floodplain development

Source: Adapted from Compagnucci et al. 2001 and Kundzewicz et al. 2007.

A recent analysis of impacts and adaptation for the water sector in Canada, for example, provides several "no regret" adaptation options, including greater emphasis on water conservation, improved weather monitoring efforts, and better planning and preparedness for floods and droughts (Lemmen and Warren 2004).

Changes in climate should be examined in association with future socio-economic trends and objectives, such as population and economic development. Climate change is but one of multiple stressors on our water-management systems. Population growth, land-use change, rising affluence in many developing countries, and pollution may have an equally or even larger impact on water in the coming years. Integrating climate change with projected socio-economic trends would help to leverage concerns about climate change to achieve broader societal goals. It would also facilitate the "mainstreaming" of adaptation through "the integration of policies and measures to address climate change into ongoing development activities" (Klein et al. 2003). Indeed, the IPCC suggests that "a high priority should be given to increasing the capacity of countries, regions, communities and social groups to adapt to climate change in ways that are synergistic with wider societal goals of sustainable development" (Adger et al. 2007).

Water managers and policymakers must start considering climate change as a factor in all decisions about water investments and the operation of existing facilities and systems. Accordingly, adaptation policy must identify the potential impacts of climate change on water resources, assess vulnerability, and begin planning for and adapting to those changes, each of which is discussed in greater detail.

Adaptation Assessment

Each adaptation option has associated social, economic, environmental, and political advantages and disadvantages that vary according to local conditions. Furthermore, climate-change impacts are highly variable. The site specificity of climate-change impacts and adaptation options highlight the need for a local or regional adaptation assessment.

An adaptation assessment is the "practice of identifying options to adapt to climate change and evaluating them in terms of criteria such as availability, benefits, costs, effectiveness, efficiency, and feasibility" (IPCC 2007b). These assessments typically rely on climate scenarios developed using GCMs. Because the spatial and temporal resolution of GCM output is typically too coarse for making detailed local water-management decisions, GCM data must be downscaled to a basin scale and evaluated using local hydrologic and water systems models.

Since completion of the previous IPCC assessment report, significant research has focused on improving downscaling methodology. Figure 3.1 provides an outline for evaluating the hydrologic impacts of climate change at a local or regional scale. Two general approaches can be taken to determine changes in temperature, precipitation, and other hydroclimatic factors at the regional scale: (1) using mathematical relationships to downscale GCM output, or (2) hypothetically modifying temperature and precipitation inputs by some arbitrary amount, e.g., a 10% increase in precipitation. These changes are then input into more detailed regional hydrologic models to simulate streamflows under altered climate conditions. Although not shown in Figure 3.1, simulated streamflows can then be input into system simulation models to determine the potential impacts on water resource systems.

This methodology was adopted by a recent regional water supply planning process in the Puget Sound region of the Pacific Northwestern United States. In February 2005, numerous stakeholders, including academics, utilities, local governments, environmental groups, Native American tribes, and business groups, initiated a planning process that would provide technical information to inform decision making in the region. The climate change component entailed an assessment of the effects of climate change on water supply, demand, and instream flows.[2] Three GCMs and two emission

---

2. Climate change was an important component of the planning process, but the process also addressed future water supply and demand and potential environmental restoration efforts.

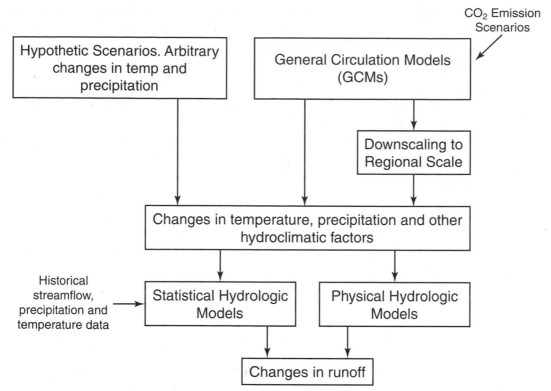

FIGURE 3.1 METHODOLOGY TO EVALUATE HYDROLOGIC IMPACTS OF CLIMATE CHANGE.
Source: Vicuna and Dracup 2007

scenarios were used to project changes in temperature and precipitation through 2075. Statistical downscaling was then applied to the GCM output to produce meteorological data for various weather stations in five water supply basins. The meteorological data, along with other land and vegetation variables, were then input into a rainfall-runoff model calibrated for each of the five basins. Results from this effort were provided to each of the participating water agencies "to assist in the management of their respective water systems and resources, and in their water supply planning activities" (Palmer 2007).

The scenario-based approach described may not be feasible in all regions or settings. The United Nations Development Programme developed an alternative approach, the Adaptation Policy Framework (APF), which may be more appropriate in regions with limited technical or financial resources. The APF starts with an evaluation of vulnerability and adaptation to *current* climate variability and extremes. While future climate risk and socio-economic trends are key components of the analysis, these factors may be expressed quantitatively or qualitatively. Thus while climate models can be used, the analysis does not necessarily rely on "high-quality data, or extensive expertise in computer-based models" (Spanger-Siegfried et al. 2004).

The APF approach has been taken by numerous developing countries through the preparation of national adaptation programs of action (NAPAs), which are designed "to address urgent and immediate needs and concerns related to adaptation to the adverse effects of climate change" (GEF 2002). As of January 2008, 29 of the 48 least developed countries that are a Party to the United Nations Framework Convention on Climate Change (UNFCC) have submitted NAPAs.

Preparation of the NAPAs helps to build adaptive capacity – which refers to the ability to adapt to climate change through moderating potential damages, taking advantage of opportunities, and coping with climate change impacts. NAPAs are initiated at the national level but are conducted in an inclusive and transparent environment. While an expert group is available to provide technical assistance, NAPAs are conducted by a team that consists of a lead agency and various stakeholder groups, particularly local groups. The project teams:

- Synthesize available information
- Assess vulnerability to the current climate regime
- Identify where climate change would increase risks
- Identify adaptation measures
- Provide a short list of projects or activities to address immediate adaptation needs

Although this approach can be particularly useful in regions that lack the technical or financial resources needed to take a scenario-based approach, it may ultimately be inadequate for addressing climate change over longer time periods. In particular, the risk of simply relying on existing climatic vulnerabilities may miss new risks imposed by climate change. Climate change may, for example, produce impacts that are larger than existing infrastructure and management systems are designed to handle. It may also produce impacts of a completely different nature than these systems can address. And, they may lead to impacts that occur at rates faster than we have seen in the past. Each of these risks must be evaluated in light of existing systems.

Community Participation

Regardless of the approach taken, the assessment and implementation of adaptation options should be based on participatory approaches that involve stakeholders throughout the entire process. This approach helps to secure community buy-in and can be particularly effective when dealing with controversial topics. Furthermore, a participatory decision-making process serves to build adaptive capacity, which is key given that impacts associated with climate change have some degree of uncertainty and that adaptation efforts will need to be adjusted periodically (UNFCC 2007a). The Living with Water strategy, adopted by the Netherlands government in 2000, illustrates the effectiveness of a participatory approach. Netherlands is a low-lying country: an estimated 60% of the Netherlands is below sea level, and economic activity in these areas accounts for 70% of the country's gross national product (Kabat et al. 2005). Nearly half of the country is bordered by the North Sea and three major European rivers (the Rhine, the Meuse, and the Scheldt) flow through the country. As a result, the Netherlands is particularly vulnerable to sea level rise, changes in runoff, and changes in demand for water upstream. In 2000, the Dutch government diverged from the traditional flood management approach of building increasingly higher levees and adopted a new strategy, Living with Water, which encourages controlled flooding in designated regions to protect more valuable areas. Because some of the flooded areas include regions that people have inhabited for hundreds of years, a participatory stakeholder process was essential for moving this controversial project forward.

TABLE 3.3   Regional Adaptation Costs (billions of USD) in 2030 Under Two Climate
            Scenarios

Region	SRES A1B	SRES B1
Africa	233	223
Developing Asia	303	230
Latin America	23	23
Middle East	151	148
OECD Europe	87	25
OECD North America	41	16
OECD Pacific	3	1
Transition economies	57	54
World total	**898**	**720**

See Box 3.1 for a description of the Special Report on Emissions Scenarios (SRES) A1B and B1.
Source: UNFCC 2007b.

Economic Cost of Adaptation

Adapting to climate change will require tremendous financial investment, as, of course, will failing to adapt. Estimates of the price vary widely, in part because few climate-specific adaptation projects have actually been completed. The World Bank estimates that "climate-proofing" investments in developing countries will cost $10–40 billion per year (World Bank 2006). While useful, this estimate ignores several key components of adaptation, including the cost of new infrastructure and adapting existing infrastructure (Oxfam 2007). Based on information in NAPAs, a limited number of community-level adaptation projects, and identifying costs that have not yet been calculated, Oxfam estimates that adaptation costs are likely to exceed $50 billion per year. Because of limited experience with adaptation, they recommend that a more thorough economic assessment of adaptation costs be done, along the lines of the U.K. government's "Stern" Review estimate of mitigation costs (Stern 2006).

Water-sector adaptation costs are some unknown fraction of overall adaptation costs. Few studies, however, have looked at costs on a sector basis. A recent report by the UNFCC provides a notable exception. The UNFCC estimates that building new infrastructure (reservoirs, wells, desalination, re-use facilities) to meet water demand in 2030 associated with socioeconomic trends and climate change will cost $720 billion to $900 billion (Table 3.3). Climate change alone accounts for an estimated 25% of the 2030 cost, or $180 billion under the B1 scenario and $225 billion under the A1B scenario (UNFCC 2007b).[3] Although substantial, even these estimates may grossly

---

3. The IPCC, through their Special Report on Emissions Scenarios (SRES), developed four storylines that differ according to demographics, social, economic, environmental, and technological factors and lead to different levels of greenhouse gas emissions. The A1B scenario has high greenhouse gas emissions associated with high population growth, income convergence, a market-oriented economy, and a balanced mix of energy sources (both fossil and non-fossil fuels). By contrast, the B1 scenario has lower greenhouse gas emissions due to slower economic growth, an emphasis on economic, social, and environmental sustainability, and clean, resource efficient technologies.

underestimate total adaptation costs for the water resource sector because they fail to include investment needed for changes in water quality, flood control, or distribution systems.

Equity Concerns

Adaptation will require multifaceted efforts, such as providing access to information, financing, and technology, and building institutional capacity. Burton et al. note that "because climate change results from human activity, rather than pure forces of nature, the question of who pays for adaptation is more complicated and contentious" (Burton et al., 2006). Greenhouse gas emissions are largely a result of economic activity in developed countries. Yet, those in developing in countries may be more vulnerable to the impacts of climate change. Because "those most vulnerable to climate change are the ones least responsible for it" (Gleick 1989, Burton et al. 2006), many argue that rich countries have a moral obligation to support adaptation efforts in developing countries. In addition to the moral obligation, Oxfam (2007) argues that assistance to the developing world provides a number of other important benefits: it demonstrates a willingness to cooperate, reduces potential climate litigation in national or international courts, and can promote global stability.

Under Article 4.4 of the UNFCCC, developed countries have committed to help developing countries meet the financial requirements of adaptation to climate change: "The developed country Parties and other developed Parties included in Annex II shall assist the developing country Parties that are particularly vulnerable to the adverse effects of climate change in meeting the costs of adaptation to those adverse effects." But the commitments are not keeping pace with need.

The UNFCCC supports two special funds: the Special Climate Change Fund (SCCF) and the Least Developed Countries Fund (LDCF). Both funds are supported by the UNFCCC and managed by the Global Environment Facility (GEF). The SCCF is designed to support adaptation planning and technology transfer, while the LDCF is intended to help immediate adaptation needs (Oxfam 2007). These funds rely on voluntary donor contributions made on a rolling basis, which provide flexibility to donor countries but do not provide a predictable and reliable funding source.

The Adaptation Fund, established under the Kyoto Protocol, is intended to support concrete adaptation projects and programs in developing countries that are Parties to the Kyoto Protocol. The fund is financed through a 2% tax on certified emission credits (CERs) issued for clean development mechanism (CDM) project activities and other voluntary sources. Although no funds have been distributed, it is estimated that the Adaptation Fund will provide $80–300 million per year for the period 2008–2012 (UNFCC 2007c), depending on the number of CERs issued. Funding is also available through various multilateral, bilateral, and regional agencies and institutions, including the World Bank, African Development Fund, and the United States Agency for International Development (USAID).

In addition, the GEF is considering funding adaptation via the GEF Trust Fund. Funding through the GEF Trust Fund, initiated in 2003 and still in pilot phase, is designed to support capacity-building adaptation measures (Oxfam 2007). Compared to mitigation, funding for adaptation is small, or less than 3% of the funds allocated by the GEF for climate change activities since 2003 (UNFCC 2007c). Despite only modest investment in adaptation, however, the GEF is mainstreaming adaptation by developing

TABLE 3.4   Global Environment Facility Adaptation Funds (millions of USD)

Fund	Resources mobilized	Resources committed	Fund remaining	Pipeline
GEF Trust Fund	50.0	29.6	20.4	14.4
Special Climate Change Fund	57.0	33.5	23.5	42.0
Least Developed Countries Fund	163.0	16.0	147.0	16.7
Adaptation Fund	0	0	0	0
Total	**270.0**	**79.1**	**190.9**	**73.1**

Source: UNFCC 2007c.

a screening tool to ensure that all GEF-supported projects be modified to increase resilience to climate change (UNFCC 2007c).

Recent funding for adaptation activities is summarized in Table 3.4. In total, $270 million have been mobilized from three sources. The Special Climate Change Fund (SCCF) has received pledges for $57 million for adaptation and has already committed funding for nine projects totaling $33.5 million. Demands for funding from this source is high, as funding for projects in the pipeline ($42 million) exceed the remaining financial resources. Available financial resources for the Least Developed Countries Fund (LDCF), which includes support for the preparation of the NAPAs, total $163 million. Of that amount, only $16 million has been allocated and an additional $17 million has been approved. Since 2003, the GEF has allocated $50 million for 11 adaptation projects, of which $30 million has been distributed.

As shown in Table 3.4, the funds distributed total $79 million. Of that amount, $67 million was dedicated to adaptation projects and $12 million for the preparation of NAPAs. The water sector has received about 20% of the funds approved for projects, or about $14 million, with the majority going to projects in Latin America and the Caribbean (see Fig. 3.2).

Oxfam, noting that "The ethical obligation upon rich countries to stop harming and start helping is extraordinarily clear," developed a methodology to evaluate who should pay adaptation costs (Oxfam 2007). The Oxfam Adaptation Financing Index estimates the financial obligations of developed countries according to four principles: responsibility, equity, capability, and simplicity. Oxfam's Index suggest that the primary contributors, with the percent of costs they should be expected to contribute shown in parenthesis, should be the United States (44%), the European Union (32%), Japan (13%), Canada (4%), Australia (3%), and the Republic of Korea (2%). Oxfam contends that this assistance is compensatory rather than aid and should be provided in addition to other long-standing aid commitments. As of April 2007, the United Kingdom, France, Germany, Denmark, and the Netherlands have been the largest contributors to international adaptation funds that accept voluntary contributions. By contrast, the United States, Japan, and Australia have made no contributions to these funds, although they may have provided money for adaptation projects in developing countries through other aid agencies, such as USAID (Oxfam 2007). We note there are also efforts of non-governmental organizations and other kinds of donors, such as foundations, to evaluate and promote adaptation options and efforts.

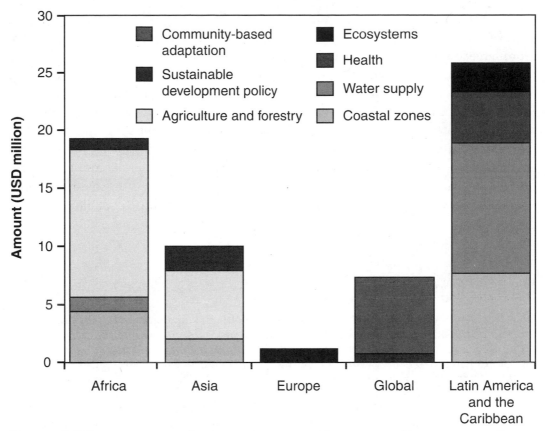

FIGURE 3.2 DISTRIBUTION OF APPROVED ADAPTATION PROJECTS, BY REGION AND BY SECTOR (2005–2007).
Source: UNFCC 2007c

Conclusion

Both developed and developing countries face a range of challenges in dealing with unavoidable climate impacts, including changes in water availability, groundwater overdraft, and water pollution from urban and agricultural activities. In 2002, an estimated 1.1 billion people lacked access to an improved water supply, and 2.6 billion people lacked access to improved sanitation. While governments have made serious efforts to address these and other water supply and quality problems, climate change will make solving these challenges even more difficult.

Climate change is already affecting our water resources. As atmospheric greenhouse gas concentrations continue to rise, significant climatic changes will continue to develop and may even accelerate. Because we have already committed to a certain degree of climate change and emissions continue unabated, adaptation must be a central element of all climate change policy. Additionally, adaptation efforts can minimize existing climate vulnerabilities and promote sustainable development.

Adapting to climate change will require tremendous financial investment and institutional attention. The World Bank estimates that "climate-proofing" investments in developing countries will cost $10–40 billion per year (World Bank 2006). A recent analysis by Oxfam suggests that adaptation costs will be much higher, exceeding $50 billion per year (Oxfam 2007). Because of limited experience with adaptation, however, the actual cost remains largely unknown. To promote adaptation, a more thorough economic assessment of adaptation costs is needed.

In addition to the financial challenges, climate change adaptation raises some important practical, ethical, and moral issues. Greenhouse-gas emissions are largely a result of economic activity in developed countries. Yet, those in developing countries are often more vulnerable to the impacts of climate change. While developed countries have committed to help developing countries meet the financial requirements of adaptation to climate change under Article 4.4 of the UNFCC, these commitments are not keeping pace with need. Greater emphasis must be placed on securing stable financial, educational, and institutional commitments from those capable of helping those in need.

REFERENCES

Adger, W.N., Agrawala, S., Mirza, M.M.Q., Conde, C., O'Brien, K., Pulhin, J., Pulwarty, R., Smit, B., and Takahashi, K. 2007. Assessment of adaptation practices, options, constraints and capacity. *Climate Change 2007: Impacts, Adaptation and Vulnerability. Contribution of Working Group II to the Fourth Assessment Report of the Intergovernmental Panel on Climate Change,* M.L. Parry, O.F. Canziani, J.P. Palutikof, P.J. van der Linden and C.E. Hanson, editors. Cambridge, UK: Cambridge University Press, 717–743.

Bernstein, L., Bosch, P., Canziani, O., Chen, Z., Christ, R., Davidson, O., Hare, W., Huq, S., Karoly, D., Kattsov, V., Kundzewicz, Z., Liu, J., Lohmann, U., Manning, M., Matsuno, T., Menne, B., Metz, B., Mirza, M., Nicholls, N., Nurse, L., Pachauri, R., Palutikof, J., Parry, M., Qin, D., Ravindranath, N., Reisinger, A., Ren, J., Riahi, K., Rosenzweig, C., Rusticucci, M., Schneider, S., Sokona, Y., Solomon, S., Stott, P., Stouffer, R., Sugiyama, T., Swart, R., Tirpak, D., Vogel, C., and Yohe, G. 2007. Summary for Policymakers. Climate Change 2007: Synthesis Report. Intergovernmental Panel on Climate Change Fourth Assessment Report. Accessed March 2008, www.ipcc.ch.

Burton, I., Huq, S., Lim, B, Pilifosova, O., and Shipper, E.L. 2002. From impacts assessment to adaptation priorities: The shaping of adaptation policy. *Climate Policy* 2:145–159.

Burton, I., Diringer, E., and Smith, J. 2006. Adaptation to Climate Change: International Policy Options. Pew Center on Global Climate Change. Accessed on December 26, 2007 at http://www.pewclimate.org/docUploads/PEW_Adaptation.pdf.

Compagnucci, R., da Cunha, L., Hanaki, K., Howe, C., Mailu, G., Shiklomanov, I., Stakhiv, E., and Doll, P. 2001. Hydrology and Water Resources. Climate Change 2001: Impacts, Adaptation and Vulnerability. Contribution of Contribution of Working Group II to the Third Assessment Report of the Intergovernmental Panel on Climate Change. Accessed March 2008, www.ipcc.ch.

Frederick, K.D. and Gleick, P.H. 1999. Water and Global Climate Change: Potential Impacts on U.S. Water Resources. Washington. D.C.: Pew Center on Global Climate Change.

Gleick, P.H. 1989. Greenhouse warming and international politics: Problems facing developing countries. *Ambio.* 18(6): 333–339.

Global Environment Facility (GEF). 2002. Operational Guidelines for Expedited Funding for the Preparation of National Adaptation Programs of Action by Least Developed Countries. Accessed on January 30, 2008 at http://napa.undp.org/about.asp.

Intergovernmental Panel on Climate Change. 2000. Special Report on Emissions Scenarios. Cambridge, UK: Cambridge University Press, pp 570.

Intergovernmental Panel on Climate Change (IPCC). 2007a. Climate Change 2007: Synthesis Report. Contribution of Working Groups I, II and III to the Fourth Assessment Report of the Intergovernmental Panel on Climate Change. Pachauri, R.K and Reisinger, A., editors. IPCC, Geneva, Switzerland, 104 pp. Accessed on March 28, 2008 at www.ipcc.ch/pdf/assessment-report/ar4/syr/ar4_syr_spm.pdf

Intergovernmental Panel on Climate Change. 2007b. Appendix I: Glossary. Accessed on March 25, 2008 at http://www.ipcc.ch/pdf/assessment-report/ar4/wg2/ar4-wg2-app.pdf.

Kabat, P., Vellinga, P., van Vierssen, W., Veraart, J., and Aerts, J.[2005] Climate proofing the Netherlands. *Nature* 438: 283–284.

Klein, R.J.T., Schipper, E.L., and Dessai, S. 2003. Integrating mitigation and adaptation into climate and development policy: three research questions. Working Paper 40. Tyndall Centre for Climate Change Research.

Kundzewicz, Z.W., Mata, L.J., Arnell, N.W., Döll, P., Kabat, P., Jiménez, B., Miller, K.A., Oki, T., Sen, Z., and Shiklomanov, I.A. 2007. Freshwater resources and their management. Climate Change 2007: Impacts, Adaptation and Vulnerability. Contribution of Working Group II to the Fourth Assessment Report of the Intergovernmental Panel on Climate Change. Parry, M.L., Canziani, O.F., Palutikof, J.P., van der Linden, P.J., and Hanson, C.E., editors. Cambridge, UK: Cambridge University Press, 173–210.

Lemmen, D.S. and Warren, F.J., eds. 2004. Climate Change Impacts and Adaptation: A Canadian Perspective. Natural Resources Canada. Ottawa, Ontario. Accessed on January 8, 2008 from http://adaptation.nrcan.gc.ca/perspective/pdf/report_e.pdf.

Meehl, G.A., Stocker, T.F., Collins, W.D., Friedlingstein, P., Gaye, A.T., Gregory, J.M., Kitoh, A., Knutti, R., Murphy, J.M., Noda, A., Raper, S.C.B., Watterson, I.G., Weaver, A.J., and Zhao, Z.-C. 2007. Global Climate Projections. Climate Change 2007: The Physical Basis. Contribution of Working Group I to the Fourth Assessment Report of the Intergovernmental Panel on Climate Change. Solomon, S., Qin, D., Manning, M., Chen, Z., Marquis, M., Averyt, K.B., Tignor, M., and Miller, H.L., editors. Cambridge, UK: Cambridge University Press.

Olmos, S. 2001. Vulnerability and Adaptation to Climate Change: Concepts, Issues, Assessment Methods. Prepared for the Climate Change Knowledge Network. Accessed on January 11, 2008 from http://www.cckn.net/pdf/va_foundation_final.pdf.

Oxfam. 2007. Adapting to climate change: what's needed in poor countries, and who should pay. Oxford, UK: Oxfam Briefing Paper 104.

Palmer, R.N. 2007. Final Report of the Climate Change Technical Committee. A report prepared by the Climate Change Technical Subcommittee of the Regional Water Supply Planning Process, Seattle, WA.

Randall, D.A., Wood, R.A, Bony, S., Colman, R., Fichefet, T., Fyfe, J., Kattsov, V., Pitman, A., Shukla, J., Srinivasan, J., Stouffer, R.J., Sumi, A., and Taylor, K.E. 2007. Climate Models and Their Evaluation. In: *Climate Change 2007: The Physical Science Basis.* Contribution of Working Group I to the Fourth Assessment Report of the Intergovernmental Panel on Climate Change. Solomon, S., Qin, D., Manning, M., Chen, Z., Marquis, M., Averyt, K.B,Tignor, M., and Miller, H.L., eds.. Cambridge, United Kingdom and New York, NY: Cambridge University Press.

Schneider, S.H., Semenov, S., Patwardhan, A., Burton, I., Magadza, C.H.D, Oppenheimer, M., Pittock, A.B., Rahman, A., Smith, J.B., Suarez, A., Yamin, F. 2007. Assessing key vulnerabilities and the risk from climate change. Climate Change 2007: Impacts, Adaptation and Vulnerability. Contribution of Working Group II to the Fourth Assessment Report of the Intergovernmental Panel on Climate Change. Parry, M.L., Canziani, O.F., Palutikof, J.P., van der Linden, P.J., and Hanson, C.E., editors. Cambridge, UK: Cambridge University Press, 779–810.

Spanger-Siegfried, E., Dougherty, B., Downing, T., Hellmuth, M., Hoeggel, U., Klaey, A., and Lonsdale, K. 2004. User's Guidebook. In: Lim, B. and Spanger-Siegfried, E., editors. Adaptation Policy Frameworks for Climate Change: Developing Strategies, Policies and Measures. Cambridge, United Kingdom and New York, NY: Cambridge University Press, Accessed on February 8, 2008 at http://ncsp.undp.org/report_detail.cfm?Projectid=151.

Stern, N. 2006. Stern Review on the Economics of Climate Change. The Edinburgh Building, Cambridge, UK: Cambridge University Press.

Trenberth, K.E., Jones, P.D, Ambenje, P., Bojariu, R., Easterling, D., Klein, A., Parker, D., Rahimzadeh, F., Renwick, J.A., Rusticucci, M., Soden, B., and Zhai, P. 2007. Observations: Surface and Atmospheric Climate Change. In: *Climate Change 2007: The Physical Science Basis.* Contribution of Working Group I to the Fourth Assessment Report of the Intergovernmental Panel on Climate Change. Solomon, S., Qin, D., Manning, M., Chen, Z., Marquis, M., Avery, K.B., Tignor, M., and Miller, H.L., eds. Cambridge, United Kingdom and New York, NY: Cambridge University Press.

United Nations Framework Convention on Climate Change (UNFCC). 2007a. Synthesis of outcomes of the regional workshops and expert meeting on adaptation under decision 1/CP.10. Accessed January 30, 2008 at http://unfccc.int/resource/docs/2007/sbi/eng/14.pdf

United Nations Framework Convention on Climate Change (UNFCC). 2007b. Investment and financial flows to address climate change. Accessed January 7, 2008 at http://unfccc.int/cooperation_and_support/financial_mechanism/items/4053.php.

United Nations Framework Convention on Climate Change (UNFCC). 2007c. The assessment of the funding necessary to assist developing countries in meeting their commitments relating to the Global Environment Facility replenishment cycle. Accessed on December 31, 2007 at http://unfccc.int/documentation/documents/advanced_search/items/3594.php?rec=j&pr iref=600004355#beg.

Vicuna, S., and Dracup, J.A. 2007. The evolution of climate change impact studies on hydrology and water resources in California. *Climatic Change* 82: 327–350.

World Bank. 2006. Clean Energy and Development: Towards an Investment Framework. Washington, D.C.

World Health Organization (WHO) and United Nations Children's Fund (UNICEF). 2000. Global Water Supply and Sanitation Assessment 2000 Report. Accessed on March 25, 2008 at http://www.who.int/docstore/water_sanitation_health/Globassessment/GlobalTOC.htm.

World Health Organization (WHO) and United Nations Children's Fund (UNICEF). 2004. Joint Monitoring Programme for Water Supply and Sanitation; Meeting the MDG drinking water and sanitation target: a mid-term assessment of progress. Accessed on April 17, 2008 at http://www.who.int/water_sanitation_health/monitoring/jmp04_1.pdf

World Meteorological Organization (WMO). 2004. Integrated Flood Management. The Associated Programme on Flood Management. APFM Technical Document No. 1.

Millennium Development Goals: Charting Progress and the Way Forward

Meena Palaniappan

Billions of people live without access to safe water and sanitation, and every year millions of children die from preventable water-related diseases. Compounding this tragedy is the fact that the global community has both the technologies and resources to provide adequate water and sanitation for all and has failed to do so. Recognizing these and many other failures of human development, the United Nations, its member countries, and non-governmental partners committed to a set of Millennium Development Goals to address the interrelated needs of the world's poorest communities.

Water and sanitation are explicitly recognized as targets in the Millennium Development Goals: the international community committed to halving the proportion of people without access to safe water and sanitation by 2015. As the target date for achieving the goals approaches, this chapter assesses how successful efforts have been in providing water and sanitation in the regions and countries that need it most. We identify trends and challenges and make recommendations for shifting course so that the Millennium Development Goals for water can be achieved—and ideally— surpassed.

Millennium Development Goals

In 2000, 189 United Nations member countries and supporting organizations established eight overarching goals for tackling persistent poverty in the global community, referred to as the Millennium Development Goals (MDGs). The MDGs are designed to address many of the interrelated causes and outcomes of poverty, including child and maternal mortality, hunger, malnutrition, lack of water and sanitation, and education. Each MDG consists of anywhere from one to seven time-bound targets (Table 4.1), and progress toward achieving these targets is measured by a set of 48 indicators.

In September of 2000, when the United Nations General Assembly adopted the Millennium Development Goals, improving access to water was recognized as a key target and included as Target 10 of Goal 7:

> "Halve by 2015 the proportion of people without sustainable access to safe drinking water" (UNDP 2003)

Because sanitation is as critical as water supply for protecting human health, the World Summit on Sustainable Development in 2002 expanded this target to include improving access to basic sanitation:

> "We agree to halve, by the year 2015, the proportion of people who are unable to reach or to afford safe drinking water (as outlined in the Millennium Declaration) and the proportion of people who do not have access to basic sanitation. . ." (United Nations 2002)

Beyond this explicit mention, adequate and safe water and sanitation are implicitly linked to the achievement of almost every other MDG. Providing water and sanitation is central to eliminating poverty and improving the lives of billions worldwide. For example, better water management increases water available for agriculture and industrial production, thereby increasing economic opportunities and food availability. Having water available closer to home increases the available time that can be spent in income-producing labor. Thus, improving access to water can help meet MDG 1, eradicating extreme poverty and hunger.

Access to water and sanitation also has direct effects on education and gender equity (MDGs 2 and 3). In the developing world, severe and repeated cases of diarrhea, often from water-related disease, are an important cause of malnutrition, which leads to short- and long-term impacts on cognitive and physical development among children, called *wasting* and *stunting*. Additionally, children miss school due to water-related illnesses like diarrhea, or when taking care of a sick family member. This in turn affects education and opportunities to get out of poverty, addressed by both MDG 1 (poverty and hunger) and MDG 2 (education). Millions of women and children in Africa travel long distances to get water, spending time that could be spent in school. Adding to gender inequity in education (MDG 3), many girls stop attending school after reaching menstrual age because they don't have a private place to use the toilet at school.

Unsafe water and inadequate sanitation can also worsen maternal health and increase childhood mortality (MDGs 4 and 5). Water-related diseases, including diarrhea, are among the largest causes of childhood mortality in the developing world: every year, diarrheal disease kills two million children, mostly under the age of 5. In addition, mothers often give birth in unsanitary conditions lacking clean water, which increases the mortality of mothers in developing countries. Efforts to improve access to an adequate water supply and basic sanitation would reduce childhood mortality and improve maternal health.

MDG 6, combating HIV/AIDs and malaria, is also affected by the management of water and sanitation in communities. Diarrheal disease increases risk of death among those suffering with HIV/AIDS. Better management of standing water and wastewater can also reduce the threat of malaria.

TABLE 4.1   The Millennium Development Goals and Targets

Goals	Targets
Goal 1: Eradicate extreme and poverty hunger	Target 1: Halve, between 1990 and 2015, the proportion of people whose income is less than one dollar a day
	Target 2: Halve, between 1990 and 2015, the proportion of people who suffer from hunger
Goal 2: Achieve universal primary education	Target 3: Ensure that, by 2015, children everywhere, boys and girls alike, will be able to complete a full course of primary schooling
Goal 3: Promote gender equality and empower women	Target 4: Eliminate gender disparity in primary and secondary education, preferably by 2005, and in all levels of education no later than 2015
Goal 4: Reduce child mortality	Target 5: Reduce by two-thirds, between 1990 and 2015, the under-five mortality rate
Goal 5: Improve maternal health	Target 6: Reduce by three-quarters, between 1990 and 2015, the maternal mortality ratio
Goal 6: Combat HIV/AIDS, malaria and other diseases	Target 7: Have halted by 2015 and begun to reverse the spread of HIV/AIDS
	Target 8: Have halted by 2015 and begun to reverse the incidence of malaria and other major diseases
Goal 7: Ensure environmental sustainability	Target 9: Integrate the principles of sustainable development into country policies and programmes and reverse the loss of environmental resources
	Target 10: Halve, by 2015, the proportion of people without sustainable access to safe drinking water and basic sanitation
	Target 11: By 2020, to have achieved a significant improvement in the lives of at least 100 million slum dwellers
Goal 8: Develop a global partnership for development	Target 12. Develop further an open, rule-based, predictable, non-discriminatory trading and financial system (Includes a commitment to good governance, development, and poverty reduction — both nationally and internationally)
	Target 13. Address the special needs of the least developed countries (Includes: tariff and quota free access for least developed countries' exports; enhanced programme of debt relief for HIPCs and cancellation of official bilateral debt; and more generous ODA for countries committed to poverty reduction)
	Target 14. Address the special needs of landlocked countries and small island developing States
	Target 15. Deal comprehensively with the debt problems of developing countries through national and international measures in order to make debt sustainable in the long term
	Target 16: In cooperation with developing countries, develop and implement strategies for decent and productive work for youth
	Target 17: In cooperation with pharmaceutical companies, provide access to affordable essential drugs in developing countries
	Target 18: In cooperation with the private sector, make available the benefits of new technologies, especially information and communications

Source: United Nations Development Programme website. www.undp.org/mdg

Measuring Progress: Methods and Definitions

The provision of safe water and sanitation are critical elements in improving lives, alleviating poverty, and promoting development. What is "safe" water and sanitation, and how can we measure progress toward this goal? The past few decades have seen shifting definitions of progress toward this goal, as well as methods of measuring progress.

Monitoring the state of water and sanitation and progress over time has been a challenging effort. The Joint Monitoring Program (JMP) of UNICEF and WHO is the defacto international standard for measuring progress toward the water and sanitation MDG. Since 1990, the JMP has changed how it measures and what indicators are used to determine how well the water and sanitation needs of communities are being met.

In the 1980s and 1990s, providers reported information on access to safe water and sanitation. National governments, regional governments, or water providers would provide information on the number and percentage of people they considered served with water and sanitation. There were some obvious problems with this system. Providers rarely had comprehensive or accurate data and had an incentive to overestimate coverage numbers. Sometimes they failed to account for those living in informal settlements or slums who did not have access. Providers also did not know and could not accurately account for private household wells or sanitation facilities built by the householder. These problems led to huge shifts from year to year in terms of the population reportedly served that did not correlate to actual changes in access but were the result of shifting definitions and monitoring methods. It also became clear in this period that definitions of safe water and sanitation varied widely from country to country. In many places where "safe" water was reportedly available, for example, there was no water testing done to determine if the water was safe for consumption.

Based on many of these limitations, the WHO/UNICEF Joint Monitoring Program shifted in 2000 to clearer definitions for water supply and sanitation and changed its data collection methods. In its Global Water Supply and Sanitation 2000 Report, WHO/UNICEF shifted from monitoring "safe" to "improved" water supply and sanitation. They also introduced definitions of what constituted "improved" water supply and sanitation technologies. The technologies that WHO/UNICEF considers improved and unimproved are listed in Table 4.2.

TABLE 4.2 Technologies Considered Improved and Unimproved by WHO/UNICEF

	Water Supply Technologies	**Sanitation Technologies**
Improved	Household connection	Connection to a public sewer
	Public standpipe	Connection to septic system
	Borehole	Pour-flush latrine
	Protected dug well	Simple pit latrine
	Protected spring	Ventilated improved pit latrine
	Rainwater collection	
Unimproved	Unprotected well	Service or bucket latrines (where excreta
	Unprotected spring	are manually removed)
	Vendor-provided water	Public latrines
	Bottled water	Latrines with an open pit
	Tanker truck-provided	

Source: WHO/UNICEF 2006.

In 2000, there was also a shift from collecting only self-reported information from providers to also collecting information from users through assessment questionnaires and household surveys. There is now more attention to actual use and maintenance of private household or community facilities. Assessment questionnaires are completed by WHO country representatives in partnership with local UNICEF staff and national agencies. The JMP also compiles results from in-country household surveys, including the USAID Demographic Health Surveys (DHS) and UNICEF's Multiple Indicator Cluster Surveys (MICS). Both DHS and MICS survey several thousand households in each country that are spread across different regions and representative of urban and rural areas. These surveys ask questions primarily related to health, a few of which are related to water, sanitation, and hygiene. Questions on hygiene practices, infant stool disposal, and diarrhea treatment help to expand the scope of information in the water sector, and can better target and track the success of interventions, although much of this broader information is not reported when measuring progress toward the MDGs related to water and sanitation.

Limitations in Water and Sanitation Data and Reporting

Numerous limitations to the existing data and its collection continue to hinder both analysis and implementation of policy. The household surveys are not conducted every year in every country. DHS surveys are conducted in seven to nine countries every year. MICS surveys are conducted about once every 5 years. There are also differences in the capacities of different national statistical organizations to conduct the survey. The typical limitations of the survey include how well the question translates to the cultural context and other challenges faced by interviewers. Thus, global water supply and sanitation assessments attempt to standardize and compare data that was collected at different times in different countries, with potentially varying accuracy in the results depending on the capacity of the supporting agencies.

One of the major limitations of the current data is that even if people have access to "improved" water sources, this water may not actually be safe to drink. In many developing countries, most types of improved water sources, including piped connections, standpipes, and wells, would not provide water that was fit for human consumption. In Bangladesh, for example, millions of people are classified as having access to an improved water supply despite the fact the water is heavily contaminated with arsenic.

Quality of life, based on the type of access afforded, varies greatly. For example, a household with access to a protected well within 30 minutes of their home is considered on par with a household that has water through a piped connection. Yet, as Bartram (2007) documents, a distant protected water source provides limited health benefits and seriously encumbers productive household economic activity compared with an in-home connection.

While current monitoring efforts provide a better assessment of the actual use of facilities, many NGOs reported that facilities counted in national estimates are broken or are no longer in use. Although the MDGs do not explicitly cover it, hygiene is an integral part of the water, sanitation, and hygiene sector. The health surveys used by the Joint Monitoring Program include several questions that could be relevant in determining the hygiene status of a community and identifying appropriate interventions.

WHO and UNICEF are making plans to unveil a new set of standards and ways of monitoring water and sanitation in developing countries that is more health based, including more information on water quality and types of access. IRC in its "Monitoring Millennium Development Goals for Water and Sanitation" (Shordt et al. 2004) have also identified recommendations to improve the system of monitoring the MDGs, including the need to strengthen in-country statistical capacity, linking MDG monitoring to action planning, and connecting the multiple in-country programs, including NGO efforts, with knowledge of the water and sanitation sector.

Progress on the Water and Sanitation MDGs

Despite the important caveats that underlie the reported numbers of people with water and sanitation access, changes in access throughout the world tell an important and compelling story. The good news is the world as a whole appears to be on track to meet the MDG for water supply. The bad news is that many regions, like sub-Saharan Africa, will fail to meet the water targets given current levels of effort, and the world as a whole, is unlikely to come close to meeting the MDG target for sanitation.

The baseline conditions for most of the MDGs, including water and sanitation targets, were set to be 1990. The water and sanitation targets sought to halve the proportion of people without sustainable access to safe drinking water and basic sanitation by 2015. In 1990, 1.2 billion people, or 22 percent of the population, did not have access to improved water, and 2.7 billion people—50 percent of the world's population—did not have access to sanitation. By 2015, according to the MDGs, the proportion of people without access to improved water and sanitation should be reduced to 11 percent and 25 percent, respectively. But these percentages can obscure the absolute number of people without access. Box 4.1 provides more information on the difference between proportion of those without access and the *number* of those without access.

Over the past 15 years, some progress has been made, although these efforts will need to be intensified in the coming years. The global community and regions of the world are considered to be "on target" if the current coverage is within 5 percent of the coverage needed to be on track in 2004 (WHO UNICEF 2006). In the 15 years between 1990 and 2004, over half of the reduction in proportion of people should have taken place for the region to be positioned to achieve the MDG target in the next 10 years, by 2015. To be on target in 2004, the percentage of those without improved water should be 15 percent. In 2004, the percentage of those without improved water was 17 percent, just two percent shy of what was needed. Thus, the global community as a whole is considered to be approximately "on target" toward reaching the water MDG (Table 4.1).

In contrast, the world has made insufficient progress toward the sanitation MDG. To reach the sanitation target of 25 percent by 2004, the proportion of the population without access to sanitation should have been reduced to 35 percent (over half way to the 2015 target of 25 percent). The actual proportion of those without access to sanitation remained high at 41 percent in 2004. Table 4.3 summarizes the progress of major regions of the globe toward the drinking water and sanitation MDGs from 1990 to 2004. Based on the rate of progress thus far, it also projects whether the region is likely to

---

Box 4.1   Proportion versus Number

The goal of efforts in the water sector is ultimately to ensure that no one lives without improved water or sanitation. What the global community committed to in the Millennium Development Goals was to halve by 2015 the *proportion* of those without safe water and sanitation. This has sometimes been mistakenly understood or reported as halving the *number* of those without safe water and sanitation. In fact, reducing by half the number of people without water and sanitation is a far more difficult achievement. The unknowns of global and regional population growth make it difficult to commit to a precise reduction at a future date in the *number* of people without water or sanitation. In fact, most of the Millennium Development targets are ratio or proportion based, thus related to the total population but not contingent on it. While the actual number of people without improved water and sanitation will actually decrease if the MDGs for water and sanitation are met, they will not be reduced by half. For example, if the water MDG is met, nearly 800 million people will remain without improved water. This is only a third less than the 1.2 billion without water in 1990, because the total population will have grown substantially.

---

reach the drinking water or sanitation target. Table 4.4 defines which countries are included in each of the worlds' regional groupings as defined by WHO and UNICEF.

Unfortunately, these gains are not enough, and efforts will need to be intensified in the coming years if the water supply and sanitation targets are to be achieved. Since 1990, about 200,000 people gained access to water services per day, and more than 200,000 gained access to sanitation. If we are to reach the MDGs by 2015, however, an estimated 450,000 people must gain access to sanitation daily, and 300,000 more per day need access to water (WHO UNICEF 2006). This is a huge additional commitment that will require increased financial resources, more focused efforts in the regions that are falling furthest behind, and greater political will among governments and partners from the developed and developing world.

In absolute numbers, even the improvements made so far, however, have been almost entirely wiped out by population growth. In 1990, 1.2 billion people did not have access to improved water, and 2.7 billion people—half of the world's population—did not have access to basic sanitation (Figs. 4.1 and 4.2). Between 1990 and 2004, 1.2 billion additional people gained access to improved water sources and sanitation. But, by 2004, because of population growth in regions with inadequate water services, 1.1 billion people still did not have improved water, and 2.6 billion people still did not have access to basic sanitation. As shown in Figures 4.1 and 4.2, even if the MDGs are achieved, a shocking 800 million people will still live without access to improved water sources, and 1.8 billion people will live without access to improved sanitation in 2015. In a 2002 study, Gleick found that even if the MDGs are met, between 34 and 76 million people will die between 2000 and 2020 from preventable water-related diseases (Gleick 2002).

TABLE 4.3 Progress Toward the Water and Sanitation MDG by Region

	Percent Access to Improved Drinking Water					Percent Access to Improved Sanitation				
	1990	2004	2004 (% needed to be on target)	2015 Target	On Target?	1990	2004	2004 (% needed to be on target)	2015 Target	On Target?
World	78	83	85	89	Almost on target	49	59	65	75	Inadequate progress
Northern Africa	89	91	93	95	Almost on target	65	77	76	82.5	On target
Sub-Saharan Africa	49	56	65	75	Inadequate progress or deterioration	32	37	52	66	Inadequate progress or deterioration
Latin America and the Caribbean	83	91	88	92	On target	68	77	78	84	On target
Eastern Asia	71	78	80	86	On target	24	45	47	62	On target
Southern Asia	72	85	80	86	On target	20	38	44	60	Making progress but insufficient
South-Eastern Asia	76	82	83	88	Almost on target	49	67	65	74.5	On target
Western Asia	85	91	90	93	On target	81	84	87	90.5	Almost on target
Oceania	51	50	66	76	Inadequate progress or deterioration	54	53	68	77	Inadequate progress or deterioration
Commonwealth of Independent States	92	92	94	96	Target nearly met in Europe but deterioration in Asia	82	83	87	91	On target in Europe, deterioration in Asia

Source: WHO/UNICEF 2006; UN 2006.

TABLE 4.4   Definitions of World Regions

Region	Countries
Northern Africa	Algeria, Egypt, Libya, Morocco, Tunisia.
Sub-Saharan Africa	Angola, Benin, Botswana, Burkina Faso, Burundi, Comoros, Cameroon, Central African Republic, Chad, Congo, Djibouti, Ethiopia, Eritrea, Equatorial Guinea, Gabon, The Gambia, Ghana, Guinea, Guinea-Bissau, Ivory Coast, Kenya, Lesotho, Liberia, Madagascar, Malawi, Mali, Mauritania, Mozambique, Namibia, Niger, Nigeria, São Tomé and Príncipe, Senegal, Seychelles, Sierra Leone, Somalia, South Africa, Sudan, Swaziland, Tanzania, Togo, Rwanda, Uganda, Zambia, Zimbabwe.
Latin America and the Caribbean	Argentina, Belize, Bolivia, Brazil, the Caribbean islands, Chile, Colombia, Costa Rica, Ecuador, French Guiana, Guatemala, Guyana, Honduras, Mexico, Nicaragua, Panama, Paraguay, Peru, Suriname, Uruguay, Venezuela.
Eastern Asia	China (including Taiwan and Hong Kong), Korea, Mongolia.
Southern Asia	Afghanistan, Bangladesh, Bhutan, India, Iran, Nepal, Pakistan, Sri Lanka.
South Eastern Asia	Cambodia, Indonesia, Laos, Malaysia, Myanmar, Philippines, Thailand, Vietnam.
Western Asia	Iraq, Israel and the Palestinian territories, Jordan, Kuwait, Lebanon, Oman, Saudi Arabia, Syria, Turkey, United Arab Emirates, Yemen.
Oceania	Pacific islands, Papua New Guinea
Commonwealth of Independent States	Armenia, Azerbaijan, Belarus, Georgia, Kazakhstan, Kyrgyzstan, Moldova, Russia, Tajikistan, Ukraine, and Uzbekistan. Turkmenistan
Developed Regions	Australia, Canada, Europe, Greenland, Japan, New Zealand, USA.

Source: WHO/UNICEF 2006.

Regional Progress in Drinking Water

Most regions of the world are on track to meet the drinking water target (Table 4.3 and Figure 4.3). Significant progress in both South Asia and Latin America has meant those regions have nearly achieved the MDG drinking water target—10 years ahead of schedule (UNICEF 2006). In these regions, the percentage and absolute number of people with access to improved water has increased. If this pace is maintained, these regions will exceed their 2015 MDG target for drinking water access.

Other regions of the world, however, are falling behind, including sub-Saharan Africa, Oceania, and the Asian countries of the Commonwealth of Independent States (CIS). In sub-Saharan Africa and Oceania, the total number of people without improved drinking water actually increased from 1990 to 2004, and these regions will need to pick up the pace and greatly increase the number of people gaining access to

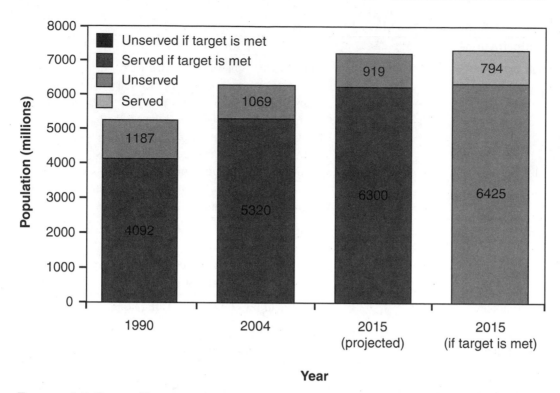

FIGURE 4.1 GLOBAL POPULATION WITH AND WITHOUT ACCESS TO IMPROVED WATER SOURCES, 1990, 2004, AND PROJECTED FOR 2015 COMPARED TO THE 2015 MDG TARGET. The number of people in each year with access to improved drinking water is shown in the bottom of the bar; the number of people without access to improved drinking water is shown in the top portion of the bar. In the 1990 and 2004, the numbers are based on UNICEF and WHO data. The 2015 projection assumes the same rate of progress for the next 10 years as for the preceding 15 years. The 2015 target shows the number of people that need to be served, as well as the number left unserved if the MDG target is met. Even if the MDGs are met, nearly 800 million people will remain without water services. If we continue at the same rate of progress, in 2015, over 900 million will live without improved water services.

Source: WHO UNICEF 2006.

drinking water to meet the drinking water MDG. Figure 4.3 also demonstrates that sub-Saharan Africa and Oceania are also the regions with the lowest percentage of people with access to improved drinking water. In 2004, only 51 percent of people in Oceania, and 56 percent in sub-Saharan Africa had access to improved water sources. By comparison, more than 75 percent of people had access to improved water in all the other regions of the world.

Figure 4.4 presents the total number of people without access to improved drinking water in each region of the world in 1990 and 2004. In 1990, the largest number of people without access to safe drinking water lived in Eastern Asia, followed closely by Southern Asia and sub-Saharan Africa. The significant progress made in Southern Asia and good progress toward improved drinking water in Eastern Asia reduced the total number of people without access in these two countries. With population growth and limited progress in sub-Saharan Africa, however, the total number of people without access increased, and by 2004, the largest number of people without safe water, 300 million, lived in this region. With 11 percent of the world's people, Sub-Saharan Africa

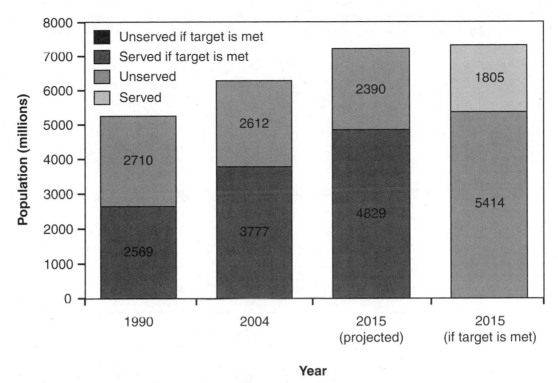

FIGURE 4.2 GLOBAL POPULATION WITH AND WITHOUT ACCESS TO IMPROVED SANITATION, 1990, 2004, AND PROJECTED FOR 2015 COMPARED TO THE 2015 MDG TARGET. The number of people in each year with access to improved sanitation is shown in the bottom of the bar; the number of people without access to improved sanitation is the top portion of the bar. In the 1990 and 2004, numbers based on UNICEF and WHO data. The 2015 projected assumes the same rate of increase for the next 10 years as for the preceding 15 years. The 2015 target shows the number of people that need to be served, as well as the number left unserved if the MDG target is met. Even if the MDGs are met, 1.8 billion people will remain without basic sanitation. If we continue at the same rate of progress, we will fail to meet the targets in 2015, and 2.3 billion will live without basic sanitation.
Source: WHO UNICEF 2006.

is home to almost a third of all the people without access to safe water (UNICEF 2006). Sub-Saharan Africa, together with East and South Asia, make up a majority of those without drinking water access, over 80 percent.

Regional Progress in Sanitation

Whereas most regions are on track to meet the drinking water MDG, many of the world's regions are not on track to meet the sanitation targets. Four of ten people in the world have to defecate in the open or in unsafe or unhygienic conditions, threatening the health of the surrounding community (WHO UNICEF 2006).

Northern Africa, Latin America and the Caribbean, and Eastern, Western, and Southeastern Asia appear to be on track to meet the sanitation target. Southern Asia has made some progress, increasing the percentage of people with access to sanitation from 20 to 38 percent in 15 years. But, to be on track to reach 60 percent of people served with sanitation by 2015, Southern Asia should have achieved 44 percent access by 2004.

As shown in Figures 4.5 and 4.6, two of the most populous regions in the world, Southern Asia and sub-Saharan Africa had sanitation coverage rates below 40 percent

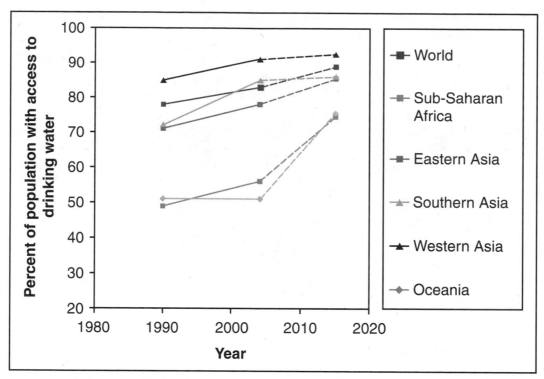

FIGURE 4.3 Percent of population by region with access to improved drinking water source, 1990, 2004, and needed to satisfy 2015 target.
Source: UN 2007.

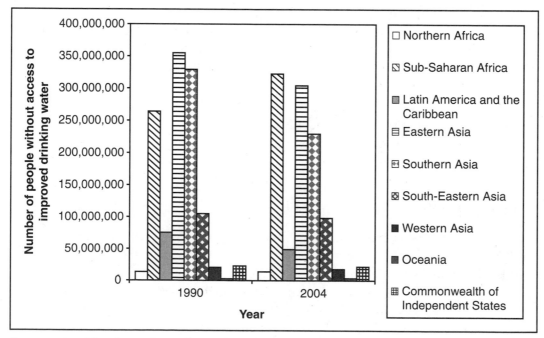

FIGURE 4.4 Number of people without improved drinking water, 1990 and 2004, by region.
Source: UNICEF WHO, 2006.

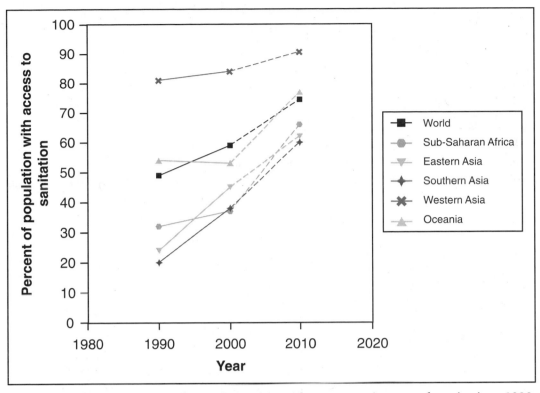

FIGURE 4.5 Percent of population by region with access to improved sanitation, 1990, 2004, and needed to satisfy 2015 target.
Source: UN 2007.

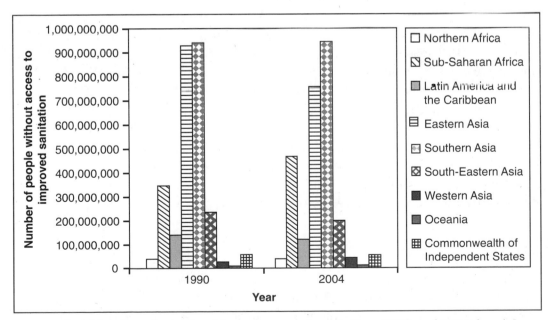

FIGURE 4.6 Number of people without improved sanitation, 1990 and 2004, by region.
Source: National Health Statistics, Ministry of Health, 2005, reprinted in Pankhurst, 2007.
Source: UNICEF WHO, 2006.

and large populations without adequate sanitation. Eastern Asia, which is on track to meet the sanitation MDG, still has the third lowest coverage in sanitation at 45 percent. In these three highly populated places, half of all people do not have a private, safe, and hygienic place to go to the bathroom. Significant increases in the number of people gaining access to sanitation every year need to be made in sub-Saharan Africa and Oceania, as demonstrated by the increased slope of change needed from 2004 to 2015 in Figure 4.5.

Over half of the total number of people that gained access between 1990 and 2004 lived in Eastern and Southern Asia. In Eastern Asia, 330 million people gained access to sanitation, and in Southern Asia, 340 million. Despite this significant progress, population growth in Southern Asia has wiped out almost all these gains. Nearly a billion people are without safe sanitation in Southern Asia, the largest population without access in any of the world's regions. Eastern Asia and sub-Saharan Africa have the second and third largest populations without access to sanitation, at 760 million and 460 million, respectively.

A Closer Look at Water and Sanitation Disparities

While the overall percentages and number of people served in each of the global regions tell an important story, there are also some important disparities that exist within countries and among countries. It might be obvious that in every country, the poor are the ones most likely to lack access to safe water and sanitation. But it is important to recognize that this poses yet another burden for the poor as they try to climb the ladder of development. A UNICEF/WHO study in developing countries found that the richest 20 percent were four times more likely to have access to sanitation than the poorest 20 percent. And while fewer than 4 in 10 of the poorest households had access to improved water, 9 in 10 of the richest households had access (WHO UNICEF 2004). Those who have the least access to water and sanitation are also often the least likely to have health care and stable jobs. Bouts of waterborne disease reduce income further, and for the most vulnerable, often lead to death.

We analyze additional inequalities in water and sanitation access. Achieving the MDG for water and sanitation in a particular country may mask significant disparities between urban or rural areas, and among different regions or groups within the country. Also, each of the technologies that satisfy the definition of improved drinking water or sanitation has different benefits. Consideration of the type of access also helps identify the challenges and decisions faced by residents as they use water and sanitation.

Urban-Rural Divide

The huge disparity among urban and rural areas has been the focus of significant analysis and reporting. Both UNICEF and the WHO acknowledge that the disparity between access to water and sanitation in urban and rural areas needs to be addressed. In the last 15 years, there has been progress in reducing this gap, but a part of this has been the result of increasing rural to urban migration. The reality is that the reduction in the percentage of those without access in rural areas is partly due to efforts to provide access to water and sanitation in these areas, but is also the result of slower population growth in rural areas as people move to cities.

The urban/rural gap in drinking water access in developing countries has been narrowing, although not enough in some regions. In the developing world in 1990, 93 percent of urban residents and 60 percent of rural residents had access to drinking water; by 2004, 92 percent of urban residents and 70 percent of rural residents had access. The largest divide in urban/rural drinking water access persists in sub-Saharan Africa and Oceania, where those living in urban areas are twice as likely to have access to improved drinking water. As noted earlier, between 1990 and 2004, 1.2 billion people gained access to improved drinking water, but two-thirds of these people lived in urban areas (UNICEF 2006).

The urban/rural gap in sanitation access has also narrowed in the last 15 years, although the divide between urban and rural access in sanitation is significantly larger than it is in drinking water. In developing countries, access to sanitation in rural areas nearly doubled, from a very low 17 percent to 33 percent, between 1990 and 2004. During this same period, access to sanitation increased from 68 percent to 73 percent in urban areas. Thus, people in urban areas are still more than twice as likely to have access to a toilet as those in rural areas. Two billion people of the 2.6 billion without access to sanitation live in rural areas. South Asia and East Asia have low rates of access to sanitation in rural areas (28 percent), while over 60 percent of the urban population has access. Western Asia has nearly universal coverage in urban areas (96 percent), while only 60 percent of those in rural areas has access to sanitation.

Regional Inequities in Sanitation and Water Access: The Case of Ethiopia

In addition to the inequities found in urban verses rural areas, there are also regional inequities in sanitation and water access. The national level estimates used to measure progress towards MDGs are valuable to compare countries and continents. But, averaged national numbers often hide extreme regional variability and inequities within countries. Understanding the enormous differences in water and sanitation access within countries requires evaluating disparities within a particular country. For this analysis, we take a closer look at Ethiopia, a country in sub-Saharan Africa with the lowest water access in the world.

In Ethiopia, only 22 percent of the population has access to improved water sources, and only 13 percent has access to improved sanitation. Within these figures, there is a high degree of regional variability. Figures 4.7 and Figure 4.8 demonstrate the inequities in both access to water and sanitation by province. In the province of Afar, for example, only 8 percent of the population has access to improved water, while the capital Addis Ababa has a reported 100 percent access to water, although, as noted earlier this complete coverage number often does not account for lack of access in informal settlements in peri-urban areas. The province of Gambella has the lowest coverage in latrines, at 7 percent, and Addis Ababa, the highest with 76 percent. Addis Ababa has highest percentage of coverage with water and latrines, whereas all of the other regions have less than 50 percent latrine coverage (Pankhurst 2007).

Regional variability is often the result of historic political neglect of areas without political or economic power and can be further exacerbated by ethnic and cultural differences. The MDGs are an important source of international attention and help focus

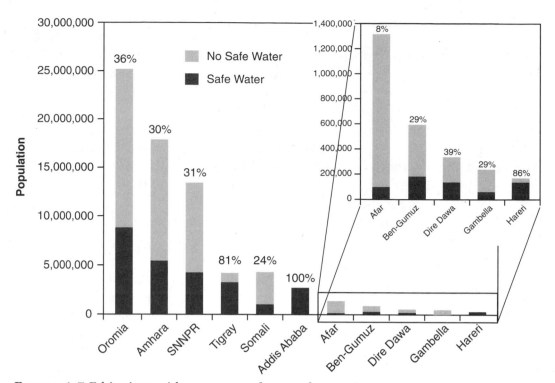

FIGURE 4.7 Ethiopians with access to safe water by province.

Source: National Health Statistics, Ministry of Health, 2005, reprinted in Pankhurst, 2007. The darker bars and percentages shown are population with access to safe water.

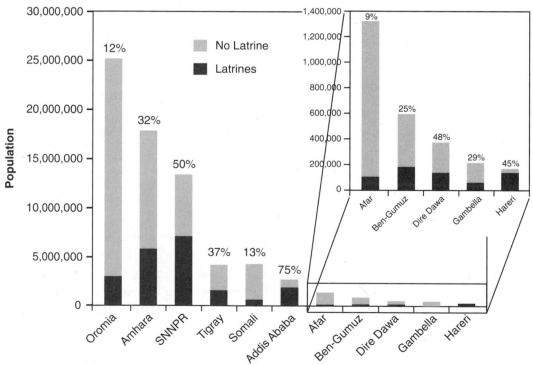

FIGURE 4.8 ETHIOPIANS WITH ACCESS TO ADEQUATE SANITATION BY PROVINCE. The access numbers reported here are significantly different from the 2004 improved water and sanitation statistics compiled by the Joint Monitoring Program. These higher percentages for access come from the Government of Ethiopia. The darker bars and percentages shown are population with access to sanitation and improved water.

Source: National Health Statistics, Ministry of Health, 2005, reprinted in Pankhurst, 2007.

the efforts of international and national organizations. As such, regional information on where the greatest inequities exist can help pressure governments to focus on regions that have been ignored or provided with fewer resources for addressing the problem.

Meeting the MDGs: The Way Forward

Meeting the Millennium Development Goals for water and sanitation is critical for providing a foundation to achieve all of the other development goals. As we near the end of the first decade of the new millennium, the need to intensify the global drive to meet these goals becomes more urgent. While we are making good progress toward meeting the water targets, there are still significant regional, urban/rural, and economic disparities. Furthermore, we are likely to far undershoot the MDG sanitation target.

Even if the MDGs for water are met, preventable waterborne diseases will still take the lives of tens of millions of people, mostly children, by 2020 (Gleick 2002). The human toll of the lack of water and sanitation cannot be overstated: for too many, it is still a matter of life or death. How then, can the global community move forward in the next decade to course correct and reach the MDG for sanitation, and far exceed it for drinking water? We offer a few important recommendations for the next 10 years.

More Resources

Recent studies estimate the economic cost of achieving the MDGs. Guy Hutton and Jamie Bartram conducted a recent analysis of these costs in a paper that reviews and updates previous studies (Hutton and Bartram 2008). They find that reaching the MDGs will require an investment of $42 billion for water and $142 billion for sanitation over the next decade, a combined annual expenditure of $18 billion. Maintaining existing services is estimated to cost an additional $322 billion for water supply and $216 billion for sanitation over 10 years, or a total of $54 billion every year. If this was spread out equally every year from 2005 to 2014, a total annual expenditure of $72 billion would be required to increase coverage to the un-served and maintain and renew existing facilities for those that are already served. *Maintaining* existing services makes up three quarters of the total projected costs of attaining the MDGs. Other estimates vary, depending on assumptions, but this amount does not seem either unreasonable or unachievable.

From where can this money come? Funding in the water and sanitation sector comes from international aid in the form of international development assistance (IDA), national governments, non-governmental organizations, and individual households that pay water bills or even build their own toilets. There are few comprehensive studies of the total amount of funding available in the sector. Funding by NGOs and household expenditures are particularly hard to document. There have been a few estimates of global spending. Hutton and Bartram (2008) review a few estimates in their paper: The Global Water Partnership estimates annual global spending of $14 billion for water and sanitation. The Joint Monitoring Program estimates that in the 1990s a total of $12.6 billion was spent on water and $3.1 billion spent on sanitation every year by government and external support agencies. This does not include private or household expenditures, which particularly in sanitation, can make up a significant part of the total. Nevertheless, the estimates of government and aid spending alone demonstrate that funding in the sector is far less than the total estimated financial need.

Further compounding these dismal numbers, there has been a decline in international financial assistance in the water sector in the past few years. According to the OECD's data, aid to the water sector actually deteriorated from 1996–1998 to 1999–2001. Gleick's (2004) analysis shows that between 1996 and 1998, an average of $3.5 billion per year was given by OECD countries and major international funding organizations. From 1999 to 2001, this decreased to $3.1 billion per year. In addition, the focus of development aid has not been the countries that have the greatest need. Between 2000 to 2004, 10 countries received half of all water-related aid, and none of these countries were in Sub-Saharan Africa.[1] In fact an increase in water aid in 2004 is almost entirely due to the huge amount of reconstruction aid from the U.S. to Iraq. From 2000 to 2004, the U.S. claims to have provided $170 million in water aid to Iraq, accounting for half of *all* U.S. international water assistance (OECD 2005).

And, within countries, water and sanitation are not winning enough of the budgeting debates. The Human Development Report (2006) advises that developing countries should aim to spend 1 percent of GDP on water. Returning to the case of Ethiopia, Pankhurst (2007) documents the consequences of its failure to rally the political will and the needed resources to meet the MDGs. In 2004, Ethiopia had among the lowest rates of access to water and sanitation in the world. In 2005, Ethiopia spent about $26.7 million on water and sanitation, while 1 percent of its GDP would have been an $80 million dollar expenditure, or three times more than what was spent (Pankhurst 2007). There are multiple demands on the budgets of developing countries, and chronic poverty, hunger, emergencies, and environmental calamities often take precedence. Yet, as one of the poorest countries in the world, Ethiopia is also one of the least assisted countries in Africa at $24.1 per capita ODA net in 2004 (Pankhurst 2007).

To meet the resource needs of attaining the MDGs, and to ensure that those with coverage continue to have water and sanitation, efforts at all levels must be intensified. International development assistance must be increased significantly, and these resources must be directed to the countries that are most in need. Within countries, water and sanitation should be given a higher priority, and efforts must be made to increase the social demand for safe water and sanitation services. Of special value may be local community processes that can develop and maintain these services. Greater efforts need to be made to provide capital for household or community-level solutions to drinking water and sanitation needs. Emerging microfinance initiatives to provide funding for water and sanitation projects should be expanded. NGO resources should be targeted to sanitation efforts (which are typically under-funded) and focused on ensuring the sustainability of services.

---

1. The top 10 countries receiving water-related aid in 2000–2004 were (in order of amount of aid received): China, Iraq, Viet Nam, Palestinian adm.areas, India, Jordan, Malaysia, Morocco, Peru, and Tunisia (OECD 2005).

Sanitation and Hygiene

The key components of the WASH challenge are water, sanitation, and hygiene. Unfortunately, for many decades, sanitation and hygiene have been largely ignored or under-funded, despite the fact that in almost every country in the world, more people live without adequate sanitation than live without safe drinking water. For years, international agencies, development banks, donors, country governments, and NGOs have put fewer resources and attention towards sanitation and hygiene as compared with water. This has left 2.6 billion without sanitation in 2004, a minor improvement in the total number of people without sanitation in 1990.

Sanitation and hygiene are less "sexy" than their far more universally favored sibling—water. It is far more satisfying for a donor to put a plaque by a new well or standpipe than next to a new toilet bank or wastewater treatment facility. Whereas the benefits of sanitation accrue to the community at large, the benefits of safe drinking water primarily benefit the household that is served. This is why studies of the willingness of consumers to pay for services find that people are willing to pay more for water services than their actual cost, while they value sanitation services far lower than their actual cost (The Millennium Project 2005).

Despite the difficulties in prioritizing sanitation, sanitation plays a critical role in stopping the spread of waterborne disease. Diagrams detailing the spread of diarrheal disease demonstrate that safe sanitation, because it stops the fecal-oral route for spread of disease at its source, is effective at closing off a number of areas of potential transmission. Secondly, the WHO has compiled studies showing that hygiene education and sanitation are significantly more effective at reducing diarrheal disease than improved water (Table 4.5).

Efforts in sanitation in the last few decades have taught the WASH community some important lessons. Past development efforts have found toilets that were built were unused or not used for their intended purpose. Generating demand for sanitation and hygiene and helping the community develop and implement sanitation that meets their needs is the most effective way of creating long-term sustainable change in the sector.

Sanitation needs to be made a priority in resources and attention. Governments and multi-lateral agencies should work to institute appropriate policies that support community-directed and sustained efforts in sanitation. The UN Millennium Project (2005) recommends the following roles of the government to drive sanitation improvements in the country:

- Commissioning (and funding) research into communities' priorities, needs, preferences, and practices, as well as into factors that motivate behavior change.

TABLE 4.5   Percent Reduction in Diarrhea Morbidity by Various WASH Interventions

Intervention	Percent Reduction In Diarrhea Morbidity
Hygiene Education	45%
Point of Use Water Treatment	35–39%
Sanitation Improvements	32%
Water Supply Improvements	6–25%

Source: WHO 2004: http://www.who.int/water_sanitation_health/publications/facts2004/en/index.html

- Funding an effective national hygiene promotion program.

- Funding an effective national sanitation marketing program.

- Supporting policies that spur expansion of services, such as the provision of microcredit and support for small-scale, independent service providers.

- Promoting and financing innovations in low-cost sanitation technologies, especially those appropriate for congested settlements.

- Requiring and financing hygiene curricula and separate sanitary facilities for girls and boys in schools.

- Targeting public funds toward elements of sanitation systems for which the public benefit is greater than the private benefit (for example, trunk infrastructure, shared facilities, environmental infrastructure, and household facilities for the small proportion of households whose effective demand is not high enough to obtain hygienic sanitary facilities.

- Supporting the development of community-based "franchising" approaches that are flexible, sustainable, and replicable on a large scale.

Focus on Sub-Saharan Africa

As the earlier sections highlighted, the greatest needs for increased effort to provide water and sanitation are in sub-Saharan Africa, as a result of rapid population growth, slow economic growth, inadequate or misdirected government efforts, and conflicting demands for limited public resources. Of all the major regions in the world, sub-Saharan Africa has the lowest coverage in drinking water access, at 56 percent.[2] Nearly all of the other regions of the world have drinking water access in excess of 75 percent.[3] Sub-Saharan Africa also has among the lowest rates of sanitation coverage in the world at 37 percent. Although Eastern Asia and Southern Asia had significantly lower sanitation coverage rates in 1990, by 2004 these regions had almost surpassed sub-Saharan Africa in access to sanitation.[4]

Other regions with significant population growth and low rates of access to water and sanitation, such as Eastern and Southern Asia, are achieving economic growth that helps to meet increased demands for water and sanitation. In sub-Saharan Africa, recent economic growth is still slower than GDP growth rates in Eastern and Southern Asia and has not been sufficient to provide an economic engine for increased development activities.

There is an urgent need to corral international funding and attention to address the dire need in sub-Saharan Africa. For the developed nations, this means greatly increasing aid dollars to improve water and sanitation in the region. Within the countries in sub-Saharan Africa, there needs to be more political pressure to increase the percentage of the government budget that is devoted to water and sanitation efforts. Other developing countries that have been successful at increasing coverage in water and sanitation should sponsor bi-lateral information exchange sessions, so that policy-

---

2. Oceania, which hosts 1/100th the number of people as Sub-Saharan Africa, has lower levels of access to drinking water at 50 percent.

3. Except Oceania, which has lower water access than Sub-Saharan Africa.

4. Southern Asia had 38 percent sanitation access in 2004; Eastern Asia had 45 percent access.

makers and practitioners can visit regions with successful water and sanitation efforts and receive support in implementing these approaches in their country or region.

Disparities within Countries

Significant differences in access to water and sanitation exist among regions and economic groups within the same country. This is also true when it comes to attaining the MDGs. One of the goals of the MDGs should not only be to increase sanitation and water access within countries, but also to reduce inequities in water and sanitation access within countries.

For political, cultural, or environmental reasons, more attention may have been paid to particular regions of a country in terms of providing water and sanitation. Those in political power will often provide increased resources and infrastructure to their home region or areas where their ethnic or cultural group predominates. As a result, other regions may suffer from historic political neglect. These regions with the least economic and political power are also the least able to demand improved services. Reversing this trend requires greater spotlight placed on regional inequities within countries in the provision of water and sanitation.

The rich often pay far less for water as a fraction of their income (or even in absolute terms) than the poor; this should be rectified. When wealthier urban consumers pay the full costs of water services, then these services can be extended to poorer peri-urban and slum areas. Ideally, as we evaluate progress toward the MDGs, we should reward increases in access to sanitation and water in countries as well as the reduction in inequities among regions, between the urban and rural, and between the rich and poor.

Conclusion

At the turn of the century, the leaders of the development community came together to call for an end to poverty and inequality and set the Millennium Development Goals in place. Among the eight goals, the goal on environmental sustainability contained targets to halve the proportion of people without access to safe drinking water and improved sanitation by 2015. The world seems to be on track to meeting the drinking water target, but individual regions such as Sub-Saharan Africa, Oceania, and some countries in the Commonwealth of Independent States are falling further behind. Progress toward the sanitation target has been far less satisfactory, and several regions seem certain to fail to meet the sanitation targets by a large amount, especially countries in Southern Asia and Sub-Saharan Africa.

The Millennium Development Goals have certainly increased international and national attention to the critical nature of the water and sanitation challenge. If we have any hope of achieving, or ideally, exceeding the MDGs by 2015, the global community needs to invest more resources, focus on sanitation, increase financial and technical assistance in sub-Saharan Africa, and direct resources to reduce disparities within countries. Access to safe water and sanitation services is a foundation upon which all other MDGs are based. Reducing poverty, improving health, and providing opportunities for advancement must begin with providing people with their most basic needs for water to drink and a safe place to go to the bathroom.

REFERENCES

Bartram, J. 2007. Improving on haves and have-nots. *Nature* 452(3):283-284.

Gleick, P. 2002. *Dirty Water: Estimated Deaths from Water-Related Diseases: 2000-2020*. Oakland, California: Pacific Institute.

Gleick, P. 2004. The millennium development goals for water. In: *The World's Water 2004–2005: The Biennial Report on the World's Freshwater Resources*. Washington, D.C.: Island Press. pp. 1–15.

Hutton, G., and Bartram, J. 2008. *Regional and Global Costs of Attaining the Water Supply and Sanitation Target (Target 10) of the Millennium Development Goals*. Geneva, Switzerland: Public Health and the Environment, World Health Organization.

Organization for Economic Cooperation and Development (OECD). 2002. *Creditor Reporting System, Aid Activities in the Water Sector: 1997–2002*. Paris, France: Development Assistance Committee.

Organization for Economic Cooperation and Development (OECD). 2005. Has the Downward Trend in Aid for Water Reversed? Measuring AID for Water. www.oecd.org/dac/stats/crs/water

Pankhurst, H. 2007. *Ethiopia Water and Sanitation Sector: Progress Towards Targets for Water and Sanitation MDGs*. United Kingdom: Department for International Development.

Shordt, K., van Wijk, C. Brikke, F., and Hesselbarth, S. 2004. *Monitoring Millennium Development Goals for Water and Sanitation*. Delft, the Netherlands: IRC International Water and Sanitation Centre.

UNICEF. 2006. *Progress for Children: A Report Card on Water and Sanitation*. Number 5, September 2006. New York, United States: United Nations.

UN. 2007. *The Millennium Development Goals Report 2007*. New York, United States: United Nations.

UN Millennium Project 2005. *Health, Dignity, and Development: What Will It Take?* Task Force on Water and Sanitation. http://www.unmillenniumproject.org/documents/WaterComplete-lowres.pdf

WHO/UNICEF. 2004. *Meeting the MDG Drinking Water and Sanitation Target: A Mid-Term Assessment of Progress*. Geneva and New York: United Nations.

WHO/UNICEF. 2006. *Meeting the MDG Drinking Water and Sanitation Target: The Urban and Rural Challenge of the Decade*. Geneva and New York: United Nations.

CHAPTER 5

China and Water

Peter H. Gleick

The remarkable growth in China's population and economy over the past several decades has come at a tremendous cost to the country's environment. China has experienced an economic growth rate averaging 10 percent per year for more than 20 years. But sustained growth and the health of the country are increasingly threatened by environmental deterioration and constraints, particularly around water. Water is critical for economic growth and well-being; conversely, economic activities have an impact on water availability and quality. When water resources are limited or contaminated, or where economic activity is unconstrained and inadequately regulated, serious social problems can arise. And in China, these factors have come together in a way that is leading to more severe and complex water challenges than in almost any other place on the planet.

China's water resources are overallocated, inefficiently used, and grossly polluted by human and industrial wastes, to the point that vast stretches of rivers are dead and dying, lakes are cesspools of waste, groundwater aquifers are over-pumped and unsustainably consumed, uncounted species of aquatic life have been driven to extinction, and direct adverse impacts on both human and ecosystem health are widespread and growing. Figure 5.1 shows the major rivers of China. Of the 20 most seriously polluted cities in the world, 16 are in China. The major watersheds of the country all suffer severe pollution. Three hundred million people lack access to safe drinking water. Desertification, worsened by excessive withdrawals of surface and groundwater, is growing in northern China (Feng 2007).

These problems are threatening to slow economic expansion and weaken political stability in a variety of ways. Significant outbreaks of illness, including cancers, are being reported in heavily polluted regions, driving up health care costs and public concern. Companies are canceling business ventures because of water concerns. There is growing internal dissent and conflict over both water allocation and water quality, raising new political pressures on the central and regional governments to come to grips with water problems. In 2005, the Chinese government acknowledged that 50,000 environmentally related protests occurred that year, many of which were related to water degradation (Turner 2006). Even the official Chinese media has reported that "The pursuit of economic growth has been the priority overshadowing the vital issues of water resources and ecological balance" (China Daily 2007a). It is not yet clear how quickly the Chinese will get their severe water challenges under control, or at what ultimate cost to human and ecological conditions.

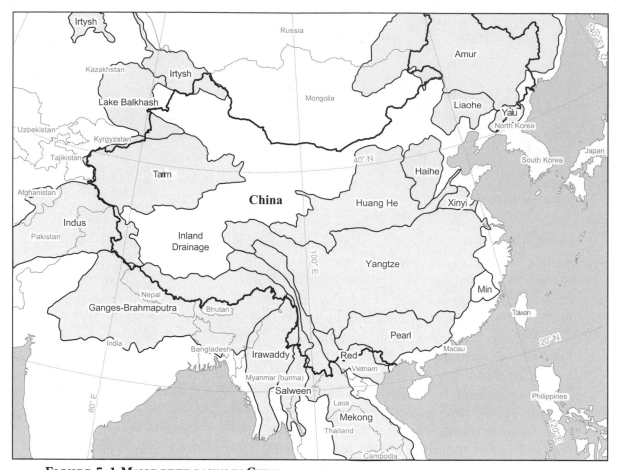

FIGURE 5.1 MAJOR RIVER BASINS IN CHINA.
Source: ESRI, USGS/WWF. Lambert conformal conic projection. Courtesy of Matthew Heberger.

Addressing China's crippling water problems is hampered by the efforts of local governments to protect local industries and jobs, government corruption, the desire to sustain rapid economic growth, and what has been described as the "crippling weakness" of the leading national environmental regulatory body, the State Environmental Protection Administration (SEPA) (Turner 2006). At the same time, these problems have encouraged public concern and efforts. Grassroots environmental efforts have grown in China and have had some success at raising awareness and spurring action, although nongovernmental organizations are still harassed and viewed with suspicion by officials. This chapter reviews the state of water problems in China and offers insights into new trends and efforts to address those problems.

The Problems

Water Quality

Comprehensive data on water quality in China are hard to find, either in English or Chinese, but there are growing indications of China's severe water contamination. The State Environmental Protection Agency (SEPA) publishes quarterly reports, but the consistency and accuracy of the data in those reports are uncertain. The World Health Organization (WHO) and the Organization for Economic Cooperation and Development (OECD) in Paris also publish data on water quality in China.

These data offer some snapshots: Chinese statistics estimate that 40% of the water in the country's surface waters was fit only for industrial or agricultural use, and even then only after some treatment. An estimated 20,000 chemical factories, half of which are along the Yangtze River, are dumping uncontrolled or only marginally controlled pollutants into China's rivers. In 2006, nearly half of China's major cities did not meet state drinking-water quality standards (OECD 2007), and a third of surface-water samples taken were considered severely polluted (Xinhua 2007f). The Tenth Five-Year Plan (2001–2006) mandated the construction of thousands of new wastewater treatment plants, yet a 2006 survey by SEPA revealed that half of new plants actually built were operating improperly or not at all (Boyle 2007). Groundwater quality is degraded by the routine and massive dumping of untreated or partially treated wastewater.

According to SEPA statistics, China experienced over 1,400 environmental pollution accidents in 2005, around half of which involved water pollution (Xinhua 2007j), and many incidents are never reported. In May 2007, the SEPA released its first quarter 2007 report indicating little improvement over time in China's seven main rivers, and it noted significant deterioration in water quality in the Songhua, Hai He, and Huai He rivers, and in Taihu, Chaohu, and Dianchi lakes, despite efforts to clean them up. According to this report, only 69 percent of key cities met national potable water standards (China Daily 2007s, Xinhua 2007p), an improvement from previous years but still problematic. Officially, Beijing tap water has been declared safe to drink under China's new national drinking water standards for 106 contaminants, but complaints in parts of the city indicate that local sources of contamination still affect quality, particularly the old distribution system that was put in place 50 to 60 years ago (China Daily 2007p).

As a consequence of these problems, the OECD estimates that hundreds of millions of Chinese are drinking water contaminated with inorganic pollutants such as arsenic and excessive fluoride, as well as toxins from untreated factory wastewater, inorganic agricultural chemicals, and leeching landfill waste (OECD 2007). In an extreme indication of the growing concern over water quality, local farmers in contaminated regions grow grain with poor quality water, sell that grain, and purchase grain from other parts of China they believe has safer water (Guo 2007). In the Huai He basin, widely acknowledged to be extremely badly polluted, there are numerous villages where no young men have been able to pass the physical examination for entering the Army, which some analysts ascribe to water-related illnesses and contamination (Economy 2004).

Poor water quality is having an impact on Chinese cancer rates. The Ministry of Health acknowledged in 2007 that air and water pollution (together with food additives and pesticides) helped make cancer the most lethal disease for urban residents in China. "The main reason behind the rising number of cancer cases is that pollution of the environment, water and air is getting worse day by day," said Chen Zhizhou, a health expert with the cancer research institute affiliated to the Chinese Academy of Medical Sciences. Reports on "cancer villages" have appeared more frequently in recent years, with clusters of cancers being linked to the use of heavily polluted water (China Daily 2007t).

These problems are increasingly well known in China, but despite rhetoric from officials, little progress has been made in reducing discharge of pollutants, according to Zhou Shengxian. Zhou is the director of the State Environmental Protection Administration (Xinhua 2007f). Efforts to clean up China's grossly polluted rivers have been

FIGURE 5.2 PROVINCES IN CHINA.
Source: ESRI, USGS. Lambert conformal conic projection. Courtesy of Matthew Heberger.

underway for more than a decade, with limited results. The Huai He remains heavily polluted despite major government investments and efforts of local and national authorities for more than a decade. Untreated wastewater volumes still exceed national standards; chemical oxygen demand remains high – 30 percent above targets, even by official statistics. Official data show that more than 4.4 billion tons of untreated or partially treated wastewater are dumped into the river annually (China Daily 2007k). Part of the problem is China's very large population. The Huai He, for example, runs through four major provinces, including Henan, China's most populous province, with more than 100 million people. Another part of the problem, however, rests with weak, incompetent, or corrupt public environmental agencies (see Fig. 5.2 for a map of China's provinces).

Water-Related Environmental Disasters in China

In the United States, much of the existing environmental legislation had its origins in environmental disasters, such as the burning of the Cuyahoga River, the Love Canal toxics catastrophe, and severe air-quality contamination events. Such disasters are now beginning to occur with disturbing regularity in China. As in the United States in the 1970s, these disasters are spurring social concern and political activism.

In 2005, a severe environmental disaster occurred on the Songhua River, when a chemical plant explosion in the city of Jilin contaminated the river with 100 tons of benzene-related pollutants. The contamination flowed downstream and forced the temporary suspension of water supply to nearly 4 million people in Harbin, the capital of Heilongjiang Province. It also led to contamination problems in the Russian city of Khabarovsk along the Heilongjiang River shared by China and Russia. This incident was covered extensively in worldwide media and led to new efforts on the part of the Chinese government to tackle water-quality problems (China Daily 2007o).

After the Songhua disaster, the government prepared plans to build over 200 "pollution control projects" along the Songhua River at a cost of nearly \$2 billion. In Heilongjiang Province, the local government inspected 4,000 commercial and industrial enterprises and shut down a small number of them in an effort to cut the worst pollution. The city of Jilin in Jilin Province built a new sewage treatment plant to process a substantial amount of previously untreated waste. And new monitoring systems have been installed in the Songhua and Heilong rivers (Xinhua 2007h).

The benzene contamination incident in Jilin is not the only water disaster in recent years. Several incidents have caused the shutdown of local or municipal water systems (Eng and Ma 2006). A mere three months after this accident, a plant in Sichuan Province spilled toxins into the upper reaches of the Yuexi River, disrupting the water supply of 20,000 people in the city of Yibin (Turner 2006). According to statistics provided by the SEPA, another 130 water pollution incidents occurred after the Songhua River spill by September 2006.

In 2007, local reservoirs around Changchun City in Jilin Province suffered a blue-green algal outbreak attributed to pollutants from both industrial and agricultural sources, including both fish and pearl farms, which rely on heavy use of fertilizer and pesticides. Such outbreaks lead to the suffocation of native fisheries as the algae consume all of the oxygen in the water. The outbreak also threatened the quality of water, and led to a reduction in drinking water supply to the city of more than seven million. Similar blue-green algae outbreaks were reported in Taihu Lake, Chaohu Lake, and Dianchi Lake, threatening local domestic water systems, leading the director of SEPA order all fish farms to be removed from the three lake areas by the end of 2008 (Xinhua 2007m, China Daily 2007q).

In mid-2007, a series of water contamination incidents in Jiangsu Province in eastern China led to the cutoff of water supplies to millions of people. A severe blue-green algae outbreak affected tap water in the city of Wuxi. A subsequent surface water incident led to severe contamination by ammonia, lead, and nitrogen. That water apparently originated from industrial sources upstream in Shandong Province (China Daily 2007o, China Ministry of Water Resources 2007, Xinhua 2007m). Also in 2007, the city of Yan'an in northwestern Shaanxi Province was forced to shift water supply after the major reservoir for the city was polluted by crude oil from a broken pipeline that contaminated the Xingzihe River. Yan'an has a population of 2.15 million (Xinhua 2007k).

Water Availability and Quantity

China faces serious water challenges from constraints on water supply as well as deteriorating quality. China's per-capita annual renewable water availability is around 2,140 cubic meters compared to 1,720 m^3/p/yr in India and over 10,000 for the United

TABLE 5.1   Per-Capita Water Availability (Total Renewable Water Resources) (m^3/person/yr): 2003–2007 Average

China	2,138
India	1,719
United States of America	10,231

Source: 2008 FAO of the UN: Aquastat database, www.fao.org.

TABLE 5.2   Major Rivers in China With Their Average Annual Runoff

River	Length (km)	Drainage Area (km^2)	Average annual runoff (km^3)
Changjiang (Yangtze)	6,300	1,808,500	951.3
Huang He (Yellow)	5,464	752,443	66.1
Heilongjiang (Amur)	3,420	896,756 *	117.0
Songhua (Sungari)	2,308	557,180	76.2
Xijiang (Pearl)	2,210	442,100	333.8
Yarlung Zangbo	2,057	240,480	165.0
Tarim	2,046	194,210	35.0
Lancangjiang	1,826	167,486	74.0
Nujiang	1,659	137,818	69.0
Liao He	1,390	228,960	14.8
Hai He	1,090	263,631	28.8 **
Huai He	1,000	269,283	62.2
Irtysh	633	57,290	10.0
Luan He	877	44,100	6.0
Minjiang	541	60,992	58.6
Total		5,224,473	2,039.0

Notes:
* Including the Songhua River Basin
** Including the Luan He River Basin

Source: http://www.eoearth.org/article/Water_profile_of_China

States (Table 5.1). The distribution of water in China, as in other countries, is highly variable in both space and time. While parts of China have abundant natural water resources, other regions are naturally arid and water scarce; for example, northern China is far drier than southern China. China has several of the world's largest rivers, bringing water from the Tibetan Plateau and western China to coastal cities. Table 5.2 shows the major rivers in China along with their average annual runoff.[1] These rivers are unevenly distributed, with large rivers and flows in the south.

China has also long suffered from extremes of floods and droughts. Some of the worst floods on record, in terms of loss of human life, have occurred in China,

---

1. Westerners know China's rivers by different names than often used by the Chinese, such as the Yellow (Huang He) or the Yangtze (Changjiang). While this chapter tries to consistently use the Chinese names, some of the more familiar western names are used for some of the major rivers. Table 5.2 lists both the Chinese transliterations and the most common English names.

including a flood in 1930 that claimed 3.7 million lives. Half a million more people died in floods in 1939 and another 2 million died in floods in 1959 (Cooley 2006). And periodic droughts are worsening China's water-supply challenges, as described.

This uneven distribution, combined with China's extensive population, inadequate urban infrastructure, and poor management, have caused more than two-thirds of the country's more than 600 cities to suffer from water shortages. Over 100 of them are seriously affected.

Guangdong Province is located in China's southern subtropical zone and is relatively water rich compared to other parts of China. Yet even here water-quantity problems are developing as a result of inefficient and wasteful use combined with growing demand and drought. Quotas on water use for industry, agriculture, and residences are being imposed in the province for a 2-year trial period (China Daily 2007a).

These concerns over supply, along with growing overdraft of groundwater, are an increasing problem for Chinese officials and water managers and are driving investments in new infrastructure and demand management efforts. To make matters worse, large numbers of Chinese do not have access to safe water and adequate sanitation – a consequence of both water quantity and quality problems. In 2004, 88.8 percent of China's urban population reportedly had access to clean drinking water and 70 percent had access to adequate sanitation, but availability of both is significantly lower in rural areas and the data are self-reported (Ministry of Foreign Affairs 2005). Government officials acknowledged in 2007 that 300 million rural Chinese had no access to safe drinking water (Xinhua 2007i, Lee 2007).

The failure to meet basic human needs for water in China, as elsewhere, leads to water-related diseases and preventable deaths, especially among children. Long-term data on water-related diseases are hard to find for China, but official statistics from the mid- to late-1990s suggested that intestinal worms, associated with the lack of safe water and adequate sanitation, are a severe problem in rural China. In 1992, for example, the Chinese Ministry of Health reported that roundworm infected nearly 200 million children under the age of 14, with additional infections from hookworm and whipworm (NPHCCO 1999). Between 1995 and 1999, typhoid incidence rates in rural Guangxi Province ranged from 27 to 153 per 100,000 (Yang et al. 2001, Yang et al. 2005). Typhoid continues to be endemic in southern China despite recent progress in meeting basic water needs (Boyle 2007). The OECD Environmental Indicators in China report issued in July 2007 estimated 30,000 rural children die each year from diarrhea caused by polluted water (OECD 2007). The World Health Organization reported an incidence of 108.4 mortalities per 100,000 persons from diarrhea-related illness in China in 2002 (WHO 2003). In comparison, Vietnam's diarrheal disease mortality rate in 2002 was under 11 per 100,000 people; Thailand's was under 5 (WHO 2004).

Groundwater Overdraft

One critical consequence of China's maldistribution of water is excessive, and ultimately unsustainable, withdrawals of water in more arid regions. As China has grown, its policy of food self-sufficiency has led to extensive agricultural production in the North China Plain, a region with relatively limited natural endowment of water. The North China Plain produces around half of all of China's wheat. Throughout this area, especially where populations have soared in recent years, groundwater is being pumped out far faster than it is naturally recharged and levels are falling fast. Some groundwater levels are now hundreds of meters below ground (Griffiths 2006). These levels of pumping cannot be sustained. "There will be no sustainable development in the future if there is no groundwater supply," acknowledged hydrologist Liu Changming of the Chinese Academy of Sciences (Griffiths 2006).

Overpumping and contamination of groundwater is forcing cities and business to dig deeper to find clean, adequate supplies. In northern Hebei province, villages are digging 120 to 200 meters to find clean drinking water; a decade ago wells were only 20 to 30 meters deep. Deep wells cost thousands of yuans – as much as half the annual income of farmers (Guo 2007).

Among the consequences of groundwater and surface water overdraft is the loss of wetlands. One survey estimated that over 80 percent of the wetlands in the North China Plain have been lost, and natural streams and creeks have dried up. Major river levels have dropped significantly due to human consumption. Northern China's largest natural freshwater lake, Lake Baiyangdian, is both disappearing and grossly polluted (Griffiths 2006). Until groundwater withdrawals are limited to sustainable levels, China's economic productivity will be threatened by rising water costs and scarcity.

Overuse of groundwater is even affecting China's most well-known cultural monument, the Great Wall. A 220-kilometer long section of the wall runs through the Minqin region of China in the Shiyang river basin. Withdrawals of water from the basin have reduced both surface and groundwater levels and led to desertification in the Minqin oasis region. Groundwater levels, for example, have dropped by 14 meters in the past half century. This in turn has led to the burial of large sections of the Great Wall by sand. Li Bingcheng, an expert on the Great Wall, said the sections of the wall in Minquin will be gone in 10 to 20 years if action is not taken to reduce the threat (China Daily 2007j).

Recent Floods and Droughts

China has experienced consecutive droughts over recent years with significant economic consequences. According to Zhang Jiatuan of China's State Flood Control and Drought Relief agency, "Since the 1990s, losses from drought have been equivalent to 1.1 percent of China's average annual gross domestic product, or about 300 billion yuan ($41 billion)" (China Daily 2007c).

Even the more relatively water-rich regions of the country appear to be experiencing increasing natural shortages. In 2007, a severe drought left well over a million people short of drinking water in southern China (Xinhua 2007a,b) and was spreading throughout the country (China Daily 2007c). The drought, which decreased rainfall between 20 and 35 percent from normal in the region, dried up hundreds of water-supply reservoirs, and thousands of wells, according to the Guangdong Provincial Hydraulics Bureau, and even the major rivers of the Yangtze (Changjiang), Yellow (Huang He), and Zhujiang are low. 2007 also saw a decrease in the water level of China's largest freshwater lake, Poyang Lake, to its lowest level in recorded history because of a combination of low rainfall and excessive human withdrawals of water (Xinhua 2007d). The low lake level led to shortages of drinking water for local residents and to cutbacks in industrial production.

China is also prone to severe flooding because of a combination of large rivers, variable climate, and vast populations living in floodplains. In 2005 more than 1,000 people were killed in China's annual flood season, while in 1998, 4,185 people lost their lives in the deadliest rainy season of the past decade (China Daily 2007l). In just a few weeks of heavy rainstorms in central China in 2007, almost 200 people died from flooding. By the end of 2007, floods nationwide had affected 180 million people, with over 1,200 deaths. The 2007 floods also ruined 12 million hectares of crops and destroyed more than one million houses, leading to a direct economic loss of over 100 billion yuan (Xinhua 2007c, l).

The economic impact of floods on China's economy is greater than that felt by most industrialized countries, and the Chinese Minister of Water Resources, Chen Lei, said in 2007 that China's annual direct economic losses from floods since the 1990s averaged 110 billion yuan (Xinhua 2007c) or nearly 2 percent of national GDP. These figures, if correct, are substantially higher than flood losses in the United States, which have been pegged at an average of less than 0.25 percent of GDP annually (Xinhua 2007c).

Part of the problem is that so many people live in areas prone to flooding, and these numbers are growing. Minister Chen projected that by 2020, forty-one percent of China's population will be exposed to flood risks (Xinhua 2007c). Almost 67 percent of the country's gross domestic product (GDP) also comes from these vulnerable regions.

Climate Change and Water in China

Climate changes will have direct and significant impacts on water availability and quality by altering precipitation patterns, increasing the intensity of extreme events, raising water temperatures, and accelerating the melting of snow and glaciers. Some Chinese experts have begun to publicly attribute increasing severity of drought to man-made climate change (Xinhua 2007n). Minister of Water Resources Chen Lei said in 2007 that China is already suffering a shortfall of water supply of around 40 billion cubic meters annually because of climate change and that there has been both increased flooding and drought (China Daily 2007b). In particular, data from the ministry suggests that rainfall in northern China is decreasing, and resources in the watersheds surrounding the Yellow River, Huai He River, Hai He River and Liao He River had dropped by 12 percent over the past decade (China Daily 2007b). "The lack of rain is mainly due to global warming," Chinese climate experts are reported to have said (China Daily 2007c). Precipitation seems to have increased in western China at the headwaters of some major rivers, but this is not translating into increased flows because evaporation rates (and unmonitored human withdrawals) are also rising rapidly (Xinhua 2007n).

One of the most significant risks to water resources from climate change is expected to be dramatic changes in snowfall and snowmelt dynamics (IPCC 2007, National Assessment 2000). In some countries, this will mean more rapid glacier melt and retreat, with impacts on long-term water availability to downstream communities. Such glacier melt is already being seen in most regions, including China, which gets as much water annually from glaciers as from the entire flow of the Yellow River. China ranks fourth in the world in terms of both area and ice volume of glaciers, after Canada, the United States, and Russia. China's glaciers cover approximately

60,000 square kilometers and have a total volume of 5,590 cubic kilometers (China Daily 2007e).

Scientists are reporting that the overall area occupied by glaciers has shrunk by about a third over the past century (China Daily 2007i). They further stated that global warming will make the trend of retreating glaciers "irreversible" (China Daily 2007e).

The shrinking of China's rivers at their mouths has long been observed and attributed to overuse and excessive withdrawal of water along those rivers. Recently, however, drying of China's major rivers has also been observed at the source and headwaters of those rivers, leading the Chinese Academy of Sciences (CAS) to conclude that climate change is already having an effect (Coonan 2006). The water resources of the Sanjiangyuan region - the headwaters of the Yangtze, Yellow, and Lancang rivers – depend on glacier melt and appear to be diminishing. This region, also known as the Qinghai-Tibetan Plateau, provides 25 percent of the water flowing down the Yangtze River, 49 percent of the flow of the Yellow River, and 15 percent of the flow of the Lancang River (China Daily 2007i). The Qinghai-Tibetan Plateau used to host 36,000 glaciers covering an area of 50,000 sq km, but their area has shrunk by 30 percent over the past century (Xinhua 2007n). In 2007, Chinese scientists warned that major glaciers in China, including the most well-known "Glacier No. 1" at the headwaters of the Urumqi River in the Tianshan mountains had decreased by over 10 percent in the past four decades and that the rate of retreat is accelerating. The loss of river flows from the dwindling Glacier No. 1 is threatening oases in the Xinjiang Uygur Autonomous Region (China Daily 2007e).

In July 2007, the CAS issued a report concluding that climate changes are also shrinking wetlands on the Qinghai-Tibetan plateau (Associated Press 2007). Aerial photos and remote sensing from satellites show that the wetlands have shrunk more than 10 percent in the past 40 years, with losses of nearly 30 percent occurring at the headwaters of the Yangtze. Even though rainfall has increased in the region, the increase in evaporation from warmer temperatures has more than compensated. Other observed changes include melting permafrost and dying vegetation.

Water and Chinese Politics

Water problems have begun to affect local and regional politics in China. President Hu Jintao in his 2007 report to the 17th National Congress of the Communist Party of China (CPC) called for a more efficient and environmentally friendly approach to development, growth, and consumption. Hu called for more "scientific development" that focused on major water issues including "securing more clean drinking water, improving water conservation, water pollution prevention, restricting excessive water resources exploitation and cutting water waste" (China Daily 2007b). Ma Jun, a water expert and the author of the book *China's Water Crisis*, publicly warns that current levels of water consumption and contamination are unsustainable (Ma 1999).

China's water problems are exacerbated by water laws that remain outdated, weak, and inadequately enforced. Because China is a heavily centralized country, governments at both the national and regional levels play a critical role in water policy and management, with traditionally little input from non-governmental organizations or individual participation in review and decision making. But there has been little comprehensive water policy development and few consistent national laws. Most water-quality laws were put in place several decades ago and lack enforcement mechanisms, with minimal fines and "vague civil liabilities" for polluters (China Daily 2007n). These laws have also traditionally limited the power of national environmental agencies in favor of local control, leading to widely differing levels of enforcement, incentives for local corruption, and confusing standards for industries.

There is growing perception that the nation's water woes result from insufficient centralized regulation – an odd problem in a country often perceived to be dominated by a strong centralized government. In fact, China – like the United States – manages water resources with a complex set of agencies at all levels of government, from the local to the central. Responsibilities for water resources, data and information, construction of infrastructure, environmental protection, agricultural development, transportation, and other water-related activities are split among competing and conflicting institutions.

China has also devolved substantial management responsibility to provincial and sub-provincial governments, undermining watershed-based management efforts (Eng and Ma 2006). Some advocates of centralized water management tools call for increasing the power of Beijing at the expense of provinces. They point to the success of centralized management in helping to restore at least some perennial flows in the Yellow River delta, which drew international attention in the late 1990s, when flows in the delta disappeared for over 200 days per year because of excessive withdrawals upstream. At that time, Beijing imposed limits to water allocations to the provinces.

Like similar historical trends in the United States and Eastern Europe, some of the first effective citizen organizations are developing around environmental issues, specifically water. While there continue to be only a small number of NGOs addressing water quality and quantity problems in China, their numbers and influence are expanding. In the past 5 years, a growing number of Chinese NGOs have begun to track water issues and to challenge projects they deem damaging. In the few cases in which public participation has been permitted, several large infrastructure projects have been successfully delayed, which some observers think is sending a signal encouraging more public participation (Eng and Ma 2006). This is leading to a struggle between existing powerful interests and environmental groups in China. Eng and Ma (2006) and Yardley (2007b) have offered examples that these efforts are having an effect:

- Local organizations and individuals worked to inform the public and media about the impacts of Yangliuhu Dam on an ancient and still functioning irrigation system that had been declared a World Cultural Heritage Site. Extensive media coverage and public dissent forced the developer to abandon the project in 2003.

- In 2004, Chinese NGOs opposed development projects on the Nujiang, one of the last two free-flowing rivers in China. Their efforts drew national attention and led Premier Wen Jiabao to halt the project pending a more comprehensive EIA.

- Environmentalists have been working to preserve the Tiger Leaping Gorge in a campaign to reduce the impact of a massive dam project on ecological and cultural diversity.

- A dam in Sichuan Province that would have inundated an ancient Qin Dynasty cultural site was canceled after local opponents called it an attack on China's heritage.

Among the obstacles NGOs have to overcome are restrictive regulations on their actions and limited internal capacity and funding. This may be changing (see Improving Public Participation).

Growing Regional Conflicts Over Water

As noted earlier, China's water resources are unevenly distributed. Because much of China's water policy revolves around massive transfers of water from one region to another, or large infrastructure projects that affect multiple political jurisdictions, there are growing regional conflicts over water-management decisions.

In one of the most serious examples of regional water conflict, there is a long history of violence over allocations of water from the Zhang River, a tributary of the Hai He that originates in Shanxi Province and flows through both Henan and Hebei provinces (see Figure 5.2 Map of Provinces). Conflicts over excessive water withdrawals and the subsequent water shortages have been worsening for over three decades between villages in Shenxian and Linzhou counties. In the 1970s, militias from competing villages fought over withdrawals. In 1976, a local militia chief was shot to death in a clash between Shexian's Hezhang village and Linzhou's Gucheng village over the damming of Zhang River. The violence escalated significantly in the 1990s: in December 1991, Huanglongkou village of Shenxian county and Qianyu village of Linzhou city actually exchanged mortar fire over the construction of new water diversion facilities. In August 1992, bombs were set off along a distribution canal collapsing part of the canal and causing flooding and economic losses (China Water Resources Daily 2002, Eng and Ma 2006). Despite efforts to mediate the dispute, violence continued in the late 1990s with confrontations, mortar attacks, and bombings, leading up to a clash on Chinese New Year in 1999 that reportedly killed nearly a hundred villagers and caused millions of dollars of damages to homes and water facilities. Some progress has been made to negotiate a settlement to this dispute, but new projects in the region may fuel new disputes (Eng and Ma 2006).

The North China Plains are also seeing growing tensions over water. As the population of Beijing has soared over the past several decades, the city has taken control of almost all of the major rivers flowing through surrounding Hebei Province. Until recently, one exception was the Juma River, a tributary of the Hai He, which flows 30 kilometers from the capital. Both Beijing and Hebei provincial officials have built major water diversions on the Juma leading to a conflict between the two governments. Withdrawals from the Juma now divert almost all of its flow, forcing downstream residents in Hebei to rely on groundwater resources. In the last few years, Beijing moved forward with new plans to tap groundwater connected to the Juma River, to raise its dam on the Juma to capture more water, and to transfer that water to Beijing's Yanshan Petrochemical Plant, the largest industrial water user in the city. Hebei officials fear that this series of new developments will cut off water to nine cities and counties downstream, affecting water supply to nearly three million people, worsen desertification in the region, and threaten the ecology of Lake Baiyangdian in north China (Eng and Ma 2006). Despite protests from top officials in Hebei Province to the Beijing water authority, no effective agreement or collaboration has occurred, and local tensions are rising.

All countries face old and new water challenges and have a variety of economic, institutional, and technological tools available for solving those challenges. But the priorities of the Chinese water management authorities have focused on a limited set of solutions and it is not yet apparent whether officials will move quickly enough to address quality and quantity problems in order to avoid more serious catastrophes in the near future.

Moving Toward Solutions

Expanding Water Supply

The standard response of the global hydrologic community to water scarcity has been, for over a century, to try to find more traditional sources of "supply" by looking farther and farther afield. China is no different. In fact, China today represents the epitome of the hard path approach to water,[2] with its intense reliance on large infrastructure projects to tap into dwindling supplies and sources and to divert water from one region to another. Many of the top leaders in China today were trained as engineers, including Hu Jintao, China's president and party chief (Griffiths 2006). It is thus no surprise that Beijing and central water agencies have typically responded to issues of scarcity with proposals for massive new infrastructure rather than new approaches to management.

This hard-path philosophy has driven work on a wide range of ambitious projects to build hydroelectric plants, dam rivers, and transfer water from one region to another. Almost half of the world's large dams (defined as dams higher than 15 meters) built since 1950 are in China (Fuggle and Smith 2000). On the Yangtze alone, there are an estimated 50,000 dams including the largest in the world, the Three Gorges Dam project (see the Water Brief, in this volume). The Chinese are now building the South-to-North Water Transfer Project, to funnel 45 billion cubic meters a year to the northern part of the country from the Yangtze River basin. That project was approved in 2002 to address water shortages in the north. Even if fully built, it will not be completed until in the middle of this century, and while several phases are already under construction, there is growing concern about both environmental and social problems.

This massive diversion consists of three major pieces: 1) an eastern route that will move water from the lower Yangtze to the north through a 1,200 kilometer long canal; 2) a middle route that will tap the Hanjiang, a major tributary of the Yangtze, and 3) a western route that will move water from the upper reaches of the Yangtze, Tongtian, Yalong, and Dadu rivers to augment water in the Yellow River Basin. As part of this project, in late 2007 China began digging a tunnel nearly 8 kilometers long under the Yellow River in Shandong Province – just one of the major rivers and physical objects the water-transfer project will have to overcome. Upon completion, water from the project will reportedly benefit about a dozen provinces, municipalities, and autonomous regions in north China including, especially, the regions around Beijing. Those areas produced one-third of the country's grain output and GDP with about 20 percent of the country's average per-capita water resource (Yardley 2007a). The Chinese government claims that as many as 300 million people could benefit from the project but it will also inevitably lead to adverse impacts in the regions where the water originates or would otherwise have flowed.

---

2. See Gleick (2003) for a discussion of the hard versus soft paths for water.

China is also looking to increase its reliance on hydroelectricity to satisfy the rapidly growing energy needs of its rapidly growing economy. Dirty coal presently accounts for two-thirds of all electricity projection, killing miners, polluting air and water, and emitting vast quantities of greenhouse gases. To ease these problems, China wants to greatly expand non-coal energy sources, including hydroelectricity, which presently provides around six percent of total electricity. Chen Deming, a governmental economic planner, stated in 2007, "We believe that large-scale hydropower plants contribute a lot to reduce [fossil] energy consumption, air and environmental pollution" (Yardley 2007b). Similarly, regional shortages lead to calls for more development of rivers and aquifers.

In late 2007, workers began damming the Jinsha River to build the Xiluodu hydropower station, which will be the second largest facility in China, and the third largest in the world when it is completed around 2015 (Xinhua 2007e). This dam had been halted previously because of the lack of a proper environmental impact assessment, but work has resumed. Moreover, the dam is being built in a national protection zone for several species of endangered fish (Yardley 2007b). Upon completion, the dam will be 278 meters high and have an installed capacity of 12.6 gigawatts. The Jinsha River is a major tributary of the upper Yangtze River and flows between Yushu in Qinghai Province and Yibin in Sichuan Province. This project is one of only dozens planned for this region along the Jinsha, Yalong, and Dadu rivers.

More traditional water-supply and treatment infrastructure is also being built rapidly, including water and wastewater treatment plants. Officials announced plans to build ten sewage disposal plants in northwest China's Shaanxi province, along the Weihe River, the largest tributary of the Yellow River. Another 30 plants are to be built by 2010. Statistics from Shaanxi province show that more than 800 million tons of uncontrolled sewage and wastewater are currently discharged into water of the Weihe basin each year, which is around 20 percent of total sewage loading in the Yellow River basin (Xinhua 2007g).

China recently announced that total investments in the water sector during the ongoing "11th Five-Year Plan" (from 2006 to 2011) could be as large as a trillion yuan (Xinhua 2007o, China Daily 2007u), with a focus on investment in water-distribution systems and the construction of a thousand water and wastewater treatment plants. Among the challenges that have hindered China's efforts to upgrade its water systems are a lack of technical expertise, a shortage of capital, and competition for resources from other sectors of the society equally in need of modernization.

As a result, China has begun to explore working with private corporations and funders. Water officials have explicitly encouraged foreign participation in China's water markets. Deputy Director of the Ministry of Construction, Qin Hong, called for foreign investment at a meeting of water business leaders from many developed countries. "Foreign investment will be encouraged especially in wastewater treatment projects," Qin said (China Daily 2007u).

In 2007, a large wastewater treatment plant developed as a "public-private partnership" opened in Guangzhou. This plant was built by the Guangzhou Wastewater Treatment Co. Ltd, a partnership of the state government and Earth Tech, a subsidiary of Tyco International. The plant was built as part of a broad effort to reduce the flow of untreated sewage into the Xijiang (Pearl) and is a build-operate-transfer (BOT) agreement in which the plant is to be transferred to the government in 17 years (China Daily 2007d). Even with the operation of this plant, total daily wastewater treatment capacity in the Pearl River Delta is only around 30% of total wastewater discharge volumes.

Another leading international water company, Veolia has been aggressively seeking joint ventures in China and now has more than 20. They have announced that their Asian business, which currently accounts for less than two percent of its global activities, could grow to as much as 20 percent in coming years. Veolia projects are being developed in Beijing, Shanghai, Tianjin, and Shenzhen. One project, a renovated water supply plant in Tianjin, now supplies drinking water to 1.8 million people (Xinhua 2007o). In another agreement, Veolia set up a joint French-Chinese venture to build a series of water projects, including urban and industrial wastewater treatment plants, desalination facilities, water-treatment equipment, and water-management services in the northern city of Teda. The joint venture, called the Tianjin Teda Veolia Water Company Limited, was sought to help the Chinese with management expertise and the provision of financial capital. Total expenditures may grow to nearly 2 billion yuan (Xinhua 2007o).

China is also beginning to explore desalination as a source of coastal water supply. Large facilities have been proposed for the county of Xiangshan in eastern Zhejiang Province, and the in northern China city of Tianjin. Xiangshan suffers severe water shortages. The plant proposed for Xiangshan is to be the largest in China, with a production capacity of 100,000 cubic meters per day. Unlike most desalination plants currently in development, which use reverse osmosis membranes, the Xiangshan facility will use multi-stage flash distillation, using heat from an existing power plant. Initial estimates are that the plant will cost 1.1 billion yuan. The cost of water will be around 6 yuan per cubic meter and will be blended with local supplies and sold for 2.5 yuan, with the government bearing the cost of the subsidy (China Daily 2007r).

Improving Efficiency

As is true elsewhere, part of China's problems with water quantity is caused by wasteful use. Absolute scarcity of water is seriously aggravated by grossly inefficient use in some sectors. There are vast opportunities to improve the efficiency of water use, and Chinese hydrologists and water managers are working to tap into this potential. In north China, projects are under way to try to learn how to reduce the water demands of winter wheat. Cities are beginning to raise the price of water as an economic signal to use it more efficiently, though many economic subsidies that encourage inefficient use remain.

Water use per unit of GDP or economic productivity is higher in China compared to many other countries, according to government statistics. In 2003, 465 cubic meters of water were used to produce 10,000 yuan worth of GDP, four times the world average and nearly 20 times that of Japan and Europe at that time (Economic Daily August 8, 2005). Similarly, to produce 10,000 yuan of "industrial added value, 216 cubic meters of water were used, 10 times more than in developed countries" (China Daily 2007a). In the relatively water-rich southern province of Guangdong, per-capita water use in the city of Guangzhou is more than double the use in Beijing, and triple the use in Paris (Zheng Caixiong 2007).

More draconian actions to curb inefficient uses or to cut demand may be required in the coming years. Some Chinese scientists have suggested that growing urban and industrial water demands may eventually lead to the elimination of winter wheat in northern China as agricultural uses give way to higher-valued uses that produce more jobs and income per unit water. This would be a dramatic change from their policy to continue to satisfy food needs as much as possible from internal production rather than international markets. The international consequences of massive Chinese purchases of grain are not well understood, but there are already serious pressures on global food markets and new imbalances could worsen the risk of shortages and famines.

Improving Environmental Protection and Enforcement

Improvements in water quality will require both new technology and new laws with two key components: clear standards and adequate enforcement. While debates about the adequacy of China's environmental standards continue, there is little dispute that enforcement of existing water-quality and monitoring laws has been grossly inadequate.

A 2007 opinion piece in the English language *China Daily* noted "We need more severe rules and penalties to change business as usual including stopping discharged waste water from further polluting our rivers, oceans and underground water supplies" (China Daily 2007a).

In 2007, Zhou Shengxian of SEPA acknowledged the country's serious and unresolved water-quality problems and called for tighter controls on pollutant discharges and better enforcement. "To contain water pollution, we should, firstly, continue to strictly control the discharge of various pollutants." He also said that tougher emission standards would be adopted by 2010 for drinking water and indicated that beginning in 2009, all new "enterprises which discharge pollutants" will have to obtain permits in order to operate or to be listed on the stock exchange (Xinhua 2007f). The use of agricultural fertilizers was also acknowledged to be a problem for water quality and SEPA called for gradual reductions in fertilizer use together with improved oversight over poultry farms. There was, however, less clarity on when existing facilities would be more tightly regulated.

In July 2007, SEPA asked local authorities in areas along the country's four major rivers to change the priority from economic development to environmental protection. Local authorities in six cities, two counties and five industrial zones—all in the Yellow, Yangtze, Huai He and Hai He river basins—were given 3 months to rectify their "environmental problems" (Xinhua 2007f). According to official sources, the campaign has led to the closure, suspension, or renovation of 700 enterprises (Xinhua 2007f), although these kinds of closures have often been lifted or ignored when the attention of the central government turns elsewhere.

The SEPA has also announced new efforts to raise drinking-water quality standards and to rehabilitate rivers and lakes. "Serious water pollution has been an obstacle to the healthy development of society," said Zhou. "We should be more determined and devoted to the rehabilitation of rivers and lakes" (Xinhua 2007i). The new standards are the first major amendment to the older one, enacted in 1985 and set drinking water limits for 106 parameters, with a deadline of full implementation by 2012. Provincial governments are able to set secondary standards (China Daily 2007q). The SEPA also announced that projects over the next decade that discharged heavy metal or organic pollutants into lakes and rivers being rehabilitated would be rejected and that new limits would be imposed on nitrogen and phosphorous discharges into closed water bodies.

Also in 2007, the government of Jiangsu Province promulgated new water-quality regulations to clean up Taihu Lake, where pollution has led to the almost complete eutrophication of the lake, severe blooms of blue-green algae, and the contamination of major drinking water supplies for the region around Shanghai. The lake is located in a densely populated area northwest of Shanghai and is home to numerous factories from six major polluting industries, including dye, chemicals, paper production, steel manufacturing, and food processing (China Daily 2007g). Clean-up plans may cost as much as $14 billion over a 5-to-10-year period. Algal blooms in June 2007 led to the shutdown of water supply in the industrial city of Wuxi and forced as many as 5 million people to rely on bottled water (China Daily 2007f). As a temporary measure, regulators ordered the mass closure of chemical plants on the margins of the lake. The new regulations will tighten standards for emissions of COD (chemical oxygen demand), ammonia, nitrogen, and phosphorus in industrial wastewater and sewage.

While these kinds of periodic campaigns have been launched by environmental agencies, consistent enforcement is still rare. The failure of the state regulatory agencies to successfully regulate, monitor, and enforce Chinese water-quality laws will ultimately require a change in approach. Standard methods, such as improving enforcement and monitoring, are being tried, but new methods are also being explored. In mid-2007, for instance, the SEPA sent a list of 30 major polluters to leading national financial institutions, including the People's Bank of China and the China Banking Regulatory Commission, in an effort to reduce their access to credit and loans for operations. The listed industries were mostly in water-intensive sectors like paper-making, coking, pharmaceuticals, iron and steel, and brewing. Most of the plants on the list are small and medium-sized facilities, which face more challenges getting bank loans, but some criticism from the Chinese Academy for Environmental Planning suggests that this approach would be more influential if larger companies and facilities were also listed. Other financial methods to promote enforcement being explored include policies on taxation, insurance, and the listing of securities (China Daily 2007m).

The Use of Smart Economics

Water policymakers are increasingly looking to economic tools, such as proper pricing and the elimination or modification of subsidies, to help in the sustainable management of limited water. In China, where water prices have long been heavily subsidized by the government, new efforts are underway to update pricing structures to encourage both improvements in efficiency and wastewater treatment. In Beijing, for example, prices for domestic water use have more than doubled to around 4 yuan per cubic meter. Water prices for certain commercial uses such as car-washing, are far higher – as much as 45 yuan per cubic meter (China Daily 2007h). In the city of Shenzhen, local government officials have been pushing for a new pricing structure to encourage the use of recycled water, rainwater, and other resources. Jiang Zhunhu, director of the Shenzhen water resources bureau said, "Increasing the price of water is an effective solution to easing the shortage."

In southern China, some regions are also imposing price-driven quotas on residential use. For urban homes, the quota means that homes that use more than 210 liters a day will have to pay a surcharge above the basic rate. This amount of water is just enough to satisfy the most basic human needs of around 50 liters per person per day (Gleick 1996) for a household of four. Use above the quota will lead to additional charges in the form of a three-tier rate structure, similar to those increasingly being used to encourage efficient use in the United States (for a discussion of the use of rate structures to encourage urban conservation and efficiency, see Chapter 6). Families who use less than 22 cubic meters a month will pay a basic rate of 1.32 yuan a cubic meter, still well below the average cost of water in most industrialized countries. Those who use between 23–30 cubic meters per month will pay a higher rate of 1.98 yuan a cubic meter; use above 30 cubic meters a month will cost 2.64 yuan a cubic meter, double the base rate (Zheng Caixiong 2007). Separate quotas are being imposed on the industrial, agricultural, and commercial sectors. China has a long way to go, however, to rationalize the use of pricing and economics as a tool to sustainable water management.

Improving Public Participation

Water problems, including recent environmental disasters, are spurring the public to action. Open debate and public participation in Chinese environmental policy have been limited and unusual, but there are signs that growing concern over water pollution and contamination is leading to efforts by citizens to change water policies and laws. A major environmental law passed in China in 2003 for the first time ostensibly encouraged public participation in environmental decision making. This law, the Environmental Impact Assessment (EIA) Law requires all major construction projects to undertake an impact assessment. Further, it states "The nation encourages relevant units, experts and the public to participate in the EIA process in appropriate ways" (Eng and Ma 2006). In addition, the law states that "the institutions should seriously consider the opinions of the relevant units, experts and the public" and "should attach explanations for adopting or not adopting the opinions." Eng and Ma (2006) note that like many other laws in China, "the EIA Law is merely a guideline and the requirement for public participation is very briefly stated. Still, it has provided an initial legal cornerstone for encouraging public participation in governmental decision making processes." In an astounding admission in 2005, the Chinese government acknowledged that 50,000 environmentally related public protests occurred that year (Turner 2006).

In fall 2007, China's National People's Congress publicized a draft of a new law on water pollution to solicit public opinion (Xinhua 2007j). The law proposes heavier punishment on both polluters and "irresponsible" officials, including fines for industrial offenders and administrative punishments or criminal charges for officials who delay reporting or hide water pollution incidents.

Associated with this growing public participation in environmental issues, central government officials have had to permit the creation and operation of non-governmental organizations concerned about the environment. Many of these NGOs are focusing on water pollution and threats to aquatic ecosystems, and are learning how to use existing environmental laws to force change. Yu Xiaogang directs the Green Watershed initiative in Yunnan, and won the prestigious Goldman Environmental Prize in 2006. Yu has worked with local villagers to help them understand the impacts of dam construction. Other citizens have sued chemical plants to force compensation for health and environmental damages or to make more environmental information accessible to the public (Turner 2006).

Public participation evokes contradictory responses by the government. New regulations have recently been issued that seem to encourage public participation in some environmental reviews, while others restrict non-governmental and non-Chinese organizations from monitoring and reporting on water issues. The difficulty of obtaining independent information on water supply, use, and quality has recently been worsened by increased government control over the hydrologic activities of non-governmental actors, and non-Chinese scientists and organizations, ostensibly to protect "national security" (Xinhua 2007q). In 2006, a dam protester was executed for what government officials claimed was his role in the death of a policeman at a protest of 100,000 people opposed to Pubugou dam (BBC 2006, Haggart 2006). New regulations took effect in mid-2007 requiring official governmental approval of any hydrological monitoring and reporting. The regulations also state that water data must only be released to the public by "relevant government department or authorized hydrological organizations," which permits total control over the release of independent assessments and monitoring (Xinhua 2007q). An additional constraint on foreign efforts to report or monitor on China water issues is the requirement that local authorities must supervise all such efforts. Only time will tell whether China develops a healthy level of public participation in addressing the country's water problems.

Conclusion

Sustainable water management has long taken a backseat to the Chinese drive for economic growth. As a result, China has developed a set of water quality and quantity problems as severe as any on the planet. Water problems are so severe now that they are having a direct impact on humans, including growing constraints on economic activities and growing adverse effects on public and ecosystem health. China's SEPA minister has acknowledged these problems: "Serious water pollution has affected people's health and social stability and become the bottleneck thwarting China's sound and rapid economic and social development" (Xinhua 2007f).

The failure of the state regulatory agencies to successfully regulate, monitor, and enforce Chinese water-quality laws will ultimately require a change in approach. Unless China moves rapidly to develop the legal, technological, and institutional tools to clean up water pollution, reduce wasteful and inefficient uses of water, restore natural ecosystems, and develop sustainable sources of supply, then environmental and human catastrophes will worsen.

In addition, growing constraints on total supply are imposing limits to the size and type of economic activities the Chinese can pursue, raising the specter of reductions in agricultural production or industrial output in coming years. New tools, approaches, and technologies will all have to be tried as China attempts to move toward long-term sustainable use of its scarce and valuable freshwater resources.

REFERENCES

Associated Press. 2007. Warming Has Shrunk China's Two Biggest Rivers. July 17, 2007. http://www.msnbc.msn.com/id/19790256/

Boyle, C.E. 2007. Water-borne Illness in China. The Woodrow Wilson International Center for Scholars. A China Environmental Health Project Research Brief. August 20, 2007. http://www.wilsoncenter.org/index.cfm?topic_id=1421&fuseaction=topics.item&news_id=272856

British Broadcasting Corporation (BBC). 2006. China Executes Dam Protester. December 7, 2006. http://news.bbc.co.uk/2/hi/asia-pacific/6217148.stm

China Daily. 2007a. Before We Run Dry. February 28, 2007. http://www.mwr.gov.cn/english1/20070228/82467.asp

China Daily. 2007b. Warming Takes Toll on Water Resources. November 5, 2007. http://www.mwr.gov.cn/english/20071105/87684.asp

China Daily. 2007c. Drought Nationwide Problem Now. December 24, 2007. http://www.mwr.gov.cn/english/20071224/88527.asp

China Daily 2007d. Guangzhou Opens New Wastewater Plant. November 14, 2007. http://www.mwr.gov.cn/english/20071114/87947.asp

China Daily. 2007e. Shrinking Glacier Threat to Rivers. November 2, 2007. http://www.mwr.gov.cn/english/20071102/87650.asp

China Daily. 2007f. China Announces $14B Lake Cleanup. October 29, 2007. http://www.mwr.gov.cn/english/20071029/87506.asp

China Daily. 2007g. Tough New Rules for Taihu Lake. October 19, 2007. http://www.mwr.gov.cn/english/20071019/87322.asp

China Daily. 2007h. Shenzhen Plans to Raise Water Charges. September 28, 2007. http://www.mwr.gov.cn/english/20070928/87015.asp

China Daily. 2007i. Key Water Source Threatened. August 16, 2007. http://www.mwr.gov.cn/english/20070816/86224.asp

China Daily. 2007j. Water Rule Will Protect Great Wall. August 9, 2007. http://www.mwr.gov.cn/english/20070809/86114.asp

China Daily 2007k. River Still Polluted After Cleanup Efforts. August 7, 2007. http://www.mwr.gov.cn/english/20070807/86077.asp

China Daily. 2007l. China Faces Twin Woes of Floods, Droughts. August 2, 2007. http://www.mwr.gov.cn/english/20070802/85971.asp

China Daily. 2007m. Blacklist of Polluters Distributed. July 31, 2007. http://www.mwr.gov.cn/english/20070731/85916.asp

China Daily. 2007n. Action on Water Crisis. July 6, 2007. http://www.mwr.gov.cn/english/20070706/85346.asp

China Daily. 2007o. Water Supply Resumes in E. China City. July 5, 2007. http://www.mwr.gov.cn/english/20070705/85305.asp

China Daily. 2007p. Beijing Tap Water Now Safe to Drink. July 3, 2007. http://www.mwr.gov.cn/english/20070703/85248.asp

China Daily. 2007q. Wen: Water Quality a Major Priority. July 2, 2007. http://www.mwr.gov.cn/english/20070702/85211.asp

China Daily. 2007r. Desalination Plant Awaiting Nod. June 13, 2007. http://www.mwr.gov.cn/english/20070613/84795.asp

China Daily. 2007s. Environment Agency Publishes Q1 Report. May 22, 2007. http://www.mwr.gov.cn/english/20070522/84206.asp

China Daily. 2007t. Pollution Makes Cancer the Top Killer. May 21, 2007. http://www.mwr.gov.cn/english/20070521/84149.asp

China Daily. 2007u. Water Market to Open Further. April 23, 2007. http://www.mwr.gov.cn/english/20070423/83600.asp

China Ministry of Water Resources. 2007. Blue-green Algae Sully Reservoir in Northeast China. September 5, 2007. http://www.mwr.gov.cn/english/20070905/86594.asp

China Water Resources Daily. 2002. True Record of the Upper Zhang River Water Conflict Resolution by Henan. March 19, 2002. Cited by Eng and Ma (2006).

Cooley, H. 2006. Floods and Droughts. In: Gleick, P.H., editor. *The World's Water 2006–2007.* Washington, DC: Island Press. pp. 91–116.

Coonan, C. 2006. Global warming: Tibet's lofty glaciers melt away. Research by scientists shows that the ice fields on the roof of the world are disappearing faster than anyone thought. *The Independent*, November 17, 2006. http://www.independent.co.uk/environment/climate-change/global-warming-tibets-lofty-glaciers-melt-away-424651.html

Economy, E. 2004. *The River Runs Black: The Environmental Challenge to China's Future.* Ithaca, NY: Cornell University Press.

Eng, M., and Ma, J. 2006. Building Sustainable Solutions to Water: Conflicts in the United States and China. Woodrow Wilson International Center for Scholars. China Environment Series. pp.155–184.

Feng, Z. 2007. More Deserts, Less Water Could Sink Rising China. March 20, 2007. http://www.mwr.gov.cn/english/20070320/82887.asp

Fuggle, R., and Smith, W.T. 2000. Large Dams in Water and Energy Resource Development in the People's Republic of China (PRC). Hydrosult Canada Inc. and Agrodev Canada Inc. Country review paper prepared as an input to the World Commission on Dams, Cape Town, South Africa. www.dams.org

Gleick, P.H. 1996. Basic water requirements for human activities: Meeting basic needs. *Water International* 21(2): 83–92.

Gleick, P.H. 2003. Global freshwater resources: Soft-path solutions for the 21st century. *Science* 302: 28 November:1524–1528.

Griffiths, D. 2006. Drought Worsens China Water Woes. British Broadcasting Corporation News (BBC), Beijing. http://news.bbc.co.uk/2/hi/asia-pacific/4754519.stm

Guo, Q. 2007. Digging deeper for cleaner water. China Ministry of Water Resources. April 24, 2007. http://www.mwr.gov.cn/english/20070424/83634.asp

Haggart, K. 2006. Dam Protester Put to Death in Secret, Rushed Execution. Three Gorges Probe news service. http://chinaview.wordpress.com/2006/12/13/china-dam-protester-put-to-death-in-secret-rushed-execution/

Intergovernmental Panel on Climate Change. 2007. *Climate Change 2007 Synthesis Report: Summary for Policymakers.* http://www.ipcc.ch/pdf/assessment-report/ar4/syr/ar4_syr_spm.pdf

Lee, D.Y. 2007. Child Mortality and Water Pollution in China: Achieving Millennium Development Goal 4. China Environmental Health Project, Research Brief. Woodrow Wilson International Center for Scholars. July 2007.

Ma, J. 1999. *China's Water Crisis.* Voices of Asia, International River Network – China Environmental Sciences Publishing House.

Ministry of Foreign Affairs of the People's Republic of China, United Nations System in China. 2005. China's Progress Towards the Millennium Development Goals. http://www.undp.org.cn/downloads/mdgs/MDGrpt2005.pdf.

National Assessment. 2000. *Water: The Potential Consequences of Climate Variability and Change.* A Report of the National Water Assessment Group, U.S. Global Change Research Program, U.S. Geological Survey, U.S. Department of the Interior and the Pacific Institute for Studies in Development, Environment, and Security. Oakland, California.

NPHCCO. 1999. Sanitation and Hygiene Promotion Kit – For Our Children and For Our Own Future, Let's Create a Clean, Healthy and Safe Environment. Ministry of Health, China. http://www.sas.upenn.edu/~dludden/WaterborneDisease2.pdf

OECD. 2007. *OECD Environmental Performance Review of China.* Paris, France: Organization for Economic Cooperation and Development.

Turner, J.L. 2006. New Ripples and Responses to China's Water Woes. Woodrow Wilson International Center for Scholars, China Brief, Volume 6, Issue 25, December 19, 2006.

World Health Organization. 2003. Children's Mortality Rates 2003. http://www.who.int/child-adolescenthealth/OVERVIEW/CHILD_HEALTH/Mortality_Rates_03.pdf.

World Health Organization. 2004. Data and Statistics: Causes of Death 2002. http://www.who.int/research/en/.

Xinhua. 2007a. Drought Leaves Nearly 250,000 Short of Drinking Water in Guangdong. December 14, 2007. http://www.mwr.gov.cn/english/20071214/88467.asp

Xinhua. 2007b. 1M People Short of Drinking Water in Guangxi. December 18, 2007. http://www.mwr.gov.cn/english/20071218/88495.asp

Xinhua. 2007c. Senior Official: 41% Chinese to Live in Flood-prone Areas by 2020. December 8, 2007. http://www.mwr.gov.cn/english/20071208/88352.asp

Xinhua 2007d. China's Largest Freshwater Lake Nears Lowest Water Level in History. December 4, 2007. http://www.mwr.gov.cn/english/20071204/88283.asp

Xinhua. 2007e. China Begins Damming River for 2nd Largest Hydropower Plant Project. October 12, 2007. http://www.mwr.gov.cn/english/20071112/87889.asp

Xinhua. 2007f. China's Environmental Chief Reiterates Measures to Combat Water Pollution. November 22, 2007. http://www.mwr.gov.cn/english/20071122/88092.asp

Xinhua. 2007g. NW China Province to Reduce Pollution on Yellow River Tributary. October 25, 2007. http://www.mwr.gov.cn/english/20071025/87457.asp

Xinhua. 2007h. China Moves to Curb Pollution in Songhua River. September 21, 2007. http://www.mwr.gov.cn/english/20070921/86864.asp

Xinhua. 2007i. China to Raise Standards for Water-polluting Industries. September 19, 2007. http://www.mwr.gov.cn/english/20070919/86811.asp

Xinhua. 2007j. China Solicits Public Opinion on Draft Law on Water Pollution. September 18, 2007. http://www.mwr.gov.cn/english/20070918/86783.asp

Xinhua. 2007k. NW City Turns to Alternative Water Supply. September 3, 2007. http://www.mwr.gov.cn/english/20070903/86555.asp

Xinhua. 2007l. 1,138 Dead, 210 Missing in China Floods This Year. August 30, 2007. http://www.mwr.gov.cn/english/20070830/86511.asp

Xinhua. 2007m. Central China Bans Pearl Farming to Restore Water Quality. August 14, 2007. http://www.mwr.gov.cn/english/20070814/86195.asp

Xinhua. 2007n. Scientists Say Climate Change Reducing Flow of Rivers. July 16, 2007. http://www.mwr.gov.cn/english/20070716/85530.asp

Xinhua. 2007o. French Veolia Expands Water Business in China. May 25, 2007. http://www.mwr.gov.cn/english/20070525/84334.asp

Xinhua. 2007p. Environmental Situation Continues to Deteriorate. May 23, 2007. http://www.mwr.gov.cn/english/20070523/84260.asp

Xinhua. 2007q. China Tightens Control Over Foreigners' Hydrological Activities. May 9, 2007. http://www.mwr.gov.cn/english/20070509/83839.asp

Yang, H.H., et al. 2001. Efficacy trial of Vi polysaccharide vaccine against typhoid fever in southwestern China. *Bulletin WHO* 79(7): 625–631.

Yang, J., et al. 2005. A mass vaccination campaign targeting adults and children to prevent typhoid fever in Hechi; Expanding the use of Vi polysaccharide vaccine in Southeast China: A cluster-randomized trial. *BMC Public Health* 5(49) doi:10.1186/1471–2458–5–49.

Yardley, J. 2007a. China tunnels through Yellow River for massive water diversion project. *International Herald Tribune*, September 27, 2007.

Yardley, J. 2007b. Chinese dam projects criticized for their human costs. *The New York Times*, November 19, 2007.

Zheng Caixiong. 2007. User pays, that's the price of wastage. *China Daily*, February 27, 2007. http://www.mwr.gov.cn/english1/20070227/82453.asp

CHAPTER 6

Urban Water-Use Efficiencies: Lessons from United States Cities

Heather Cooley and Peter H. Gleick

A transition to a "soft path" to water, discussed in earlier editions of *The World's Water* (Wolff and Gleick 2002), involves a wide range of changes in water management, policies, technologies, and approaches. One part of such a path is a major effort to improve the efficiency of water use, to continue to provide the goods and services society demands, while reducing the pressure on water resources. Such improvements have been underway for many years, as evidenced by the stabilization and even decline in per-capita water use in the United States and other countries (Gleick 2003a, b).

Although municipal water consumption is a relatively small percentage of overall water use in the United States and elsewhere, it warrants close attention, in part because municipal uses have often been the drivers for new water acquisitions and for significant capital expenditures for water infrastructure. In the western United States, where water is scarce and largely allocated, new municipal demands for water, or even the projections that such demands will increase, result primarily from rapid population growth. In regions where the availability of untapped water is constrained by absolute limits on supply or by the prior reservation of water rights by other users, new sources of water can have high costs, both economically and environmentally. As a result, identifying ways to use existing water resources more effectively has become increasingly attractive. Yet even cities in regions with more abundant water, like New York and Seattle, have put serious effort into improving water-use efficiency because of the social, economic, and environmental benefits of these measures, with successful results. This chapter reviews a segment of these improvements, focusing on the experience of a wide range of urban areas in the United States in identifying and capturing efficiency improvements in residential, commercial, and industrial water use.

Use of Water in Urban Areas

Urban water use in the United States varies from region to region, but is typically split into five major categories: residential, industrial, commercial, institutional, plus overall distributional losses that occur while meeting these needs. Residential water

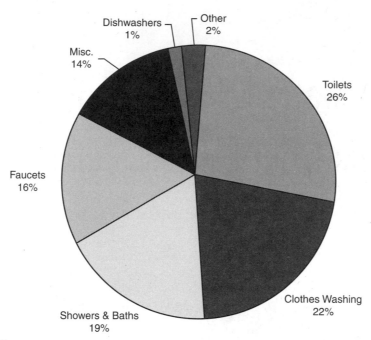

FIGURE 6.1 TYPICAL HOUSEHOLD INDOOR WATER USE.
Note: miscellaneous uses include leaks.
Source: Mayer et al. 1999

use refers to water used in the home for both indoor and outdoor needs, including cleaning, bathing, cooking, and maintaining gardens and landscapes (see Fig. 6.1 for a breakdown of typical indoor water use in the U.S.). Commercial and industrial water uses are extremely diverse depending on the goods and services being produced, but include water for cooling systems, landscaping, commercial washers, the production of industrial goods ranging from metals and chemicals to semiconductors, clothing, and finished commercial products, and much more. Institutional water use typically includes water used by schools, municipalities, prisons, and government agencies. Distributional losses include water loss due to system leakage, hydrant flushing, theft, and un-metered connections, and are typically about 10–15% of total withdrawals, although they can exceed 25% of total water use in older systems.

All of these water uses can also be evaluated separately as indoor and outdoor demands. Indoor water use is often recaptured as wastewater, treated, and then returned to the hydrologic cycle or reused. Outdoor demand is largely consumptive – used in a manner to prevent any additional immediate downstream use – and sensitive to regional and seasonal variation. In the arid Southwest, for example, outdoor water demand can account for more than 90% of total water use during the summer months, whereas in humid regions, outdoor water demands can be met largely by precipitation events. Outdoor demand is also subject to significant annual variation in response to local weather conditions.

Projecting and Planning for Future Water Demand

Although municipal water use is a relatively small percentage of overall water use in the United States and elsewhere, rapid population growth is driving urban water agencies to identify and develop new supplies. Yet, in many cases, the existing supply is already over allocated, and new supplies have significant social, economic, and environmental cost. In response, water managers are beginning to look seriously at new ways to balance supplies and demands, and are rethinking approaches to managing demand to ensure that sufficient water resources are available to meet anticipated needs. In this next section, we examine how three U.S. cities (Las Vegas, Seattle, and Atlanta) are facing this challenge.

Las Vegas, Seattle, and Atlanta appear to have very little in common: Las Vegas sits in the midst of a vast, dry valley where fewer than 125 mm (five inches) of rain fall each year and temperatures regularly exceed 40°C (over 100°F) in the summer. Seattle, on other hand, is situated in a cool, wet region that receives an estimated 860 mm (34 inches) of rainfall annually. And metropolitan Atlanta is situated in the warm, humid Southeast, where annual rainfall averages 1250 mm (50 inches) and occurs throughout the entire year.

Despite these climatic differences, all three regions are experiencing significant population growth (Fig. 6.2). Atlanta is one of the fastest-growing metropolitan areas in the United States. Between 1990 and 2000, the overall population of the counties within metropolitan Atlanta grew by 43%, from 2.8 million in 1990 to 4.0 million in 2000 (US Census 1990, 2000). The population is projected to nearly double again over the next 30 years, reaching an estimated 7.8 million by 2030 (MNGWPD 2003).

Like Atlanta, Las Vegas is experiencing rapid population growth. In absolute terms, growth in Las Vegas is far less than in Atlanta; but as a percentage of the current population, Las Vegas' growth is unparalleled. In 1990, fewer than 800,000 people called the Las Vegas Valley home (Clark County 2005). By 2000, Las Vegas' population swelled to 1.4 million, an increase of nearly 80%. These trends are expected to continue into the future: population in the region is projected to increase by an additional 2 million people, or 140%, between 2000 and 2030 (Center for Business and Economic Research 2005).

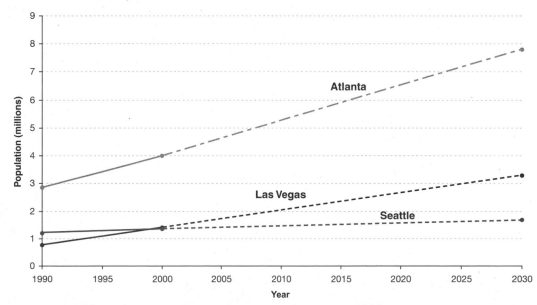

FIGURE 6.2 HISTORIC AND PROJECTED POPULATION IN ATLANTA, LAS VEGAS, AND SEATTLE.
Source: Las Vegas: Historic population estimates from Clark County. Population projections from the Center for Business and Economic Research (2005). Atlanta: Historic estimate from the US Census 1990 and 2000. Population projections from MNGWPD 2003. Seattle: Dietemann, A. personal communication.

Population in Seattle is also growing, but at a more modest pace than in either Las Vegas or Atlanta. Between 1990 and 2000, Seattle's population increased by 11%, compared to 43% in Atlanta and nearly 80% in Las Vegas. By 2030, Seattle anticipates serving an estimated 1.7 million people, a 26% increase over 2000 levels.

The traditional approach to meeting the water demands of growing populations has been to develop new sources of supply, and Las Vegas and Atlanta are taking this approach. But agencies are increasingly turning to water conservation and efficiency improvements because they are often the cheapest, easiest, and least destructive ways to meet current and future water supply needs. Although Atlanta, Las Vegas, and Seattle are all pursuing water conservation and efficiency improvements to help reduce future demand, these cities differ in the degree to which they have implemented efficiency improvements in the past and their plans for achieving efficiency improvements in the future. This section examines trends in per-capita water demand and compares the residential per-capita demand estimates in Las Vegas, Atlanta, and Seattle. We then examine each agency's water conservation and efficiency efforts in an attempt to understand the underlying factors driving differences in demand.

Per-Capita Demand

Cross-city comparisons of per-capita water demand provide a metric for evaluating the strengths and weaknesses of a city's water conservation efforts and can thus be an extremely valuable tool in gauging an agency's performance in promoting water conservation and efficiency. Cross-city comparisons, however, also have limitations. Per-capita demand is affected by a variety of factors, including the level and type of industry, income, climate, and mix of single-family and multi-family homes. Thus, a city with a high degree of water-intensive industrial or commercial development would tend to have a higher per-capita demand than a largely residential city. Likewise, a city in a hot, dry climate, like Las Vegas, typically has higher outdoor demand requirements than a city in a cool, wet climate, all other things being equal. Simple cross-city comparisons can gloss over these differences.

Although cross-city comparisons are imperfect, they can offer valuable information. Our approach in this analysis is to minimize their limitations and identify the differences where they exist. We focus on single-family residential water demand to remove the effect of the level and type of industry in a given area. We examine indoor and outdoor use separately and provide data on climatic variables to give the reader some information about regional differences.

The cities examined in this analysis exhibit significant variation in per-capita water demand, both indoors and outdoors (Fig. 6.3). At 165 gallons per person per day

TABLE 6.1 Average Temperature and Precipitation

	Average Temperature (°C)	Average High Temperature (°C)	Average Summer* High Temperature (°C)	Average Annual Precipitation (cm)	Average Monthly Summer[1] Precipitation (cm)
Seattle, WA	12	15	22	86.4	3.0
Atlanta, GA	17	22	31	127.0	10.7
Las Vegas, NV	19	27	39	10.2	0.8

*Calculated based on averages for June, July, and August.

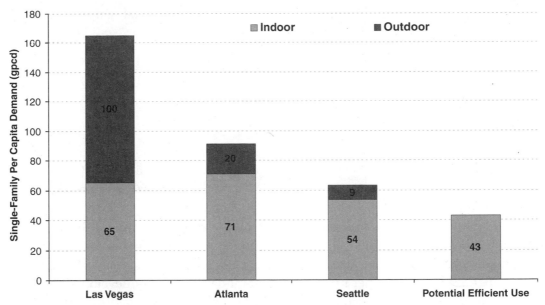

FIGURE 6.3 SINGLE-FAMILY RESIDENTIAL PER-CAPITA WATER DEMAND. Las Vegas and Seattle data for 2004; Atlanta data for 2001. Data shown for Seattle is for the retail service area. (One gallon equals 3.785 liters.)

(gpcd), per-capita demand in Las Vegas is substantially higher than in Atlanta and Seattle.[1] But indoor use in Atlanta is higher than in either Las Vegas or Seattle. We discuss indoor and outdoor per-capita demand separately.

Outdoor Per-capita Demand

Outdoor water use is a function of temperature, precipitation, and landscape type and is thus subject to significant regional variation.[2] The majority of homes in the United States are equipped with vast expanses of water-intensive turf. While precipitation is sufficient for meeting turf water requirements in some regions, lawns must be watered in others, particularly during the summer months. As shown in Table 6.1, Seattle, Atlanta, and Las Vegas are situated in very different climatic zones, resulting in tremendous variation in outdoor water demand. The average single-family resident in Las Vegas, for example, uses 100 gallons of water outdoors each day,[3] accounting for about 70% of residential water use (SNWA 2004). In Seattle and Atlanta, however, outdoor demand is considerably lower. In Seattle, with its relatively cool temperatures and modest precipitation during the summer months, outdoor demand not satisfied by rainfall is only 9 gpcd. In Atlanta, where summer temperatures are much warmer but summer precipitation is higher, outdoor demand is slightly higher, averaging 21 gpcd.[4] Because

---

1. One gallon equals 3.785 liters. We use gallons consistently here because all U.S. municipalities report in these units; we apologize to our global readers who must convert to more familiar and sensible metric units.

2. When we refer to outdoor water use, we refer to water intentionally applied outdoors, not water provided by nature. Thus, one reason Atlanta's "measured" outdoor water use is lower than Las Vegas' use is because a substantial amount of water required to grow lawns is provided by seasonal rainfall in Atlanta, while that same water has to be provided with artificial irrigation in hotter climates.

3. Note that outdoor water use can vary significantly from year to year based on local weather conditions.

4. While outdoor demand is relatively modest in Seattle and Atlanta, it could become problematic during a drought. High temperatures combined with a lack of precipitation could dramatically increase outdoor demand.

water-intensive turf is common in homes throughout the United States, it is not surprising that outdoor demand is high in hot, dry environments like Las Vegas. This suggests that replacing turf with more regionally appropriate landscapes would produce drastic reductions in outdoor, and total, demand, as has been observed.

Indoor Per-capita Demand

While outdoor demand is subject to significant regional variation, per-capita indoor demand should be fairly consistent across the United States. The Energy Policy Act of 1992 established national efficiency standards for toilets, faucets, and showerheads, and there are no major regional differences in behaviors like frequency of toilet, dish-washer, or washing machine use per person. According to the U.S. efficiency standards, residential toilets sold in the United States after January 1994 must use 1.6 gallons per flush or less. Likewise, all faucets and showerheads sold after January 1994 must use 2.5 gallons per minute or less. Standards were not developed for dish-washers and clothes washers, but there are efforts underway to establish performance standards for these appliances. Studies suggest that an "efficient" home, i.e., one that is equipped with devices that meet current national standards, along with efficient dish and clothes washers, could reduce indoor single-family residential demand to 40–45 gpcd (AWWA 1997; Mayer et al. 2000; Vickers 2001). Since these studies were completed, newer, more efficient appliances, such as dual-flush toilets and front-loading clothes washers, have been introduced to the market. Installing these devices would reduce per-capita demand even further.

Although these national standards were implemented nearly 15 years ago, a significant number of homes are still equipped with old, wasteful appliances and fixtures. As a result, we see significant variation in indoor water use (Fig. 6.3). Among the cities examined in this analysis, Atlanta has the highest indoor demand, at 71 gpcd. Indoor water use in Las Vegas is also high compared to Seattle and the average "efficient" indoor use. This is surprising given that both Atlanta and Las Vegas have experienced significant growth since 1994, and homes in these regions should be equipped with appliances and fixtures that meet current efficiency standards. By contrast, Seattle has the lowest single-family residential water use among the cities surveyed here despite the fact that Seattle has a larger proportion of older homes and thus a greater potential for homes with old, inefficient appliances and fixtures. One key explanation for the difference is that Seattle has implemented an aggressive indoor conservation program to replace old fixtures with more efficient models.

Water Conservation and Efficiency Efforts

All of the cities evaluated are pursuing water conservation and efficiency measures as a means to offset projected demand increases. Water agencies promote water conservation and efficiency improvements by developing programs that combine economic incentives and disincentives, regulatory policies, education, and voluntary actions. In this next section, we examine the conservation programs geared toward indoor and outdoor uses, focusing on similarities and difference among these communities and how they manage supply and demand.

Las Vegas

Water conservation efforts in Las Vegas have largely – and intentionally – focused on reducing outdoor water use. Ordinances limit the extent of turf in new residential and non-residential developments. To improve the efficiency of existing developments, rebates are available for pool covers, irrigation timers, evapotranspiration (ET) controllers, and rain sensors. But Las Vegas is best known among water conservation specialists for its turf removal program, which has been shown to be an effective mechanism for reducing water use in existing developments. While the quantity of water saved by turf replacement depends largely on local climate conditions, a study conducted in Las Vegas indicates replacing turf with more water-efficient landscapes reduces water use by an average of 76%, an annual savings of 56 gallons per square foot of turf replaced (SNWA 2004).

Participation in the landscape conversion program has been strong in the past and was the primary driver for recent efficiency improvements. But participation has declined markedly since 2004 (Fig. 6.4). The SNWA recently increased the incentive, from $1.00/ft^2 to $1.50/ft^2, which will likely result in higher participation levels. But the sudden decline in interest calls into question the ability of this narrow conservation program to effectively offset future demand.

Indoor water conservation and efficiency programs in Las Vegas, particularly those targeting existing homes, are far more limited (Cooley et al. 2007). For homes built before 1989, the SNWA offers free fixture retrofit kits that include faucet aerators, leak detection tablets, toilet flapper, and low-water use showerhead. Many homes never request such kits, and many homes that do never install the water-savings options. No other appliance rebates are available for single-family residents.

Programs targeting commercial and industrial users are also limited. The Water Efficient Technologies (W.E.T.) program, for example, provides rebates for a set menu of technologies as well as custom technologies that target both indoor and outdoor

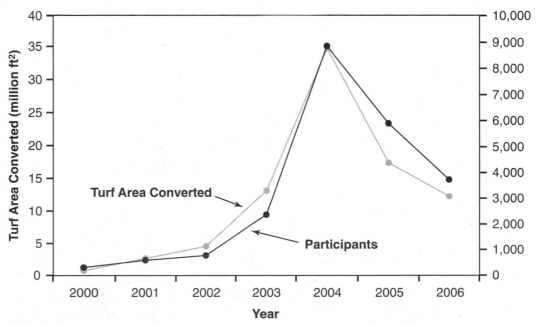

FIGURE 6.4 Participation in the Landscape Conversion Program, 2000–2007
Source: Cooley et al. 2007

uses for institutional, commercial, multi-family, and industrial customers. When implemented, these programs can be extremely effective in capturing water waste. The Mirage Hotel and Casino, for example, upgraded a cooling tower, reducing their annual water use by more than 18 AF (SNWA 2007). Unfortunately, though 2007 the W.E.T. program had provided rebates for only 30 projects since its inception, a mere 0.2% of the existing commercial and industrial accounts (Cooley et al. 2007).

Atlanta

Prior to 2001, water conservation and efficiency efforts in the Atlanta area were limited, with most efforts focused on short-term emergency measures implemented during a drought, such as restrictions on lawn watering or car washing. Half of the local districts in the region lacked conservation programs altogether. Those districts with programs emphasized school and public education, and only one district (the City of Atlanta) distributed low-flow fixtures to residents. Furthermore, rate structures that encouraged water waste, such as declining block rates, were still common throughout the region.

In 2003, however, in response to growing drought, water planners in the Atlanta region developed a more aggressive water conservation plan than in the past, but which still left significant untapped conservation potential. And even these efforts were weakened; most notably, a retrofit-on-resale ordinance, which would have required the replacement of old, inefficient appliances with more efficient models when a home is sold. This ordinance was dropped due to political opposition from the real-estate community. In its place, regional water planners opted to allow water suppliers to adopt incentive-based programs to facilitate the replacement of older, inefficient fixtures.

Atlanta's current conservation program is based on implementation of the following measures:

1. Implementing conservation pricing
2. Offering rebates to replace older, inefficient plumbing fixtures
3. Offering a "pre-rinse spray valve" education program for restaurants
4. Installing rain sensor shut-off switches on new irrigation systems
5. Installing sub-unit meters in new multi-family buildings
6. Reducing distribution system losses
7. Conducting residential and commercial water audits
8. Distributing Low-flow retrofit kits (showerheads and faucet aerators) to residential users
9. Developing education programs

Since the water conservation plan was adopted, water agencies in the Atlanta region have made considerable progress in implementing these water conservation programs (Fig. 6.5). For example, nearly 100% of the population in the Atlanta region is subject to conservation-oriented rate structures. But these efficiency efforts still fail to capture significant untapped potential. As of December 2007, even in the midst of one of the most severe droughts the region has experienced, only a handful of agencies had actually established toilet rebate programs, and regional water managers were still developing a regional rebate program. In addition, incentive programs had not yet

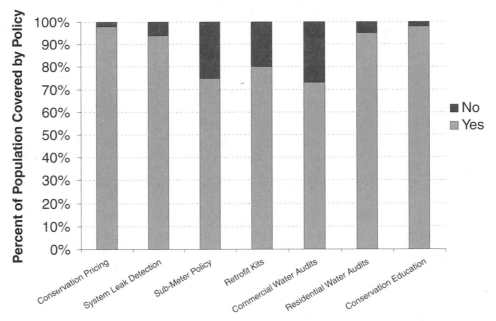

FIGURE 6.5 PERCENT OF METROPOLITAN ATLANTA'S POPULATION COVERED BY WATER CONSERVATION POLICIES AND PROGRAMS, 2007.
Source: MNGWPD 2007.

been established to replace other appliances and fixtures, such as showerheads, clothes washers, and dishwashers.

Seattle

Seattle has made some remarkable progress in improving water-use efficiency over the past 20 years. In the past, Seattle's water use was much like other cities across the United States: increasing as the population grew. Beginning in the early 1990s, however, Seattle launched a successful comprehensive water conservation program. By 2007, Seattle had reduced its total water use to 126 million gallons per day (mgd), lower than it had been at any time during the previous 30 years (Fig. 6.6). Furthermore, system-wide per-capita demand declined from 152 gpcd in 1990 to 97 gpcd in 2007.

Seattle achieved success by developing a comprehensive water conservation program that included a broad mix of pricing policies, education, regulations, rebates and incentives, and improved system operation. Beginning in 1989, Seattle introduced conservation-oriented rate structures for water and sewer rates. In 1992, a drought prompted mandatory restrictions and an aggressive conservation program. A new state plumbing code was adopted in 1993 for toilets, showerheads, and plumbing fixtures. In 2000, the Saving Water Partnership (SWP), a consortium of agencies that includes the City of Seattle and its retail and wholesale customers, launched a regional conservation program – the 1% Program. The goal of the Regional 1% Program was to reduce per-capita demand by 1% every year between 2000 and 2010. These efforts have proven successful, and the region is on track to meet its goal. SWP is now looking beyond the Regional 1% Program; they are developing a new program with a conservation goal of 15 mgd average annual savings from 2011–2030 "as low-cost insurance for meeting potential future challenges from climate change, as a low-cost way to manage and protect water resources, and as a low-cost way for customers to manage their bills" (SWP 2007).

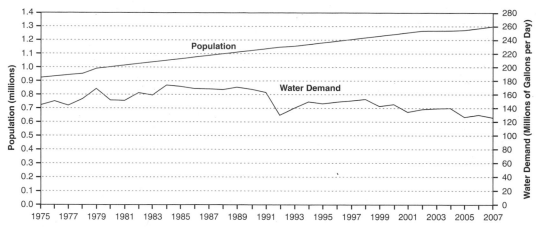

FIGURE 6.6 WATER DEMAND AND POPULATION IN THE SEATTLE REGIONAL WATER SYSTEM, 1975–2007.

Source: SWP, 2007.

Comparison of Water Conservation Programs

Rebates and Incentives

For more than two decades, agencies throughout the United States have developed water efficiency and conservation programs to accelerate the adoption of more efficient appliances and fixtures. These efforts have resulted in real water savings. Numerous communities, including Los Angeles, Seattle, El Paso, and Tucson, have stabilized or even reduced water demand while supporting population and economic growth. Despite these improvements, inefficient fixtures and appliances remain in common use, particularly in homes built prior to 1994,[5] and in a range of commercial, institutional, and industrial settings. We compare indoor and outdoor conservation efforts of Las Vegas, Seattle, and Atlanta.

Indoor Conservation Efforts

Water conservation programs throughout much of the United States have traditionally targeted indoor residential demand because savings can be achieved by installing a set of simple, cost-effective, widely available technologies, such as low-flow toilets, washing machines, and showerheads. Despite the demonstrated benefits of indoor conservation, efforts to promote efficient indoor use throughout the western United States vary considerably. Whereas Seattle provides incentives for nearly all indoor appliances and fixtures for single-family and multi-family residential customers, Atlanta only provides incentives for toilets for its single-family residential customers (Table 6.2). In Las Vegas, residents can get free fixture retrofit kits that include faucet aerators, leak detection tablets, toilet flappers, and low-water use showerheads for homes built before 1989, but provides no other rebates are available to its single-family residential customers. Rebates to multi-family customers for indoor fixtures and appliances are available through the Water Efficient Technologies (W.E.T.) program, but few have actually been provided.

---

5. National water-efficiency standards for some fixtures were signed into law in 1992; implementation began in 1994.

TABLE 6.2 Indoor Conservation Measures Provided by Las Vegas, Seattle, Atlanta

Water Conservation Measure	Las Vegas	Seattle Public Utilities	City of Atlanta
Audits			
Audits	MF	C	X
Targeted sector water audits			
Rebates			
Ultra-low-flush toilet		X	SF
High-efficiency or dual-flush toilet	MF,C	X	SF
Clothes washer		X	
Retrofit kit giveaways[1]	SF,MF	SF,MF	SF,MF
Hot water re-circulating system		SF,MF	
Appliances in new homes that exceed standards[2]			
Real-time water monitoring system			
High-efficiency urinal		C	
Waterless urinal	MF,C		
Laundry water ozonation or recycling system	C	C	
Dishwasher	C	C	
Cooling tower conductivity controller	C	C	
Replace once-through cooling systems	C	C	
Connectionless food steamers	C	C	
Medical air and vacuum systems	C	C	
Restaurant low-flow spray nozzles	C	C	
Pressurized waterbrooms	C		
Process improvements: performance based	MF,C	C	
Air-cooled refrigeration systems	C	C	
Steam sterilizer retrofit	C	C	
Hospital X-ray water recycling unit	C	C	
Regulatory Program			
Regional or City Plumbing Codes[3]			
Submetering			MF
Retrofit on resale			
Retrofit on reconnect			X
Educational Program			
School programs	X	X	X
Water Smart Home[4]	SF,MF		
Water Upon Request[5]	X	X	
Advertising/Community Events	X	X	X

There are many water agencies in the Atlanta region. The programs included in this table are for the City of Atlanta.

1. Low-flow nozzles, aerators, dye tablets, showerheads.
2. Distribution of rebates to builders for model and some production homes for appliances that exceed current efficiency standards.
3. Can include showerheads, urinals, etc.
4. Branding/labeling program for new homes.
5. Available at restaurants.
X= program available to all customer classes.
SF= program available to single-family residential customers.
MF = program available to multi-family residential customers.
C = program available to commercial and industrial customers.

The age of homes can have an important effect on indoor water demand. Homes built after 1993 should have appliances and fixtures that meet current national plumbing code standards, whereas older homes are more likely to be equipped with old, wasteful appliances and fixtures. Thus we would expect that a region with a higher percentage of older homes to have higher indoor demand. Of the three regions examined here, however, Seattle has the highest percentage of older homes but the lowest indoor water demand. As shown in Table 6.2, however, Seattle has implemented a much more aggressive water conservation program that targets older homes, resulting in lower indoor demand. Atlanta and Las Vegas, however, have much weaker indoor conservation programs. Thus despite having a higher percentage of new homes, old, inefficient appliances and fixtures are still common. In addition, some water-using appliances, such as clothes washers and dishwashers, are not covered by the national plumbing codes. There are even widely available fixtures that exceed the current plumbing code requirements, such as dual-flush toilets. An effective water conservation program can accelerate the installation of these products, thereby reducing indoor demand further.

Designing effective programs that target the commercial, industrial, and institutional sectors can be more challenging because businesses and industries use water in different ways. Yet, conservation assessments suggest that existing, cost-effective technologies can reduce demand from this sector by 25% to 40% (Gleick et al. 2003; Pollution Prevention International 2004). To capture these savings, many agencies provide defined rebates for specific technologies. Increasingly, agencies have developed performance-based programs that provide incentives for nearly any technology that reduces water use, with the financial incentive based on the quantity of water saved, e.g., $2.50 for every 1,000 gallons conserved. In Las Vegas and Seattle, business owners can get both performance-based and defined rebates. Such programs are not available in Atlanta. Because, as shown in Table 6.2, nearly any water-saving technology is covered under the performance-based programs, implementation is a key issue.

Outdoor Conservation Measures

Water conservation efforts in Las Vegas have largely, and intentionally, focused on curbing outdoor demand. As a result, it has been an innovator in developing certain outdoor conservation programs, particularly those aimed at reducing turf area but also including incentives for other measures, such as rain sensors, irrigation controllers, and pool covers (Table 6.3). Seattle also provides incentives for a variety of technologies to reduce outdoor water demand. Atlanta, however, does not yet provide these incentives. While outdoor water demand in the Atlanta region is relatively low (although we note it is substantially higher than Seattle), it could become problematic during an extended dry period. Encouraging residents to install low-water use landscapes or rain sensors could minimize this risk.

Rate Structures

Effective rate structures can play an essential role in communicating the value of water to customers and promoting long-term efficient use. Smartly designed rate structures permit an agency to cover costs of: 1) the utility's operation and maintenance; 2) procurement and development of additional water supplies and treatment for meeting future demands; and 3) the social and environmental "opportunity costs" of losing or protecting other benefits of the water and natural waterways

TABLE 6.3   Outdoor Conservation Measures for Las Vegas, Seattle, and Atlanta

Water Conservation Measure	Las Vegas	Seattle Public Utilities	City of Atlanta
Audits			
Audits	MF	C	X
Large landscape	X		
Rebates			
Artificial turf incentive	C		
Garden sprayer with shut-off valve			
Grant program[1]			
Irrigation timer/controller[2]	X	X	
Irrigation ET controller[3]	X	X	
Irrigation upgrades: performance based	C	MF,C	
Irrigation water budget			
Landscape conversion	X		
Pool covers	X		
Pressure regulating valves		MF,C	
Rain sensor	X	X	
Rainwater harvesting		X	
Rotating sprinkler nozzle		C	
Soil moisture sensor		X	
Sprinkler to drip/micro conversion		MF,C	
Regulatory Program			
Landscape efficiency codes	X		X
Seasonal watering schedule	X		
Time of day restrictions	X		
Water waste ordinance	X		
Educational Program			
School programs	X	X	X
Water Smart Home[4]	X		
Demonstration gardens	X		
Landscape training for public	X	X	X
Landscape training for irrigation professionals	X		
Plant labeling program/plant list	X	X	X
Published irrigation schedules		X	X

There are many water agencies in the Atlanta region. The programs included in this table are for the City of Atlanta.

1. Grant reward based on a request for proposal process.
2. Capable of multiple programming schedule.
3. Determines irrigation based on current or historical weather conditions.
4. Branding/labeling program for new homes.

X= program available to all customer classes
SF= program available to single-family residential customers
MF = program available to multi-family residential customers
C = program available to commercial and industrial customers

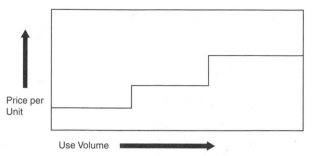

Price per Unit

Use Volume

FIGURE 6.7 GRAPHICAL REPRESENTATION OF AN INCLINING BLOCK RATE STRUCTURE

(e.g., ecological and recreation values of river basins, local/community economies, and values of river flows for diluting pollutants). Integrating all of these social and environmental values into rate structures better approximates the true cost of water and provides an economic incentive for customers to use it efficiently.

Communities throughout the west have adopted a variety of rate structures and pricing policies, including flat, uniform, seasonal, and increasing block rates. With flat rates, the customer pays a fixed amount independent of the amount of water used. With uniform rates, the unit price of water remains constant, e.g., $3.00 per 1000 gallons used, but the customer pays more as use increases. Through an increasing block rate design, the unit price for water and/or wastewater increases as the volume used increases, with prices set for each "block" of water (Fig. 6.7). Customers who use low or moderate volumes of water are charged a modest unit price and rewarded for conservation; those using significantly higher volumes pay higher unit prices. Properly designed increasing block rate structures provide incentive to conserve and ensures that lower-income consumers are able to meet their basic water needs at an affordable cost.[6]

Pricing is an important tool that allows water managers to reduce wasteful water use. The responsiveness of water demand to changes in water price is referred to as the price elasticity of water demand and is commonly expressed as a positive or negative decimal. If the price doubles and water use drops by 20 percent, for example, the price elasticity of water is −0.20. The price elasticity can vary by region, water use (indoor vs. outdoor), customer type, etc. A recent survey of price-elasticity factors by the Pacific Institute found that typical California price elasticities of water are around −0.20 for single-family homes, −0.10 for multi-family homes, and −0.25 for the non-residential sector (Gleick et al. 2005). Similarly, a Seattle study found price-elasticities of water of −0.20 for single-family homes, −0.10 for multi-family homes, and −0.23 for the non-residential sector (Seattle Public Utilities 2006).

Water Rates

While all of the agencies reviewed here have implemented inclining block rates for water, there is tremendous variation in the design of the inclining block rate structures, including the level of the fixed charge, the number of blocks (ranging from two to five), the block volume thresholds, and the block prices (Table 6.4 and Fig. 6.8). Of the three

---

6. Seasonal rate structures also provide a conservation price signal from one season to the next. This structure charges a higher unit price in the summer months when outdoor water use is more prevalent. However, within each season the seasonal rate structure does not provide an incentive to conserve because the unit price remains constant. Some cities overcome this by implementing a uniform rate in winter months and an inclining block rate throughout the irrigation season.

TABLE 6.4   Water Rate Structures

Municipality [Water Provider]	Region	Rate Structure Type	Fixed Monthly Service Charge	Consumption Rate: Rate per 1,000 Gallons of Water Consumed
Big Bend Water District	Las Vegas	Increasing Block Rate (two blocks)	$7.10	$2.70 - up to 15,000 $3.38 - over 15,000
Boulder City	Las Vegas	Increasing Block Rate (three blocks)	$7.50	$1.37 - up to 60,000 $1.73 - 60,001 to 550,000 $1.98 - over 550,000
Henderson	Las Vegas	Increasing Block Rate (four blocks)	$7.45	$1.46 - up to 6,000 $1.90 - 6,001 to 16,000 $2.47 - 16,001 to 30,000 $3.46 - over 30,000
Las Vegas Valley Water District	Las Vegas	Increasing Block Rate (four blocks)	$4.04	$1.16- up to 5,000 $2.08 - 5,001 to 10,000 $3.09 - 10,001 to 20,000 $4.58 - over 20,000
North Las Vegas	Las Vegas	Increasing Block Rate (four blocks)	$7.50	$1.37- up to 6,000 $1.78 - 6,000 to 15,000 $2.31 - 15,000 to 24,000 $3.00 - over 24,000
Seattle	Seattle	Seasonal and Increasing Block Rate (three blocks)	$8.05	Sept. 16 - May 15th: $3.38 May 16th - Sept. 15th: $3.85 - up to 3,740 $4.48 - 3,741 to 13,464 $11.43 - over 13,464
Gwinnett County	Atlanta	Seasonal and Increasing Block Rate (two blocks)	$7.30	November - May: $3.66 June - October: $4.58 - up to 10,000 $7.32 - over 10,000
City of Atlanta	Atlanta	Increasing Block Rate (four blocks)	$3.63	$1.90 up to 2,990 $3.96 - 2,990 to 5,236 $4.56 - over 5,236
Cobb County	Atlanta	Increasing Block Rate (five blocks)	$7.00	$2.47 - up to 8,000 $2.85 - 9,000 to 15,000 $3.22 - 16,000 to 29,000 $3.78 - 30,000 to 49,000 $5.40 - over 50,000
DeKalb County	Atlanta	Increasing Block Rate (four blocks)	$1.33	$1.01 - up to 2,000 $1.44 - 2,001 to 10,000 $2.17 - 10,001 to 20,000 $3.78 - over 20,001

Customers in the Atlanta region are served by 56 entities, each of which has a different pricing structure. Because of the large number of agencies in the region, we have included pricing data for the larger agencies in the region — DeKalb, Cobb, and Gwinnett Counties, as well as the City of Atlanta. Likewise, we have included pricing data for the three largest agencies in the Las Vegas region — Las Vegas Valley Water District, the City of Henderson, and the City of North Las Vegas.

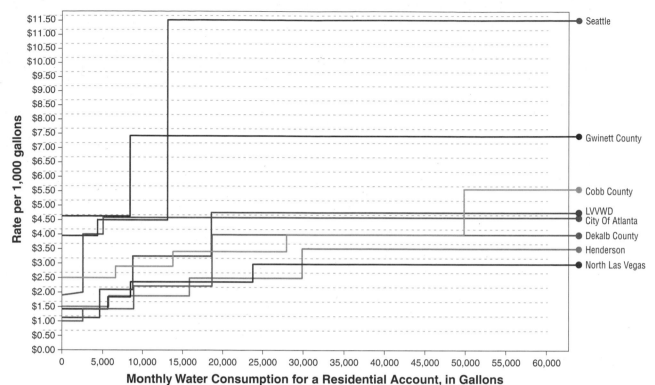

FIGURE 6.8 MARGINAL PRICE CURVE FOR WATER AGENCIES IN LAS VEGAS, SEATTLE, AND ATLANTA. Customers in the Atlanta region are served by 56 entities, each of which has a different pricing structure. Because of the large number of agencies in the region, we have included pricing data for the larger agencies in the region — DeKalb, Cobb, and Gwinnett Counties, as well as the City of Atlanta. Likewise, we have included pricing data for the three largest agencies in the Las Vegas region — Las Vegas Valley Water District, the City of Henderson, and the City of North Las Vegas.

regions, Seattle has adopted a water rate structure that sends the strongest price signal to its customers. Customers in Seattle pay a modest unit price for the first block to cover essential indoor uses such as cooking, cleaning, and bathing, at a relatively low cost. All subsequent tiers have per-unit prices that increase substantially, sending a strong conservation price signal to consumers that the more they use the more they will pay per unit.

Water rate structures in the Atlanta region are varied but are a significant improvement over past rate structures. As mentioned previously, rate structures that encouraged water waste were common in the region. Today, rates in Gwinnett County, the City of Atlanta, and Cobb County send a modest price signal to customers, with a relatively sharp increase between tiers. Rates in DeKalb County, however, are still low and send a weak price signal to customers. Because large price increases can induce significant reductions in water use and jeopardize the utilities revenue stability, rate adjustments must be phased in. Water utilities in the Atlanta region appear to be moving in the direction of conservation pricing and should continue their efforts over the coming years.

By comparison, two of the three largest water agencies in Las Vegas — the City of Henderson and the City of North Las Vegas —have adopted inclining block rate structures that send a weak price signal to their customers. The unit price increase that customers in each of these cities experience when they move from one block to the next is relatively insignificant, especially with customers who are accustomed to using (and paying for) large volumes of water. For example, the City of Henderson charges $1.46 per 1,000 gallons for the first 6,000 gallons of water and $1.90 per 1,000 gallons for the next 10,000 gallons. Thus a customer using 12,000 gallons (average household use in Las Vegas) would pay a monthly bill of $27.61, which is only $2.64 more than if all units had been priced at a flat rate of $1.46. In a consumer's mind, the difference between $27.61 and $24.97 is likely to be too low to produce a significant change in behavior. The Las Vegas Valley Water District recently adopted a more robust conservation-oriented rate structure, although one that still has significant room for improvement.

Average Price

While consumption charges are an important component of an effective water rate structure, they are not the only factor affecting the price paid by the customer. The customer's water bill includes consumption charges (discussed earlier) as well as any fixed service charge used to cover operating and maintenance costs. The customer then "sees" the average price for water, defined as the monthly service charge plus the total consumption charges, divided by the total volume used. The average price curve provides an indication of the effectiveness of water rate structures. Typically, average price curves initially trend downward, as the fixed costs are distributed. The curve can then trend upward, downward, or remain flat. If the block price increases are too small and/or the fixed monthly service charges are too high, the average price curve often declines. Curves that trend downward indicate that the unit price decreases as use increases, thereby providing a disincentive for conservation.[7] Curves that trend upward, by contrast, indicate that the unit price increases as use increases, thereby penalizing customers for wasting water. The steepness of the curve provides an indication of the strength of the price signal, with a steeper line sending a stronger price signal.

Figure 6.9 indicates that the effectiveness of the rate structures of the three cities is highly variable. Seattle's average price curve initially declines as the fixed service charge is spread out but then sharply increases at around 11,000–15,000 gallons per month, conveying the message that "the more you use, the more you pay." Georgia's Gwinnett County's average price curve is less aggressive than those in Seattle, but does increase modestly at 10,000 gallons per month. The average price curves of the other water providers in the Atlanta region, however, are much weaker and fail to convey an effective conservation message.

Likewise, the average price curves in the Las Vegas region offer only weak conservation incentives. The average price curves of two of the three Las Vegas communities remain relatively flat until about 30,000 gallons where there is a slight rise. The Las Vegas Valley Water District has the greatest rate of increase of these communities, and a lower volume at which the curve begins to rise. The moderate price signal, however, only targets high volume users that exceed 20,000 gallons a month. The median home in the Las Vegas Valley uses 12,151 gallons per month and thus does not feel this signal. In fact, only the highest 20% of homes exceed 20,000 gallons per month (LVVWD 2006).

---

7. "A rate structure with increasing marginal prices while the average price is declining sends mixed signals to consumers about their economic incentives to conserve water. Rate structures with any service charges, and in particular relatively large service charges in relation to the per unit cost and total water bill, are apt to create these mixed price signal conditions," AWWARF, at 13–14.

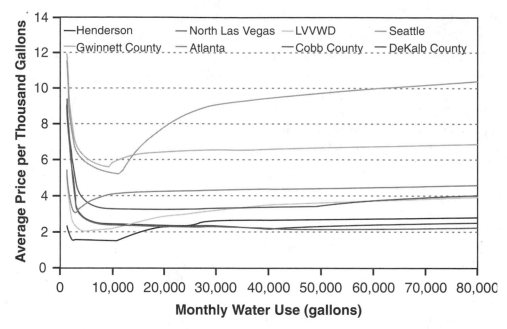

FIGURE 6.9 AVERAGE PRICE CURVES. The average cost is defined as the monthly service charge plus the total consumption charges, divided by the total volume used. The average price curve shows how the unit cost of water changes as water use increases. In all cases, the average cost declines as the fixed charge is spread out.

Wastewater Rates

In addition to paying for water, customers also pay to dispose of wastewater. Like water rates, wastewater rates can be flat, uniform, or increasing or decreasing block rates. While water managers have given an increasing amount of attention to developing conservation-oriented water rate structures in recent years, they have given far less attention to designing wastewater rates that encourage efficiency. In many regions, customers pay a flat rate for wastewater service. In Boulder City, for example, customers pay $9.65 per month for wastewater service, regardless of whether they generate 2,000 or 20,000 gallons of wastewater.

Wastewater rate structures vary tremendously in Las Vegas, Seattle, and Atlanta (Table 6.5). Four of the five wastewater providers in the Las Vegas area charge a flat rate, ranging from $9.65 per month in Boulder City to $19.00 per month in the City of Las Vegas. With these rate structures, there is no incentive to reduce water use and efficient customers are essentially subsidizing wasteful customers. By charging based on the amount of water treated, however, North Las Vegas has a much more effective wastewater rate structure that provides some incentive to conserve.

In Atlanta and Seattle, all customers are charged based on the volume of wastewater they generate, which provides a much better incentive to conserve water. In the Atlanta region, three of the four wastewater providers surveyed here charge a uniform rate, ranging from $4.52 to $5.31 per thousand gallons. Rate structures in the City of Atlanta send an even stronger conservation message: the city has implemented an inclining block rate with sharp increases between blocks; customers pay $7.21 per thousand gallons for the first tier, $10.09 for the second tier, and $11.60

TABLE 6.5 Wastewater Rate Structures

Municipality [Wastewater Treatment Provider]	Region	Rate Structure Type	Fixed Monthly Service Charge	Consumption Rate: Unit Rate per 1,000 Gallons of Water Consumed
Boulder City	Las Vegas	Flat	$9.65	
Henderson	Las Vegas	Flat	$18.63	
North Las Vegas	Las Vegas	Increasing Block Rate (two blocks)	$3.02	$3.66 - up to 3,000 $4.00 - over 3,000
City of Las Vegas	Las Vegas	Flat	$19.00	
Clark County Water Reclamation District	Las Vegas	Flat	$14.91	
Seattle	Seattle	Uniform		$10.36
Gwinnett County	Atlanta	Uniform		$4.52
City of Atlanta	Atlanta	Increasing Block Rate (three blocks)	$3.63	$7.21 - up to 2,990 $10.09 - 2,990 to 5,236 $11.60 - over 5,236
Cobb County	Atlanta	Uniform		$4.58
DeKalb County	Atlanta	Uniform	$3.23	$5.31

Rates for Clark County Water Reclamation District are an average for the communities served.

for the third tier. And in Seattle, customers do not pay an inclining block rate, but they do pay a hefty $10.36 per 1,000 gallons for wastewater service. Note that in these regions, customers pay considerably more for wastewater than they do for water service. Customers in DeKalb County, for example, pay between $1.01 and $3.78 per thousand gallons for water but $5.31 per thousand gallons for wastewater service.

Thus, when comparing water rates in different communities, it is important to consider wastewater rates as well. Failing to do so may hide some important differences. In Las Vegas, relatively weak inclining block rates for water combined with a wastewater charge that does not vary according to the volume of waste generated fails to send a conservation signal to customers. In Atlanta, wastewater rates are much more effective in promoting conservation than in Las Vegas, although water rates could be improved there as well. And in Seattle, robust water and wastewater rates combine to send a strong price signal to their customers about the value of water.

Summary of Water Rates

Each utility has a different water-supply situation and different costs associated with providing and managing these supplies; thus we would expect water prices and rate structures to vary among agencies. No surprisingly, however, per-capita water use and the aggressiveness of the water-rate structures are negatively correlated. Seattle, for example, has the most aggressive conservation rate structure and the lowest per-capita use rate, while Las Vegas has the weakest conservation incentives and the highest per-capita use rate. This conclusion highlights the potential effectiveness of rate structures as a tool for promoting efficient use.

Conclusion

Urban water-efficiency improvements are a key tool for satisfying water demands imposed by population and economic growth without tapping increasingly expensive or environmentally damaging sources of supply. While some water managers have mounted effective conservation efforts, others are lagging behind. Comparisons among cities in different regions and climatic zones of the United States can be useful for evaluating which policies and practices are the most effective.

This chapter reviews the different experiences and results from three major cities in the United States (Las Vegas, Seattle, and Atlanta). All three have implemented a range of water-efficiency and conservation programs in response to increasingly constrained supplies and rapid population growth. Yet their results vary considerably. On closer look, the more effective programs rely on strong economic signals, rebates and incentives, education, and appropriately designed and implemented regulatory programs. In particular, the more comprehensive indoor and outdoor efforts in Seattle, accompanied by an effective water-pricing program, have been successful in reducing residential water use far more than in many other cities, even accounting for differences in climate and other factors. Such comprehensive reviews of successful programs around the world can offer insights for water managers, planners, and individuals as the importance of conservation and efficiency programs grows.

REFERENCES

AWWA WaterWiser. 1997. Residential Water Use Summary – Typical Single Family Home. Denver, Colorado: American Water Works Association.

Center for Business and Economic Research. 2005. Clark County Nevada Population Forecast 2005–2035. Las Vegas, Nevada: University of Nevada.

Clark County Department of Comprehensive Planning. 2005. Historical Population by Place. www.co.clark.nv.us/Comprehensive_planning/05/Demographics.htm.

Cooley, H., Hutchins-Cabibi, T., Cohen, M., Gleick, P.H., and Heberger, M. 2007. *Hidden Oasis: Water Conservation and Efficiency in Las Vegas.* Oakland, California: Pacific Institute.

Gleick, P.H., Cooley, H., and Groves, D. 2005. *California Water 2030: An Efficient Future.* Oakland, California: Pacific Institute for Studies in Development, Environment, and Security.

Gleick, P.H. 2003a. Global freshwater resources: Soft-path solutions for the 21st century. *Science* 302(11): 1524–1528.

Gleick, P.H. 2003b. Water use. *Annual Review of Environment and Resources* 28: 275–314.

Las Vegas Valley Water District (LVVWD). 2006. *Las Vegas Valley Water District Rate Proposal Fact Sheet,* November 2, 2006.

Mayer, P.W., DeOreo, W.B., Opitz, E.M., Kiefer, J.C, Davis, W.Y., Dziegielewski, B., and Nelson, J.O. 1999. Residential End Uses of Water. Final Report. Denver, CO: AWWA Research Foundation.

Mayer, P.W., DeOreo, W.B., and Lewis, D.M. 2000. *Seattle Home Water Conservation Study: The Impacts of High Efficiency Plumbing Fixture Retrofits in Single-Family Homes.* Boulder, Colorado: Aquacraft, Inc. Water Engineering and Management.

Metropolitan North Georgia Water Planning District (MNGWPD). 2007. *2007 Activities and Progress Report.* http://www.northgeorgiawater.com/files/MNGWPD_Annual07.pdf

Metropolitan North Georgia Water Planning District (MNGWPD). 2003. Water Supply and Water Conservation Management Plan.

Saving Water Partnership (SWP). 2007. *Seattle Water Supply System Regional 1% Water Conservation Program.* 2006 Annual Report. Seattle, Washington.

Seattle Public Utilities. 2006. Official Yield Estimate and Long-Range Water Demand Forecast. Seattle, WA. http://www.seattle.gov/util/stellent/groups/public/@spu/@csb/documents/webcontent/cos_003828.pdf

Southern Nevada Water Authority. 2007. Water Efficient Technologies Success Stories. http://www.snwa.com/html/cons_wet_success.html. Website accessed July 13, 2007.

Southern Nevada Water Authority (SNWA). 2006. 2006 Water Resource Plan. www.snwa.com/html/wr_resource_plan.html.

Southern Nevada Water Authority. 2004. SNWA Water Conservation Plan 2004–2009. Las Vegas, Nevada, p. 12.

United States Census (US Census). 1990. Census 1990 Summary Tape File 1 (STF 1) - 100-Percent data, Detailed Tables.

United States Census (US Census). 2000. Census 2000 Summary File 1 (SF 1) 100-Percent Data, Detailed Tables.

Vickers, A. 2001. *Handbook of Water Use and Conservation.* Amherst, Massachusetts: Waterplow Press.

Wolff, G. and Gleick, P.H. 2002. The Soft Path for Water. In: *The World's Water 2002–2003: The Biennial Report on Freshwater Resources.* Gleick, P.H., editor. Washington, D.C.: Island Press.

Tampa Bay Desalination Plant: An Update

Heather Cooley

In the 2006 edition of *The World's Water*, we provided a case study of the Tampa Bay seawater desalination plant. Like many communities around the United States, the Tampa Bay region is experiencing significant growth combined with increasing constraints on water availability. Around Tampa Bay, overpumping of local groundwater resources forced communities to think about alternative water solutions; some water managers and planners thought desalination might be the key to the region's long-term water future.

Starting in the mid-1990s, regional water authorities in Florida partnered with Poseidon Resources, a private company, to plan and build what was to be the largest seawater desalination plant in the United States. The initial design was for a plant to produce 25 million gallons per day (mgd) through reverse osmosis, at a price far below the historical price of desalinated water (Wright 1999, U.S. Water News 1999). When the plant was announced, it was hailed as a breakthrough for desalination in the United States and sparked enormous interest and excitement.

The reality has been far different. While the plant is now running, technical problems, unanticipated costs, and political controversy have dampened unbridled enthusiasm for desalination as a silver-bullet solution for municipalities facing water-supply constraints.

The planning process for the plant began in October 1996. Several vendors submitted initial proposals in December 1997, and binding offers in the competitive bidding process were received in October 1998. In early 1999, Tampa Bay Water (formerly the West Coast Regional Water Supply Authority) selected S&W Water, LLC, a consortium of Poseidon Water Resources and Stone & Webster. Their proposal called for construction of a 25 mgd plant to commence in January 2001 on the site of the Big Bend Power Plant on Tampa Bay. Operation was to begin in the second half of 2002 (Heller 1999, Hoffman 1999).

Communities in the United States and abroad saw the plant as the beginning of a new era in desalination in the United States. "It is a historic moment," said Pasco County Commissioner Ann Hildebrand in 1999, "a kind of cornerstone of the new water projects" (AP 1999). Many were intrigued by the contract structure between

Poseidon Resources and Tampa Bay Water, a public-private partnership, and the incredibly low cost of the product water claimed by project advocates. The Prime Minister of Singapore, for example, sent a delegation to Florida to examine the plant and learn about the project contract (Johnson 2001).

S&W Water, LLC made a binding commitment to *deliver* desalinated water in the first year of operation at an unprecedented wholesale cost of $1.71/kgal ($0.45/m^3), with a 30-year average cost of $2.08/kgal ($0.55/m^3) (Heller 1999). Even the highest of the four bidders offered a price between $2.12 to $2.54/kgal ($0.56 and $0.67/m^3), well below the cost of water from other recent desalination plants. By comparison, the same year, the Singapore Public Utility Board announced plans to build a 36 mgd (136,000 m^3/d) desalination plant, to *produce* water at an estimated cost of between $7.50 and $8.74/kgal ($1.98 and $2.31/m^3) (U.S. Water News 1999) with an additional cost to deliver water to customers.

Unique conditions, difficult to reproduce elsewhere, contributed to the expected low cost of water. Energy costs in the region were low (around $0.04 per kWh) compared to other coastal urban areas. Co-locating the desalination plant on the site of the power plant also lowered the cost because the power plant provided infrastructure, supporting operations, and maintenance functions. Salinity of the source water from Tampa Bay is substantially lower than typical seawater: only about 26,000 parts per million (ppm) instead of 33,000 to 40,000 ppm typical for most seawater. In addition, financing was to be spread out over 30 years, and the interest rate was only 5.2 percent (Wright 1999).

Despite the initial optimism, the Tampa Bay desalination plant has been plagued with problems. Three contractors declared bankruptcy, forcing Tampa Bay Water to assume the responsibility and risk of building the plant. The original pretreatment method was unable to adequately remove sediment and organic matter, leading to a greater incidence of membrane fouling, decreasing the life of the membranes, and ultimately requiring a redesign of major project components. Additional chemicals were required to clean the membranes, which in turn caused the plant to exceed contaminant levels established in its sewer discharge permits. Furthermore, many of the water pumps developed rust and corrosion problems because cost cutting resulted in the use of inappropriate materials (Pittman 2005).

In November 2004, Tampa Bay Water agreed to a $29 million, 2-year contract with American Water-Pridesa (both owned by Thames Water Aqua Holdings, a wholly owned subsidiary of RWE) to fix the plant. Major improvements were made to the pretreatment, reverse osmosis, and post-treatment system. Although American Water-Pridesa missed several deadlines, the desalination plant finally passed its performance tests in November 2007. The plant became fully operational in December 2007, more than 5 years behind schedule. The final cost of the plant was $158 million, more than 40% over the original cost of $110 million. In a press release issued in early 2004, the new cost was estimated at $2.54/kgal ($0.67/m^3), up from an initial expected cost of between $1.71 and $2.08/kgal ($0.45 to $0.55/m^3) (Business Wire 2004). By early 2008, the cost of water had risen again to an estimated $3.38/kgal ($0.89/ m^3) (Kranhold 2008), nearly double the initial estimate.

To understand and mitigate the environmental impacts, the Tampa Bay plant and surrounding vicinity is subject to a monitoring requirement. Departmental budget cuts, however, have limited the monitoring plan. The Hillsborough County Environmental Protection Commission collected pre-operational data on water quality and continuously monitors temperature, pH, salinity, and dissolved oxygen at three sites near the desalination plant. Pre-operational sediment and benthic sampling were also conducted, although testing has been suspended due to budget cuts and are not expected to resume until next year (D'Aquila 2008).

The Tampa Bay desalination plant is not yet out of the woods. In August and November 2007 and January 2008, the plant violated its wastewater discharge permit by dumping too much cleaning solution into the sewer system. In November, the plant was also cited for failing to notify officials from Hillsborough County about the excessive discharge in a timely manner. Officials from American Water-Pridesa blame the violations on operator error and a faulty meter (Pittman 2008). Until the problem is fixed, cleaning solution will be placed into a tanker and hauled to the wastewater treatment plant. Now that the plant is running, Tampa Bay Water officials hope it will continue to provide reliable water. If operations are successful, there are plans to expand the capacity to 35 mgd.

The experiences at Tampa Bay should caution water managers against excessive optimism on the price, feasibility, and appropriateness of seawater desalination. Desalination is not a silver bullet that will solve all regional water woes, as many desalination advocates claim. If combined with a balanced water resource plan that includes implementing cost-effective supply and demand management measures and controls on growth; however, properly designed, built, and operated seawater desalination facilities can improve system reliability and may allow water districts to restore overexploited surface and groundwater systems.

When the barriers to desalination are overcome (see, for example, NRC 2008), carefully regulated and monitored construction of desalination facilities should be permitted (Cooley et al. 2006). Regulators should develop comprehensive, consistent, and clear rules for desalination proposals so inappropriate proposals can be swiftly rejected and appropriate ones identified and facilitated. Private companies, local communities, and public water districts that push for desalination facilities should do so in an open and transparent way, encouraging and soliciting public participation and input in decision making.

REFERENCES

Associated Press (AP). 1999. Tampa Bay area to build huge desalination plant. *Miami Herald*, July 21, 1999, page 5B.

Business Wire. 2004. American Water-Pridesa secures contract to remedy and operate the Tampa Bay Seawater Desalination plant. November 16. http://www.highbeam.com/library/docfree.asp

Cooley, H., Gleick, P.H., and Wolff, G. 2006. *Desalination: With a Grain of Salt. A California Perspective*. Oakland, California: A Report of the Pacific Institute for Studies in Development, Environment, and Security.

D'Aquila, T. 2008. Personal communication. Hillsborough County Environmental Protection Commission. March 19.

Heller, J. 1999. Water board green-lights desalination plant on bay. *Tampa Bay Business News*. March 16. http://www.tampabay.org/press65.asp.

Hoffman, P. 1999. Personal communication. Stone and Webster Company.

Johnson, N. 2001. World taking notes on desalination. *The Tampa Tribune*, December 31.

Kranhold, K. 2008. Water, water everywhere. Seeking fresh sources, California turns to the salty Pacific. *The Wall Street Journal*, January 17, 2008, page B1.

National Research Council (NRC). 2008. *Desalination: A National Perspective*. Washington, D.C.: The National Academies Press.

Pittman, C. 2008. Clean water, dirty waste. *St. Petersburg Times,* March 1, 2008.

Pittman, C. 2005. Desal delays boil into dispute. *St. Petersburg Times,* August 16. http://pqasb.pqarchiver.com/sptimes/882831921.html

U.S. Water News. 1999. Tampa Bay desalinated water will be the cheapest in the world. *U.S. Water News Online,* March. http://www.uswaternews.com/archives/arcsupply/9tambay3.html

Wright, A.G. 1999. Tampa to tap team to build and run record-size U.S. plant. *Engineering-News Record,* March 8. http://beta.enr.com/news/enrpwr8.asp.

Past and Future of the Salton Sea

Michael J. Cohen

Off in the hinterlands of remote southeastern California, the Salton Sea – an oddity created by the vagaries of human behavior and nature – shrinks slowly, perhaps irreversibly, a hazard in the making. The Salton Sea is a creeping environmental problem (Glantz 1999) that may not attract the attention and investment needed to avoid catastrophic impacts to public health and to the millions of birds the Sea supports. Rapid municipal growth and rising demand for water in urban Southern California, coupled with projected declines in future water supply, impose great pressure on the imperiled Sea. Located predominantly in one of the poorest counties in California, the Sea suffers from the indifference and outright hostility of distant communities and water users. Protecting and rehabilitating the Sea will require years of determined effort and billions of dollars, yet neither is assured.

The Salton Sea lies more than 70 m below sea level, a vast, incongruous salty lake amidst the harsh Colorado Desert. The Sea is a terminal lake – water flowing into the lake has no escape except through evaporation. Maximum temperatures in the basin exceed 40° C 136 days per year, and exceed 45° C more than 10 days per year. Fewer than 7 centimeters (cm) of precipitation fall annually in the basin, generating an annual net evaporation of about 1.8 m. The Salton Sea watershed covers 21,700 km^2, yet more than 85% of its inflows come from surface and subsurface agricultural runoff from the ~2,500 km^2 of irrigated fields of the Imperial, Mexicali, and Coachella valleys (Cohen et al. 1999). See Figure WB2.1.

Designated as an 'agricultural sump' by President Coolidge in 1924, the Sea is often dismissed as unnatural or worse, seen by some as an aberration in the desert that should be allowed to disappear (Nijhuis 2000). Yet the nutrient-rich agricultural drainage fuels a remarkably productive ecosystem, sustaining a great abundance and diversity of micro-organisms and more than 270 species of migratory and resident birds. Shuford et al. (2002) note "the Salton Sea is of regional or national importance to various species groups – pelicans and cormorants, wading birds, waterfowl, shorebirds, gulls and terns," as well as to a large number of individual species. In the winter of 1999, extensive surveys recorded 24,974 white pelicans and 18,504 double-crested cormorants at the Sea (Shuford et al. 2000). Jehl and McKernan (2002) estimated that 3.5 million eared grebes were at the Sea on March 5, 1988. The loss of 90–95% of California's predevelopment wetlands and a similar percentage of the former Colorado River delta (Cohen 2002) has left migrating birds with few other stopovers along the Pacific Flyway. See Figure WB2.2.

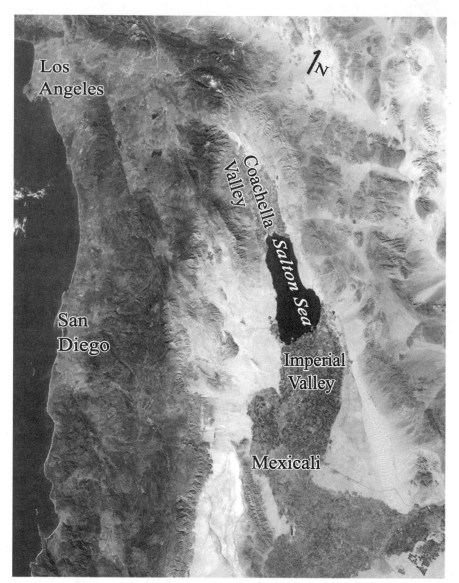

FIGURE WB 2.1 SALTON SEA LIES IN REMOTE SOUTH-EASTERN CALIFORNIA.

Source: Image ISS004-E-6119.JPG taken January 10, 2002, courtesy of Earth Sciences and Image Analysis Laboratory, NASA Johnson Space Center, available at http://eol.jsc.nasa.gov.

   The Salton Sea defies expectations. California's largest lake, it extends 56 km by about 24 km at its widest, with a total surface area of roughly 950 km². Yet it is relatively shallow, less than 15 m at its deepest, with an average depth of only 9 m. The lake currently holds some 8.56 km³ of hypersaline water, almost half of the mean annual flow of the Colorado River. But the Salton Sea, with a current salinity of about 48 g/L, is already 37% saltier than the Pacific Ocean and 67 *times* saltier than the Colorado River at Imperial Dam (the initial source of most of the Sea's inflows). With evaporation the only exit for incoming waters, the Sea inexorably concentrates incoming salts, nutrients, and contaminants in its waters and sediments. For more than 40 years, prognosticators have predicted the Sea would die within a decade (Pomeroy and Cruse 1965), certain that rising salinity signaled the impending demise of its fishery. Others dismiss the ecological value of a hypereutrophic[1] lake fed by agricultural drainage, especially one limited to a

---

1. Characterized by high levels of primary productivity, low concentrations of dissolved oxygen, and low visibility.

single species of fish – a hybrid freshwater species originally from Africa. Yet year after year, millions of birds visit the Sea, feeding on its abundance, nesting and roosting on its shores and islands, or just passing through (Shuford et al. 2000, 2002, 2004), indifferent to how natural the Sea might be, or how tenuous its future.

Background

Prior to the early years of the 20th century and the dams of the Age of Reclamation (Reisner 1993), the Colorado River meandered about its delta (see Fig. W.B2.2), periodically discharging north into the Salton Basin before shifting once again to flow south into the Upper Gulf of California (Sykes 1937). Previous incarnations of the Salton Sea, known as Lake Cahuilla, grew to several times the size of the present lake before being abandoned by the river and left to evaporate under the relentless desert sun. In early 1905, unexpected Colorado River floods tore through an unprotected headgate cut by Imperial Valley irrigators in the river's right bank, diverting the entire flow of the river into the bed of the old Lake Cahuilla for more than 18 months. After the Colorado was forced back into its original bed, this new lake, dubbed the "Salton Sea," would have evaporated if not for the agricultural drainage that continues to feed it. The taming of the Colorado River, by means of massive dams, incised and armored channels, and carefully released flows, now insulates the Salton Sea from these previous cycles of filling and drying.

These dams and associated water rights and delivery agreements mean that more than 23% of the total average annual yield of the Colorado River currently flows into the Salton Sea basin each year, regardless of the river's actual discharge. Although irrigators' senior water rights and return flows into the Sea have protected it from drying completely, inter-annual and seasonal variability, reflecting seasonal evaporation rates and farmers' irrigation and cropping patterns, still cause the Sea's surface elevation to vary about 0.3 m annually. Figure WB2.3 shows calculated inflows to the Salton Sea from 1967–2006. Prior to the 2003 signing of a large agricultural-to-urban water transfer agreement, total annual inflows to the Sea averaged about 1.6 km^3. This has since declined to about 1.5 km^3, and is projected to decline further, to about 0.88 km^3/yr within 25 years (DEIR 2006). A variety of factors account for these reductions, including reduced flows from Mexico, changes in cropping patterns, and, after 2017, the water transfer itself. Climate change impacts on evaporation from the Sea's surface, and on evapotranspiration from the irrigated fields in its watershed, are also expected to have a marked effect on the Sea's size and water quality (Cohen and Hyun 2006).

California Water Transfers

In the mid-1990s, the federal government and the other six U.S. states that share the Colorado River began to exert increasing pressure on California to reduce its consumption of Colorado River water. This pressure drove California state and local water agencies' discussions and negotiations over the "Quantification Settlement Agreement (QSA)."[2] Central among these discussions were the terms of an Imperial Valley-San Diego

---

2. The QSA and related agreements quantified the water rights of some California Colorado River contractors, enabled acquisition and transfer of conserved water, and obligated environmental impact mitigation. For texts of selected QSA documents, see: http://www.crss.water.ca.gov/crqsa/index.cfm

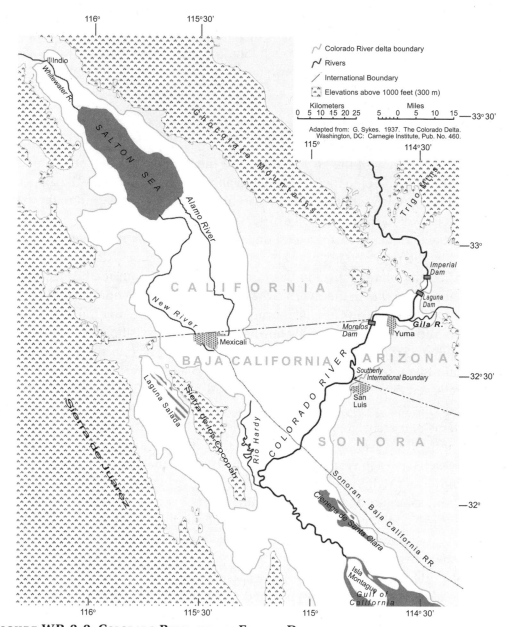

FIGURE WB 2.2 COLORADO RIVER AND ITS FORMER DELTA.

Modified from Sykes, G. 1937. *The Colorado Delta.* Publication no. 460. Washington, DC: Carnegie Institution.

water transfer. San Diego sought to invest in Imperial Valley water-efficiency improvements (such as lining canals), in exchange for receiving the conserved water over time. In 2001 and 2002, these discussions foundered over costs, liability, potential impacts to state and federal threatened and endangered species at the Salton Sea, and the costs of conveyance. In early 2003, the Bureau of Reclamation initiated proceedings to unilaterally decrease deliveries to the Imperial Irrigation District (IID), citing unreasonable water use.[3] To facilitate the water transfer and the signing of the QSA and to avoid unilateral reductions, California state negotiators agreed, among other things, to cap the water agencies' liability for QSA-related impacts to the Sea at $133 million; the

---

3. See Colorado River, Notice of Opportunity for Input Regarding Recommendations and Determinations Authorized by 43 CFR Part 417, Imperial Irrigation District, 68 Fed Reg 22738 (April 29, 2003).

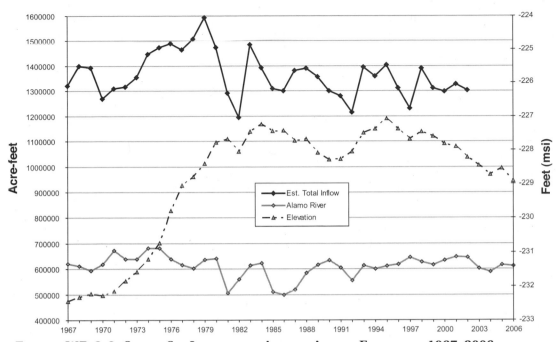

FIGURE WB 2.3 SALTON SEA INFLOWS AND AVERAGE ANNUAL ELEVATION., 1967–2006.
Source: Estimated total inflows from U.S. Bureau of Reclamation, Elevation and Alamo River data from U.S. Geological Survey gages.

state entered into contracts and adopted legislation to assume liability for costs in excess of this amount. In 2007, the state estimated capital costs for simply managing air quality and endangered species at the Salton Sea at more than $800 million (PEIR 2007).

The QSA also requires IID to offset the impacts of declining inflows due to the water transfers by delivering "mitigation" water directly to the Sea, through 2017, providing a brief window in which restoration can be designed and implemented. Without a restoration project, starting in 2018, the size and water quality of the Salton Sea will begin a period of very rapid decline, with a roughly 60% loss of volume, a tripling of salinity, and exposure of nearly 300 km² of lakebed within a dozen years (Cohen and Hyun 2006).

Restoration

The Salton Sea will change dramatically in the near future, whether or not state and federal officials take action on its behalf. For the Salton Sea, successful restoration will not mean a healthier lake of similar size. Instead, it will mean a completely unrecognizable combination of various infrastructure-heavy project elements, possibly including massive dams, multiple pumps, sedimentation basins, and hundreds of miles of berms and canals. Unlike more typical, science-based restoration projects, Salton Sea restoration requires a policy-level or political determination of a preferred set of conditions that bear no resemblance to any pre-disturbance state of the lake. These challenges underscore the differences between Salton Sea restoration and more typical projects focusing on restoring or promoting the recovery of damaged or degraded ecosystems (SER 2004).

Salton Sea restoration has been proposed for more than 40 years. Yet the goals of restoration have changed over time, and continue to differ based on the location and objectives of restoration advocates. Four general types of actions could be taken: 1) full-Sea restoration; 2) partial-Sea restoration; 3) shallow-habitat construction; and 4) the legal minimum of air quality and desert pupfish management. Additionally, the state could fail to meet its legal obligations, due to legislative inaction or the staggering costs of meeting such obligations, and not fund any significant action at the lake. The estimated $800+ million price tag for meeting these obligations (DEIR 2006) and the general lack of political will to protect the Salton Sea (San Diego Union Tribune 2006) suggest that legislative action might be deferred and delayed for many years, until litigation and court orders require it.

Full-Sea Restoration Alternative

Historically, when inflows to the lake were thought to be relatively secure, restoration proponents envisioned a full Sea, preserving the Sea's shoreline at roughly its elevation of the time, and stabilizing its salinity at marine levels (~33–35 g/L TDS). Now, with the decrease in inflows to the Sea, the only way to maintain a full Sea would be to import sufficient volumes of water to offset projected declines. Increasing demand for the over-allocated Colorado River and federal legislation prohibiting diversions of additional Colorado River flows into the lake rule out new sources of fresh water. The only other nearby water source with sufficient volume is the ocean. To maintain current elevation and a salinity of 44 g/L would require pumping some 4.2 km^3 of water up roughly 20 m and a distance of 286 to 350 km (depending on the route) from the Upper Gulf of California through Mexico to the lake. The size of the Sea could be readily managed just by importing additional water, but the ocean water would carry a huge salt load, quickly spiking the Sea's salinity beyond acceptable ranges. Stabilizing the Sea's salinity would require removing an additional 3.3 km^3 of highly saline Salton Sea water up 90 m and 286 to 350 km back to the ocean, each year, to create a flow-through system necessary to avoid the accumulation of additional salts. The required infrastructure and energy requirements of such a project would be exorbitant – costs could exceed $70 billion – and such a project would require the approval of the Mexican government (DEIR 2006). Additionally, the time required for designing, permitting, acquiring rights-of-way, and constructing such a project would delay benefits to the Sea for decades, during which the Sea would degrade at a rapid rate. Despite these problems, many people living near the Sea continue to advocate such a binational pipeline, because it is the only way to maintain the Sea as people now know it.

Partial-Sea Restoration Alternative

The exorbitant costs and institutional obstacles associated with full-Sea restoration make it, at best, a theoretical option only. Most serious observers and analysts no longer consider it a realistic possibility. Partial-Sea restoration proposals appear more feasible. Generally, these proposals seek to preserve some extent of existing shoreline, create a smaller lake with approximately marine salinity, include air-quality control measures to limit emissions of dust from an exposed lakebed, and would construct large shallow ponds to create habitat for many of the species of birds that use the Sea.

California legislation adopted in 2003 required the California Resources Agency to craft and submit to the state legislature a Salton Sea ecosystem restoration plan, in consultation with a broad range of stakeholders. This legislation (California Fish and Game Code § 2931(c)) directs the Resources Agency to submit a preferred alternative that provides the maximum feasible attainment of the following objectives:

- Restoration of long-term stable aquatic and shoreline habitat for the historic levels and diversity of fish and wildlife that depend on the Salton Sea.

- Elimination of air quality impacts from the restoration projects.

- Protection of water quality.

During consultation, the state reviewed several partial-Sea proposals that include various permutations of impoundments on the north or south side of the existing lake (see DEIR 2006), ultimately developing a preferred alternative that drew from the general concepts driving these partial-Sea proposals. The Resources Agency's preferred alternative, submitted to the legislature in 2007 and shown in Figure WB2.4, calls for the creation of: 251 km^2 of shallow saline habitats, primarily at the south end of the Sea; a 182 km^2 horseshoe-shaped marine lake impounded by an 84 km-long dam; 304 km^2 managed for air quality and 429 km^2 of total exposed lakebed; and associated canals, pumps, sedimentation ponds, and related infrastructure. The plan as designed would cost an estimated $8.9 billion, with an additional $142 million each year in operations and maintenance, at build-out (PEIR 2007). The plan would require the importation and placement of an estimated 145 million cubic meters of rock and gravel, for the dam and other barriers, and the dredging or excavation of 81 million cubic meters of lakebed. To give an idea of the scale of this construction, the PEIR estimates that 3,000 truck trips would be required each day for several years, just to transport the needed rock and gravel. However, existing air-quality restrictions limit diesel emissions, suggesting that 3,000 daily truck trips may not be feasible, potentially lengthening the time of construction and the time required to complete the project well beyond the state's projected dam closure in 2022. See Figure WB2.4.

To complicate matters, some local agencies dispute the state's estimates of the size of future inflows to the Salton Sea. The Imperial Irrigation District, source of the agricultural water being transferred, has adopted resolutions stating it will not enter into future water-transfer agreements. Pointing to these resolutions, these agency directors claim that the state underestimates future inflows. These inflow estimates determine the scope and scale of proposed restoration projects: if inflows are different than estimated, infrastructure such as dams and canals will be under- or over-built and might not function as designed. Other project elements, such as air-quality control measures or shallow habitat areas, may not receive sufficient water if flows are too low, stranding them. Insufficient flows could also render proposed marine lakes too salty. On the other hand, if inflows are larger than projected, local officials have expressed concern that this will encourage urban water agencies to demand additional water transfers, threatening the local economy. Further challenging these estimates is the historic variability in flows (see Fig. WB2.3): scaling the restoration project to estimated median annual inflows would mean that roughly 50% of the time, the project would have insufficient water to function as designed. Even designing the project to function at 20th percentile inflows would mean that the project would fail to perform as designed one year in five. This suggests that the restoration project should be sufficiently flexible to adapt to variable inflows, perhaps by varying the elevation of the lake or by creating or desiccating shallow habitat ponds based on water availability.

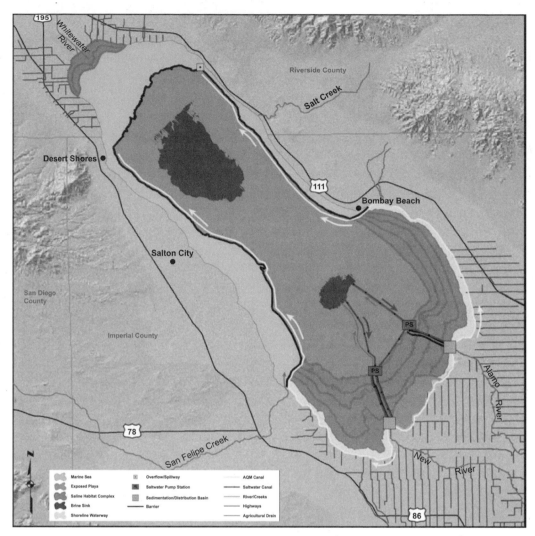

FIGURE WB 2.4 RESOURCES AGENCY'S PREFERRED SALTON SEA RESTORATION PROGRAM ALTERNATIVE.

The Salton Sea Authority (SSA), a joint powers authority comprised of representatives of the two counties, two water agencies, and one tribe bordering the lake, proposed a less ambitious plan that assumes average annual inflows will be about 12% greater than the state's estimates.[4] The SSA Plan does not account for climate change-driven increases to evaporation from the >500 km^2 of open water. Their plan seeks to maximize economic development and recreational use of the lake. It includes a large dam across the width of the Sea, plus various conveyances, pumping more than two million cubic meters of saltwater per day, and two, 1.1+ million cubic meter/day water treatment plants intended to improve water quality in the impounded north lake. The plan also designates shallow saline habitat areas for wildlife, and would attempt to manage exposed lakebed by impounding hypersaline water to create salt crusts. The state estimated the cost of its version of this plan at ~$5.2 billion, with an additional $82 million in annual operations and maintenance costs.

Despite their differences, the state and SSA plans share several significant logistical challenges. Among these is the seismic activity in the region, one of the most tectoni-

4. The SSA Plan is available at http://www.saltonsea.ca.gov/ and was reviewed in slightly modified form in the DEIR (2006).

cally active in North America (Monroe 2007).[5] The Sea itself lies atop the seismically active zone between the Pacific and North American plates; the former moves away from the latter at a rate of about 4 cm/year (Monroe 2007, Elders et al. 1972). The San Andreas and San Jacinto faults both run near or beneath the Sea. The San Andreas fault historically experiences a major earthquake in this area roughly every 200 years, though the most recent occurred 335 year ago (Monroe 2007). This seismic activity requires that proposed structures be designed accordingly, dramatically increasing costs (Reclamation 2007). Other logistical challenges include the harsh summer climate, occasional days of large waves that will limit water-based construction, staging difficulties as the shrinking lake requires frequent dredging or relocation of harbors, the restrictions on diesel emissions noted above, the absence of a proximate quarry with sufficient rock in the size and quantity needed, and the limited number of construction firms capable of handling a project of this scale.

Multi-billion dollar restoration proposals also face the harsh fiscal realities of California's projected multi-billion dollar budget deficit, and limited prospects for any federal or local funding. Although the project timeline extends through 2077, the reality is that the bulk of financing would be required during construction, in the first 10–15 years of the project. Adding to the projected high capital costs are the equally daunting annual operations and maintenance expenses; for the state's Preferred Alternative, these expenses would be equivalent to more than 35% of the California Department of Fish and Game's total FY 08–09 budget.

Shallow Habitat Alternative

In 1998, Congress adopted P.L. 105–372, directing the Bureau of Reclamation to conduct a feasibility study for maintaining the Sea as an agricultural sump; stabilizing its salinity and elevation; reclaiming fish and wildlife and their habitats; and enhancing the potential for recreation and economic development. In 2004, Congress further directed Reclamation to complete a feasibility study on a preferred Salton Sea restoration alternative (P.L. 108–361). In January, 2008, the Bureau of Reclamation released an appraisal level report on Salton Sea restoration that declined to recommend any of the five action alternatives it reviewed, due to their extreme costs and their significant risks and uncertainties (Reclamation 2007). Reclamation estimated costs for alternatives similar to those evaluated by the state: these costs ranged from $3.5 to $14 billion, with an additional $119 to $235 million in estimated annual costs. The report also noted "substantial uncertainties and risks associated with engineering, physical, and biological elements of the alternatives."

Instead, the report recommends a "Progressive Habitat Development Alternative." This concept-level alternative would develop and study the performance of 809 ha of shallow saline habitat complexes, constructed over a period of 7–10 years. Such complexes would impound water with salinities of 20 g/L and higher, to depths of up to 2 m, in a variety of configurations designed to increase habitat variability (see Fig. WB2.5.) California's Preferred Alternative contains a very similar component, known as "Early Start Habitat," though the state's plan calls for the 809 ha acres of shallow saline habitat to be constructed by 2011, fully 7 years earlier than the federal proposal. The state's plan also includes early start habitat as an interim measure, to provide habitat value for birds while the larger preferred alternative is developed and constructed.

---

5. For a map of recent earthquakes in the region, see http://www.data.scec.org/index.html

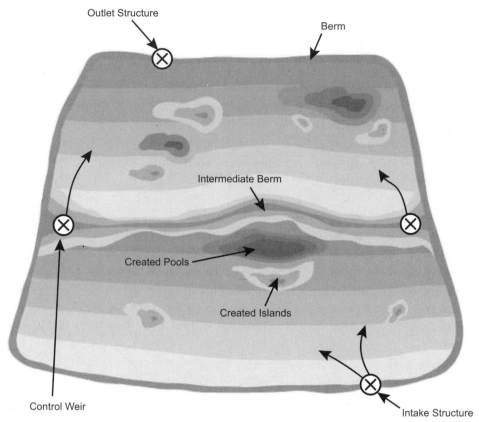

FIGURE WB 2.5 CONCEPTUAL DESIGN FOR SALTON SEA SHALLOW SALINE HABITATS.
Source: U.S. Bureau of Reclamation.

Reclamation's report, on the other hand, suggests that federal restoration would not include any major infrastructure elements.

The benefit of the federal approach is that it could be phased in over time, and would be far less expensive. Additionally, such shallow habitat ponds could be constructed relatively quickly, creating interim habitat in the near future, and could be relatively resilient in the face of earthquakes – some of the ponds would likely survive even if others were lost. The most significant drawback is that even tens of thousands of hectares of shallow saline habitat would almost certainly fail to replicate the existing habitat and recreational values of the Sea's deep open water. Such a project would bear no resemblance to the existing Sea, and will not enjoy local support.

Legal Requirements

Whether a restoration project is implemented, California is required to undertake two actions at the Salton Sea: 1) monitor land exposed as the lake recedes and control such dust-emitting soils as may be exposed due to the 2003 water transfer, and 2) promote the recovery of the endangered desert pupfish, by ensuring that pupfish populations in the various drains and rivers do not become isolated (DEIR 2006). The latter task could simply involve connecting drainage canals and rivers around portions of the Sea. Air-quality monitoring and management, however, could be very expensive, given the magnitude of lands exposed. As the surface elevation of the Sea falls to a new elevation based on declining inflows over the next 20–30 years, some 350 km^2 of lakebed will be exposed. It is not known how much of this exposed lakebed will actually emit dust. California is currently in the process of developing a monitoring network to measure such emissions. A variety of methods could be used to control dust emissions, including the use of sand fences, shallow flooding, planting and irrigating salt-tolerant vegetation, and creating salt crusts. Most of these methods require water. This water would come from the remnant Sea itself, or by diverting some of the lake's inflows. Either way, the Sea would shrink further, exposing additional lakebed.

In 2003, California assumed responsibility for air-quality monitoring and mitigation for impacts due to the water transfer (the transfer parties are responsible for the first $133 million in environmental mitigation costs; the state assumed liability for all costs exceeding this threshold). However, the transfer itself represents just over half of the state's estimated reduction in flows to the Sea. Landowners will be responsible for lakebed exposed due to factors other than the transfer, such as declines in flows from Mexico or changes in cropping patterns, or due to increases in evaporation. Determining what factors actually lead to land exposure will be very contentious, given the responsibility to control emissions from such lands. This uncertainty could lead to extensive litigation between landowners and the state (Cohen and Hyun 2006).

No Action Alternative

As of mid-2008, California was suffering from a $14.4 billion budget deficit and the legislature had demonstrated little interest in funding Salton Sea restoration. The federal government has invested millions of dollars in Salton Sea studies (Cohen et al. 1999), and has funded the construction of 40 ha of shallow ponds near the southeastern edge of the Sea, but has yet to authorize the hundreds of millions of dollars that will be required for restoration. The extremely high costs estimated for Salton Sea restoration, the high degree of uncertainty regarding future conditions, the lack of consensus on a restoration plan, the other environmental problems clamoring for state and federal intervention (such as the San Francisco/Sacramento Bay-Delta, eroding delta levees, and the Klamath) and the Salton Sea basin's limited political leverage relative to the Bay-Delta region and to urban Southern California, combine to suggest that any large-scale action at the Salton Sea may be deferred for many years, if it is ever implemented at all. These challenges make the "no-action alternative"—intentional or not – a real possibility.

Conclusion

For the next decade, change at the Salton Sea will continue to be gradual. Salinity will slowly rise to about 60 g/L, the surface of the Sea will drop another meter, more dust will blow, and fish and invertebrate populations will be stressed by worsening water quality. But the Sea will look much the same as it does now. Starting in 2018, however, the rate of change will increase dramatically as inflows drop precipitously. After 2018, the shrinking Sea will quickly become an environmental catastrophe, threatening public health with massive dust storms and potentially threatening the survival of the large populations of many species of birds that currently depend on the Sea (Cohen and Hyun 2006). The greatest challenge facing restoration advocates is convincing decision-makers that action needs to be taken now, to avert the catastrophe.

Fortunately, there have been some positive developments. The U.S. Geological Survey's Salton Sea Science Office is currently operating and monitoring a 40-ha pilot project of shallow habitat on the southeast end of the Sea, while the Torres-Martinez Desert Cahuilla Indian tribe has constructed a 34-ha wetland on their land at the northwest end of the lake (Kelly 2008). These two projects have both attracted large numbers of birds, with more than 135 different species recorded at the sites. Building upon these successes could demonstrate the benefits of state and federal commitment to the Sea, generating momentum while providing real benefits on the ground.

REFERENCES

Cohen, M.J. 2002. Managing across boundaries: the case of the Colorado River delta. In Gleick, P.H, editor. *The World's Water 2002–2003.* Washington D.C.: Island Press, pp. 133–147.

Cohen, M.J., and Hyun, K.H. 2006. *Hazard: The Future of the Salton Sea with No Restoration Project.* Oakland, CA: Pacific Institute.

Cohen, M.J., Morrison, J.I., and Glenn, E.P. 1999. *Haven or Hazard: The Ecology and Future of the Salton Sea.* Oakland, CA: Pacific Institute.

Draft Environmental Impact Report (DEIR). 2006. *Salton Sea Ecosystem Restoration Program Draft Programmatic Environmental Impact Report.* Prepared for State of California, The Resources Agency, by California Department of Water Resources and California Department of Fish and Game. State Clearinghouse #2004021120. Available at http://www.saltonsea.water.ca.gov/PEIR/draft/.

Elders, W.A., Rex, R.W, Meidou, T., Robinson, P.T., and Biehler, S. 1972. Crustal spreading in southern California. *Science* 178:15–24.

Glantz, M.H. 1999. Sustainable development and creeping environmental problems in the Aral Sea region. In Glantz, M.H., editor. *Creeping Environmental Problems and Sustainable development and in the Aral Sea Basin.* Cambridge: Cambridge University Press. pp. 1–25.

Jehl, J.R., and McKernan, R.L. 2002. Biology and migration of eared grebes at the Salton Sea. *Hydrobiologia* 473: 245–253.

Kelly, D. 2008. Part of Salton Sea's desolate shore made into a lush oasis. *Los Angeles Times,* February 24. Available at http://www.latimes.com/news/printedition/california/la-me-wetlands24feb24,0,5237505.story

Monroe, R. 2007. The shaky future of the Salton Sea. *Explorations* 14(9). Available at http://explorations.ucsd.edu/Features/Salton_Sea/

Nijhuis, M. 2000. Accidental refuge: Should we save the Salton Sea? *High Country News,* June 19. Available at http://www.hcn.org/servlets/hcn.Article?article_id=5865

PEIR. 2007. *Salton Sea Ecosystem Restoration Program Final Programmatic Environmental Impact Report.* Prepared for State of California, The Resources Agency, by California Department of Water Resources and California Department of Fish and Game. Available at http://www.saltonsea.water.ca.gov/PEIR/.

Pomeroy, R.D., and Cruse, H. 1965. *A Reconnaissance Study and Preliminary Report on a Water Quality Control Plan for Salton Sea.* Prepared for the California State Water Quality Control Board. Available at http://projects.ch2m.com/webuploads/ssRestoration/182_1965_Recon_Study_and_Prelim_WQC_Plan_1_of_5.pdf

Reclamation U.S. Department of the Interior, Bureau of Reclamation. 2007. *Restoration of the Salton Sea: Summary Report.* Available at http://www.usbr.gov/lc/region/programs/saltonsea.html.

Reisner, M. 1993. *Cadillac Desert: The American West and its Disappearing Water.* NY: Penguin Books. 2nd ed.

San Diego Union-Tribune. 2006. On Salton Pond; Man can't save this desert lake. *San Diego Union-Tribune,* October 23.

Society for Ecological Restoration (SER). 2004. International Science & Policy Working Group. *The SER International Primer on Ecological Restoration.* Tucson: Society for Ecological Restoration International. 15 pp. Available at www.ser.org

Shuford, W. D., Warnock, N., Molina, K.C., Mulrooney, B. and Black, A.E.. 2000. *Avifauna of the Salton Sea: Abundance, distribution, and annual phenology.* LaQuinta, CA: Contribution No. 931 of Point Reyes Bird Observatory. Final report for EPA Contract No. R826552-01-0 to the Salton Sea Authority, 78401 Highway 111, Suite T, La Quinta, CA 92253.

Shuford, W. D., Warnock, N., Molina, K.C., and Sturm K. 2002. The Salton Sea as critical habitat to migratory and resident waterbirds. *Hydrobiologia* 473:255–274.

Shuford, W.D., Warnock, N., and McKernan, R.L., 2004. Patterns of shorebird use of the Salton Sea and adjacent Imperial Valley, California. *Studies in Avian Biology* 27:61–77.

Sykes, G. 1937. *The Colorado Delta.* American Geographical Society. Baltimore, Maryland: Lord Baltimore Press, 19:193.

Three Gorges Dam Project, Yangtze River, China

Peter H. Gleick

Introduction

The Three Gorges Dam (TGD) and associated infrastructure is the largest integrated water project built in the history of the world. It has also been one of the most controversial due to its massive environmental, economic, and social impacts. The very first volume of *The World's Water*, published more than a decade ago, reviewed the plans underway at that time to build the Three Gorges Dam, along with many of the expected benefits and costs (Gleick 1998). A decade later, the physical dam itself has largely been completed, although work is continuing on electrical generating systems and a wide range of peripheral projects. This chapter offers an update on the project and a timeline of major events. It is crucial to note that while extensive information on the project is available from authorities and government officials, reliable independent information on environmental and social costs is harder to find (Dai 1994, 1998; Heggelund 2007). This update draws on official materials, as well as information available from non-governmental and non-Chinese sources, to get a clearer snapshot of the project's complex implications.

There are growing indications that very serious problems have started to develop. In the summer of 2007, major western media began to report on growing threats from landslides, pollution, and flooding, as well as growing social and political unrest and dissatisfaction associated with relocating millions of people (Oster 2007, Yardley 2007). Even officials in China have begun to be increasingly outspoken about unresolved challenges associated with the project. Weng Lida, secretary general of the Yangtze River Forum was quoted as saying "the problems are all more serious than we expected" (Oster 2007). In September 2007, Chinese officials "admitted the Three Gorges Dam project has caused an array of ecological ills, including more frequent landslides and pollution, and if preventive measures are not taken, there could be an environmental 'catastrophe' " (Xinhua 2007c). The complex and massive effort to relocate millions of displaced and affected people has also caused a range of social, political, and economic problems.

It is impossible to try to judge whether the TGD project will have net costs or benefits. All major water projects have complicated combinations of both costs and benefits that vary over a project's lifetime and are difficult to evaluate and quantify in a consistent, comparable way. As is typical with such large water projects, the benefits are typically far easier to identify and quantify than the costs, which often only manifest themselves over many years, in complex ways. Calculating actual costs and benefits accurately may never be possible because of the difficulty of putting monetary values on many of the complex environmental, social, and cultural impacts of the project (Tan and Yao 2006). Nevertheless, enough time has gone by, and enough information is available, to begin the process of evaluating the overall implications of the project.

The Project

The Three Gorges Dam stretches more than two kilometers across one of the greatest rivers in the world, the Yangtze. The dam was built in a stretch of the Yangtze known as Three Gorges because of the canyons formed by immense limestone cliffs. These gorges— the Xiling, Wu, and Qutang—offer some of the most scenic landscape anywhere in the world and have long been a destination spot for tourists from around the world. In recent years, tourism has boomed as people have rushed to see some of the sights to be destroyed by the dam and reservoir (China View 2008). The beauty of the region has inspired Chinese poets and artists for centuries including much of the work of Li Bai (701–762 AD), considered by many Chinese to be the world's greatest poet (Fearnside 1988).

The idea of building a gigantic dam on the Yangtze River in the Three Gorges area was proposed more than 80 years ago by Sun Yat-sen. After severe flooding along the river in the 1950s, Chairman Mao Tse Tung vowed to speed up construction of a massive dam but nothing significant happened for several more decades. In 1986, the Chinese Ministry of Water Resources and Electric Power asked the Canadian government to finance a feasibility study to be conducted by a consortium of Canadian firms. The consortium, known as CIPM Yangtze Joint Venture, included three private companies (Acres International, SNC, and Lavelin International), and two state-owned utilities (Hydro-Quebec International and British Columbia Hydro International). The World Bank was asked to supervise the feasibility study to ensure that it would "form the basis for securing assistance from international financial institutions" (Adams 1997). On April 3, 1992, the National People's Congress officially approved the construction of the project. On December 14, 1994, the Chinese government formally began construction. The first electricity was produced in 2003, and the physical dam was mostly completed in 2006.

The Three Gorges Dam is nearly 200 meters high, has a volume of 40 million cubic meters, and has created a reservoir 600-kilometer long with a total storage capacity approaching 40 billion cubic meters. Maximum storage of water behind the dam is expected to occur sometime in 2008. The 14 generators in the north side of the dam have already been installed and they reached full capacity (9,800 MWe) on October 18, 2006 after the water level in the reservoir had been raised to 156 meters. Installation of seven generators in the south side of the dam was completed by the end of 2007, bringing the total power capacity to 14,800 MWe, surpassing the generating capacity of the Itaipu Dam (14,000 MWe) in Brazil (Government of China 2006). At its completion, sometime after 2010, the project is expected to have a total installed hydroelectric capacity exceeding 22,000 MWe. This power capacity is higher than originally proposed because of an expansion initiated in 2002. In 2007, the turbines generated around 62 billion kWhr of electricity – about two-thirds of the maximum level expected by the completed project. Other benefits of the project claimed by project designers include flood protection on the historically dangerous Yangtze River and improvements to river navigation for thousands of kilometers.

Major Environmental, Economic, Social, and Political Issues

Economic, environmental, social, and political concerns have been raised about the TGD project, both before the project was launched and in recent years. One of the strongest and most consistent arguments made by project proponents has been that the electricity produced by the dam would otherwise be produced by dirty Chinese coal-burning power plants, with their serious environmental impacts. One of the strongest and most consistent arguments made by project opponents has been the vast scale of the environmental and social transformations of the watershed of the Yangtze both upstream and downstream of the dam itself. These major questions are addressed here.

Economic and Financial Costs

The total cost of the Three Gorges Dam and associated projects will be enormous, but it is no longer possible to produce any definitive quantitative estimate. Even the financial costs of the infrastructure alone cannot be known because of the magnitude of the expenditures, the related development projects in the region, and expenditures made unofficially. Estimates of the construction costs made during the mid-1990s for the major parts of the project ranged from a low of $25 billion to a high of $60 billion (Dai 1994, China 1996, JPN 1996, McCully 1996, Reuters 1997). The most recent estimates have fluctuated around the upper end of these figures. The TGD is being funded by a complex mix of both internal and external sources. China has identified four internal sources of funds: the State Three Gorges Construction Funds, power revenues from existing hydropower facilities, power revenues from the Three Gorges Project itself, and loans and credits from the Chinese State Development Bank (SDB), now renamed the Chinese Development Bank (CDB).

External sources of funding have been critical for the project. International organizations have tried to maintain a list of international financiers and companies supplying equipment and services to the project through the China Three Gorges Project Development Corporation, a state-owned entity set up to finance and build the project (see, especially, Probe International 2008). Canada's Export Development Corporation, Germany's export-import bank, and other international export credit agencies provided early loan guarantees for the project totaling hundreds of millions of dollars (Financial Times 1997). Commercial banks and investment firms have offered significant financing assistance. The SDB of China signed a loan package with Germany's Kreditanstalt Fur Wiederaufbau, Dresdner Bank, and DG Bank in 1997 for the purchase of turbines and generators. Hundreds of millions of dollars in SDB bonds were underwritten at the beginning of the project by a virtual who's who of the international financial community, including Lehman Brothers, Credit Suisse First Boston, Smith Barney Inc, J.P. Morgan & Co, Morgan Stanley & Co Incorporated, and BancAmerica Securities Inc. In 1997 and 1999, SDB issued more than a billion dollars of new bonds underwritten and managed by Merrill Lynch & Co. and Chase Manhattan Bank, with contributions from Chase Securities, J.P. Morgan, Morgan Stanley Dean Witter, Credit Suisse First

Boston and Goldman Sachs. Morgan Stanley continued to participate through a joint venture with the China International Capital Corporation (CICC), which is the lead advisor on raising overseas capital. In 2004, the CDB and the Chinese Export Import Bank (CEIB) hired Goldman Sachs, UBS, HSBC, Citigroup, and others to help raise another €1.5 billion from bonds (Carrell 2004, Probe International 2008).

Far more difficult to compute than the financial costs of building infrastructure are the non-traditional costs associated with social disruption, political corruption, massive relocation, ecological losses, and unquantified geological threats associated with landslides and earthquakes.

Environmental Impacts

Much of China's electricity is produced by thermal power plants burning one of the dirtiest fossil fuels – coal. The Chinese government estimates that if the electricity generated by the Three Gorges project were produced instead with Chinese coal, 50 million more tons of coal would be burned annually, producing 100 million tons of carbon dioxide, 1.2–2 million tons of sulfur dioxide, 10,000 tons of carbon monoxide, and large quantities of particulates (China 1996, Xinhua 2007a).

Government officials also point to efforts to remove polluting enterprises from the edge of the river or reservoir, and their construction of sewage treatment facilities to improve water quality in the Three Gorges reservoir region, although they note that eutrophic conditions and algal blooms continue to occur throughout the basin (People's Daily Online 2007). In addition, while massive funding has been committed to the dam itself, much of the proposed spending on pollution control has not yet occurred. Officials estimate that about 40 billion yuan will be spent to build at least 150 sewage treatment plants and 170 urban garbage disposal centers, but many of these are not yet complete (China Daily 2007a). The city of Chongqing alone still releases nearly one billion tons of untreated wastewater into the Three Gorges reservoir every year (Hodum 2007).

Fisheries Impacts

Ecological problems have been projected to occur as a result of the construction of the dam and modification of the watershed, including impacts on the fisheries of the Yangtze River basin. This basin has 36 percent of all freshwater fish species in China, with more than 360 fish species belonging to 29 families and 131 genera (Xie 2003). Twenty-seven percent of all of China's endangered freshwater fish are in the Yangtze basin, and there are as many as 177 endemic fish species (Yue and Chen 1998).

Major changes in fish populations have been anticipated because the project is altering the dynamics of the river, the chemical and temperature composition of the water, and the character of the natural habitat and food resources available for these fish species. The dam itself blocks migration of fish and access to spawning grounds, and these impacts will be imposed on top of other significant modifications to the Yangtze that have already caused declines in fisheries. In 1981, the Chinese completed the construction of the Gezhou Dam 40 kilometers downstream from the TGD site. That was followed by rapid and sharp declines in the populations of three of China's famous ancient fish species, the Chinese sturgeon, River sturgeon, and Chinese paddlefish, each of which is now listed as endangered (Xie 2003). Of special concern is the Chinese freshwater dolphin, which may already be extinct (Hance 2008). Fisheries in the upper watershed are also at risk. A study in 2003 identified six species at high risk of complete extinction, another 14 with an uncertain future, and two dozen more that may only survive in tributaries to the Yangtze (Park et al. 2003).

Fisheries are already beginning to show the effects of altered river ecology below the dam. Data released to a Three Gorges Dam monitoring program website indicate that commercial harvests of four species of carp are well below pre-dam levels (Xie et al. 2007). Annual harvest of these commercial fish below the dam from 2003 to 2005 was 50–70% below a 2002 pre-dam baseline and even more dramatic declines are being seen on larvae and eggs below the dam (see Table WB 3.1).

Han Qiwei, a member of the Chinese Academy of Sciences, has argued, correctly, that many of the ecological problems in the Yangtze predate the building of the dam (China Daily 2007a). But many of these problems have also been worsened, not improved by the project. Despite early calls from environmentalists, limited scientific investigations were done to prepare a baseline assessment of plant and animal communities threatened or destroyed by the Three Gorges Dam project. Only in late 2007 was a formal comprehensive assessment of plant communities initiated by the Chinese Academy of Sciences (Xinhua 2007d). Prior to this, research institutes from the region had conducted small-scale investigations, but no large-scale systematic search or collection was done. A small private botanical garden set up to support rare regional plants went bankrupt in June 2007 (Xinhua 2007d). Some gene and seed banks have been set up by the Chinese Academy of Sciences to maintain genetic stocks of plants that may end up going extinct in the wild (China Daily 2007b).

River Sediment Flow

The dam is also having a significant impact on sediment loads in the Yangtze. The Yangtze River has traditionally carried a vast load of sediment from its upper reaches of the watershed to the East China Sea, supporting ecological processes in the river delta and the productivity of fisheries in the Sea. This sediment load has varied with annual climatic factors, and more significantly, with the level of deforestation, and subsequent reforestation in the upper watershed. The completion of the Three Gorges Dam, however, has led to a rapid and significant decrease in downstream sediment load. Sediment volumes have been declining from the late 1990s due to reforestation efforts and the construction of many small- and intermediate-sized dams on Yangtze River tributaries. In 2003, the closure of the Three Gorges Dam caused a further severe decrease. Sediment load at Datong, near the Yangtze's delta dropped to only 33 percent of the 1950–1986 levels (Xu et al. 2006). Among the consequences of this drop in sediment are growing coastal erosion and a change in the ecological characteristics and productivity of the East China Sea (Xu et al. 2006). Based on estimates of the historical sediment budget and erosion data from the river's delta, scientists estimate that

TABLE WB 3.1 Annual Commercial Harvest (x1,000 metric tons) of Four Species of Carp (Silver, Bighead, Grass, Black) and Numbers (Millions) of Drift-sampled Carp Eggs and Larvae Below the Three Gorges Dam Before (1997 and 2002) and After the River Was Impounded (2003–2005)

Year	1997	2002	2003 (Dam Closure)	2004	2005
Commercial Harvest (1000 metric tons)	NA	3360	1350	1010	1680
Eggs and Larvae (millions)	250	190	40.6	33.9	10.5

Data from Xie et al. 2007.

the delta will be increasingly eroded during the first five decades after full operation and then approach a balance during the next five decades as sediments start to move through the TGD reservoir (Yang et al. 2006).

Flood Protection

A major anticipated benefit of the project is improved flood protection on the middle and lower reaches of the Yangtze River. Historically, people living along the Yangtze River have suffered tremendous losses from flooding. In 1931, 145,000 people drowned, and over 300,000 hectares of agricultural land flooded. In 1954, 30,000 more died in Yangtze floods or the subsequent diseases (Boyle 2007). In 1998, a flood in the same area caused billions of dollars of damage. More than two thousand square kilometers of farm land was flooded, and over 1,500 people were killed (CTGPC 2002).

The Chinese government has already claimed flood benefits to the dam. According to Li Yongan, the general manager of the Yangtze Three Gorges Project Development Office, the project averted floods in late July 2007 by storing waters that would have exceeded flood levels below the dam (People's Daily Online 2007), though the overall long-term flood-control benefits provided by the Three Gorges Dam are only likely to be determined over the next several decades as a wider range of high flows are experienced.

Shipping Benefits

The Yangtze River, China's "golden waterway," plays an important role in the economy of the upper river area. In that region, river navigation is almost the only means of long-distance, cost-effective transportation of freight. For Chongqing, the major port city in Sichuan province, 90 percent of goods are transported by water, and navigation on the upper Yangtze has been difficult in the past. The Three Gorges reservoir dramatically increases the depth of water and improves navigation up to Chongqing, more than 600 kilometers upstream of the dam.

Three Gorges has been built with one of the largest systems of ship locks in the world, permitting large quantities of cargo to move into the upper reaches of the Yangtze. In 2006, 50 million tons of cargo passed through the new lock system up to Chongqing, up from 18 million tons before the dam, and the 2007 estimate exceeds 50 million tons (Peoples Daily Online 2007).

Reservoir-Induced Seismicity and Geological Instability

Large reservoirs can cause seismic events as they fill and as the pressure on local faults increases (ICE 1981). Such reservoir-induced seismicity was predicted for the Three Gorges region, which is already seismically active and indeed, there has been an increase in reported seismic activity in the region following construction of the dam and the filling of the reservoir. Official statements minimize the importance of this, saying that "no unusual phenomena which could disrupt the stability of Three Gorges Dam have occurred" – a far cry from saying that there have been no significant damages to individuals, homes, or businesses (People's Daily Online 2007).

Related to the risk of increased seismic activity is the risk of increased landslides in the regions around Three Gorges with steep slopes. Landslide activity associated with the filling of the reservoir appears to be on the rise. Very soon after the closing of the dam and the filling of the reservoir, a major landslide occurred near the town of

Qianjiangping on the Qinggan River near its confluence with the Yangtze mainstream. Early on the morning of July 13, 2003, 24 million cubic meters of rock and earth slid into the Qinggan River, completely blocking its flow, capsizing 22 boats, and destroying four factories, 300 homes, and more than 67 hectares of farmland. Official reports say that 14 people were killed and 10 more were listed as missing (Wang et al. 2004). In 2007, thirty-one people died when a landslide on a tributary to the dam in Hubei province crushed a bus (Stratton 2007).

The risk of such disruptions appears to be far more severe than anticipated and is leading to new resettlement efforts as the danger zones around the margins of the reservoir expand. In the fall of 2007, officials and experts admitted the Three Gorges Dam project had caused more frequent landslides (Xinhua 2007b,c). Tan Qiwei, vice-mayor of Chongqing, told a forum in Wuhan that the shore of the reservoir had collapsed in 91 places and a total of 36 km of shoreline had caved in. In some cases, landslides around the reservoir had produced massive waves as high as 50 meters, causing even more damage along the reservoir's edge.

Relocation and Resettlement

Every large dam built in China has led to the resettlement of local people because of the high populations and the density of towns and villages along the major rivers. Even early in the debate over Three Gorges, the Chinese Academy of Sciences (1988, 1995) acknowledged that large-scale resettlement and inundation of population centers would be among the most devastating aspects of the project.

Initial estimates of the populations to be displaced varied from around one million to almost two million. Far more than a million people have already been resettled during the project's construction – official estimates typically say "at least 1.2 million" or "1.13 million" (Yardley 2007). Other estimates range from 1.3 million to almost 2 million (Dai 1998, Chao 2001, Tan and Yao 2006). More than 100 towns are ultimately to be submerged, including the major population centers of Fuling, Wanxian, and parts of Chongqing. Chongqing is the central municipality in the Three Gorges reservoir area and recently received approval to become a centrally administered municipality – only the fourth in the country after Beijing, Shanghai, and Tianjin. Fourteen thousand hectares of agricultural land will be submerged, as will more than 100 archeological sites, some dating back over 12,000 years. The cities of Wanxian and Fuling have cultural histories extending back more than 1,000 years.

In fact, it now appears possible that as many as *six million people* in total will have to be resettled because of the dam and surrounding impacts. In late 2007, a stunning announcement vastly increased the scale and scope of the relocation effort. Vice-Mayor Tan announced that "at least 4 million people from the Three Gorges Reservoir area are to be relocated to cities in the next 10 to 15 years" (Xinhua 2007b). As part of this newly announced massive relocation, more than 4 million people currently living in northeast and southwest Chongqing are to be resettled in the outskirts of Chongqing city in new settlements. Officials dispute that these new relocations are related to the dam, arguing instead that they are part of a national experiment in economic reform. Other reasons given for the resettlement include regional overpopulation, limited opportunities for industrial development, and growing ecological and geological problems along the reservoirs edge, including massive landslides (Xinhua 2007b).

Among a growing number of scholars, there is increasing concern that people displaced due to construction projects face long-term risks of becoming poorer and are also threatened with landlessness, food insecurity, joblessness, and social margin-alization (World Commission on Dams 2000, Li et al. 2001, Heggelund 2004, 2006). Certainly, the early efforts at resettlement at Three Gorges led to a worsening of condi-tions for many of the already relatively poor rural communities in the region. There has been some unprecedented discussion of these problems in scientific and policy journals, as well as the news media in China.

A factor that contributed to some of the early challenges with TGD resettlement was local government corruption, which led to significant resettlement funds ending up in the pockets of government officials, rather than passing to the refugees (Chao 2001, Heggelund 2006). Poor local planning also left many relocated people with bad land, homelessness, loss of jobs and social status, and other social ills. To make matters even worse, the resettled populations often receive farmland taken from the population who already live in the resettlement areas, raising tensions and conflicts between the host population and the new migrants (Qiu et al. 2000, Heggelund 2007). Recent research also suggests that women displaced by the project are more severely affected than men. They are more likely to become impoverished and less likely to find new work in the new areas (Yan et al. 2005). Forced migration is also apparently linked to worsening depression (Hwang et al. 2007).

Other Issues

The long-term implications of the TGD will only be understood fully over the coming decades. But it is likely to have some unanticipated implications, beyond the signifi-cant effects already predicted or observed. Some of these are already beginning to appear: The magnitude of the dam and reservoir are so large that it is already playing a role in military planning and in affecting local climatic conditions.

In 2004, the U.S. Pentagon released their annual report to Congress on military issues related to China. In that report, the Pentagon reported that Taiwanese leaders were considering the concept of targeting the Three Gorges Dam militarily as a deterrent against Chinese military action against Taiwan. They wrote:

> "Taipei political and military leaders have recently suggested acquiring weapon systems capable of standoff strikes against the Chinese mainland as a cost-effective means of deterrence. Taiwan's Air Force already has a latent capability for airstrikes against China. Leaders have publicly cited the need for ballistic and land-attack cruise missiles. Since Taipei cannot match Beijing's ability to field offensive systems, proponents of strikes against the mainland apparently hope that merely presenting credible threats to China's urban population or high-value targets, such as the Three Gorges Dam, will deter Chinese military coercion." (U.S. Depart-ment of Defense 2004).

This comment was taken by mainland Chinese media and political leaders as a direct threat, or as an effort to encourage Taiwan military to develop such capability, and provoked an angry response (Hogg 2004).

While the Chinese are increasingly concerned about the implications of climatic change for the water resources of China (see Chapter 5), there is now evidence that the TGD itself is affecting climate on a far larger scale than initially suggested. Early assessments raised the possibility that the massive new reservoir might affect temperatures and other climatic variables locally, on the scale of tens of kilometers. Now a study suggests that the effects are

TABLE WB 3.2 Chronology of Events: Three Gorges Dam Project

1919	First mention of the Three Gorges project in Sun Yat-sen's "Plan to Develop Industry."
1931	Massive flooding along the Yangtze River kills 145,000 people.
1932	Nationalist government proposes building a low dam at Three Gorges.
1935	Massive flooding kills 142,000 people.
1940s	The U.S. Bureau of Reclamation helps Chinese engineers identify a site.
1947	Nationalist government terminates all design work.
1949	Communist revolution in China.
1953	Mao Zedong proposes building a dam at Three Gorges to control flooding.
1954	Flooding along the Yangtze leave 30,000 people dead and one million people homeless.
1955	Soviet engineers play a role in project planning and design.
January 1958	Mao appoints Zhou Enlai to begin planning along Yangtze.
May 1959	Yangtze Valley Planning Office (YVPO) identifies Sandouping site for dam.
1966	All work halted by the Cultural Revolution (1965–1975).
1976	Planning recommences.
February 1984	Ministry of Water Resources and Electric Power recommends immediate commencement of construction.
Spring 1985	The National Peoples Congress delays a decision until 1987 because of economic difficulties.
1986	The Chinese Ministry of Water Resources and Electric Power asked the Canadian government to finance a feasibility study
August 1988	Canadian-World Bank "Three Gorges Water Control Project Feasibility Study" is completed and recommends construction at "an early date."
February 28, 1989	*Yangtze! Yangtze!* released.
April-June 1989	Democracy movement sweeps through China.
February 1992	Politburo Standing Committee agrees to the construction of the project.
April 3, 1992	China's National People's Congress (NPC) formally approved the "Resolution on the Construction of the Yangtze River Three Gorges Project." 177 delegates oppose the project, 644 abstain, 1,767 approve.
April 27, 1992	The Canadian government cancels development assistance for the project.
May 1992	179 members of the Democratic Youth Party reportedly detained in connection with their protests against the Three Gorges project in Kai County, Sichuan (HRW 1995).
January 1993	An armed fight involving over 300 persons occurred in the vicinity of the dam (HRW 1995).
December 14, 1993	The U.S. Bureau of Reclamation terminates agreements for technical services because of economic and environmental impacts.
Early 1994	The full resettlement program begins in earnest.

continues

TABLE WB 3.2   *continued*

Mid-1994	Excavation and preparation of the dam's foundations are underway at Sandouping, the chosen dam site.
December 14, 1994	Premier Li Peng formally declared the project under construction.
May 1996	The US Ex-Im Bank's board votes unanimously to withhold support for the project and voices serious reservations about the dam's environmental and social impacts and its economic viability.
August 1997	China awards a contract for 14 power generating units to GEC Alsthom, ABB, and an industrial consortium formed by Germany's Voith and Siemens and General Electric Canada (VGS).
September 1997	The State Development Bank of China signs a loan package with Germany's Kreditanstalt Fur Wiederaufbau, Dresdner Bank, and DG Bank that includes both export credits and a $200 million commercial loan.
November 1997	Yangtze River dammed
1998–2002	Concrete pouring on left bank
2003–2006	Concrete pouring on right bank
May/June 2003	Dam is finished and the first water is impounded. The level of water in the reservoir rises to 135 meters in June.
July 2003	The first electricity is produced.
June 2006	Dam is completed.
October 2006	North side generators reach full capacity.
2007	14 700-MW turbine generators in operation.
2008	New estimated completion date.
2009	Original estimated completion date. Reservoir estimated to be raised to 175 meters.

For a more detailed chronology of the internal political debates between 1955 and 1992, see Qing (1989, 1994).

occurring on a regional scale – a hundred kilometers (Wu et al. 2006). New research must be conducted to track and assess these kinds of unexpected consequences.

Conclusion

The Three Gorges Dam in China is rapidly approaching completion. This project, along with a vast array of peripheral projects, constitutes the largest water-supply development in the history of humanity. As with any major construction project that substantially modifies or alters a watershed, the Three Gorges Dam will have significant costs and benefits. Among the most significant benefits are the generation of electricity without greenhouse gas emissions, improvements in navigation, and potential reductions in flood risk. Among the most significant costs are massive dislocations of millions of Chinese to make way for the dam and reservoir, further ecological degradation of the Yangtze River ecosystem and fisheries, a reduction in sedimentation reaching the East China Sea, and a growing risk of new landslides and reservoir-induced seismicity. Over decades, the overall implications of the project will become more evident, but before the full benefits have begun to be delivered, the environmental, social, political, and economic costs are beginning to accumulate. Even official government spokesmen are beginning to question the substantial human and environmental costs of the project, while other officials are moving rapidly forward on new massive water infrastructure elsewhere in China, without having learned the lessons from Three Gorges. Long-term sustainable water management in China will require a better balancing of the true costs and benefits of their water choices.

REFERENCES

Adams, P. 1997. Planning for disaster: China's Three Gorges Dam. http://www.multinational-monitor.org/hyper/issues/1993/09/mm0993_08.html

Boyle, C.E. 2007. *Water-borne Illness in China.* China Environmental Health Project, Research Brief. Washington, D.C.: Woodrow Wilson International Center for Scholars. August 2007.

Carrell, S. 2004. HSBC under fire for its role in £870m bond sale to finance China's megadams. *The Independent,* July 25.

Chao, J. 2001. Relocation for giant dam inflames Chinese peasants. *National Geographic News,* May 15, 2001. http://news.nationalgeographic.com/news/pf/58264160.html.

China Daily. 2007a. Dam impact 'less than predicted.' November 27, 2007. http://www.mwr.gov.cn/english/20071127/88153.asp

China Daily 2007b. Race is on to protect Three Gorges flora. September 11, 2007. http://www.mwr.gov.cn/english/20070911/86702.asp

Chinese Academy of Sciences. 1988. Environmental Impact Statement for the Three Gorges Project. Chinese Academy of Sciences, Nanjing Geography Institute.

Chinese Academy of Sciences. 1995. Environmental Impact Statement for the Yangtze Three Gorges Project, (A Brief Edition). Environmental Impact Assessment Department, Chinese Academy of Sciences and the Research Institute for Protection of Yangtze Water Resources. Beijing, China: Science Press.

Chinese Three Gorges Project Corporation (CTGPC). 2002. Flooding on the Yangtze in 1998. April 20, 2002. Retrieved on February 8, 2008. (Chinese). http://www.ctgpc.com.cn/sxslsn/ index.php?mClassId=003000

China. 1996. The Three Gorges Project: A Brief Introduction. Official government fact sheet. Washington, D.C., Chinese Embassy.

China View. 2008. Tourist arrivals to Three Gorges Dam hit record high in 2007. http://news.xinhuanet.com/english/2008-01/06/content_7374888.htm#

Dai Q. 1994. *Yangtze! Yangtze!* English edition, 1994. Adams, P and Thibodeau, J., eds. United Kingdom: Probe International, Earthscan Publications Limited.

Dai, Q. 1998. *The River Dragon Has Come! The Three Gorges Dam and the Fate of China's Yangtze River and Its People.* Translated by Ming Yi; Thibodeau, J.G., and Williams, P.B., eds. Armonk, NY: M.E. Sharp Publishing.

Fearnside, P.M. 1988. China's Three Gorges Dam: Fatal project or step toward modernization. *World Development*16(5): 615–630.

Financial Times. 1997. Germans cement water links. *Global Water Report* 31(9):7.

Gleick, P.H. 1998. The Status of Large Dams: The End of an Era? In Gleick, P.H. *The World's Water 1998–1999: The Biennial Report on Freshwater Resources.* Washington, D.C.: Island Press, pp. 69–104.

Government of China. 2006. Three Gorges Dam. October 18, 2006. Retrieved on February 8, 2008. (Chinese). http://www.gov.cn/jrzg/2006-10/18/content_416256.htm

Hance, J. 2008. New expedition seeks evidence for survival of the 'extinct' Baiji. mongabay.com. April 16, 2008 http://news.mongabay.com/2008/0416-baiji.html

Heggelund, G. 2004. *Environment and Resettlement Politics in China: The Three Gorges Project.* Hampshire, United Kingdom: Ashgate Publishing.

Heggelund, G. 2006. Resettlement programmes and environmental capacity in the Three Gorges Dam Project. *Development and Change* 37(1):179–199.

Heggelund, G. 2007. Running into dead ends: Challenges in researching the Three Gorges Dam. Washington D.C.: Woodrow Wilson International Center for Scholars. *China Environment Series* 7:79–83.

Hodum, R. 2007. China's need for wastewater treatment, clean energy grows. Washington D.C.: Worldwatch Institute. [Online]. Available: http://www.worldwatch.org/node/4889.

Hogg, C. 2004. Storm across the Taiwan Strait. http://news.bbc.co.uk/2/hi/asia-pacific/3825927.stm

Hwang, S.S., Xi J., Cao, Y., Feng, X., and Qiao, X. 2007. Anticipation of migration and psychological stress and the Three Gorges Dam project, China. *Social Science and Medicine* 65(5):1012–1024.

Institution of Civil Engineers (ICE). 1981. *Dams and Earthquake.* Proceedings of a Conference at the Institution of Civil Engineers, London, England: Thomas Telford Ltd.

Japan Press Network (JPN). 1996. MITI and Ex-Im Bank consider aid for Three Gorges Dam. (J. Tofflemire, May 20).

Li, H., Waley, P., and Rees, P. 2001. Reservoir resettlement in China: Past experience and the Three Gorges Dam. *The Geographical Journal* 167(3):195–212.

McCully, P. 1996. *Silenced Rivers: The Ecology and Politics of Large Dam.* London, United Kingdom: Zed Press.

Oster, S. 2007. In China, new risks emerge at giant Three Gorges Dam. *The Wall Street Journal* August 29, 2007; Page A1.

Park, Y-S., Chang, J., Lek, S., Cao, W., and Brosses, S. 2003. Conservation strategies for endemic fish species threatened by the Three Gorges Dam. *Conservation Biology* 17(6):1748–1758.

People's Daily Online. 2007. Full views of Three Gorges Project. November 30, 2007. http://www.mwr.gov.cn/english/20071130/88209.asp

Probe International. 2008. Who's behind China's Three Gorges Dam: List of international 3G companies and financiers. August 2007 update. Canada, Probe International.

Qiu, Z., Lizhi, W., and Jinping, D. 2000. Sanxia kuqu nongcun yimin anzhi moshi tantao. ('Exploring the Three Gorges Rural Resettlement Pattern'), [People's] Yangtze River (*Renmin Changjiang*) 31(3):1–3 (Cited in Heggelund 2007).

Reuters. 1997. China firms win $807 million contracts on Three Gorges. *Reuters News Service* (August 19).

Stratton, A. 2007. World's largest hydroelectric dam is safe, Beijing declares. *The Guardian,* November 27, 2007.

Tan, Y. and Yao, F. 2006. Three Gorges Project: Effects of resettlement on the environment in the reservoir area and countermeasures. *Population and Environment* 27:351–371.

U.S. Department of Defense. 2004. Annual report on the military power of the People's Republic of China. FY04 Report to Congress on PRC Military Power. Washington D.C. http://www.defenselink.mil/pubs/d20040528PRC.pdf.

Wang, F.W., Zhang, Y.M, Huo, Z.T., Matsumoto, T., and Huang, B.L. 2004. The July 14, 2003 Qianjiangping landslide, Three Gorges Reservoir, China. *Landslides* 1(2):157–162.

World Commission on Dams. 2000. *Dams and Development.* London: Earthscan Publishers, 448 pages. http://www.dams.org/

Wu, L., Zhang, Q., and Jiang, Z. 2006. Three Gorges Dam affects regional precipitation. *Geophysical Research Letters* 33:L13806.

Xie, P. 2003. Three-Gorges Dam: risk to ancient fish. *Science* 302(5648):1149–1151, November 14, 2003.

Xie, S., Li, Z., Liu, J., Xie, S., Wang, H., and Murphy, B.R. 2007. Fisheries of the Yangtze River show immediate impacts of the Three Gorges Dam. *Fisheries* 32(7):343–344.

Xinhua. 2007a. Three Gorges offers clean energy, no major geological threats. November 28, 2007. http://www.mwr.gov.cn/english/20071128/88168.asp

Xinhua. 2007b. 4 million more people to be moved from Three Gorges area. October 12, 2007. http://www.mwr.gov.cn/english/20071012/87188.asp

Xinhua. 2007c. China warns of environmental "catastrophe" from Three Gorges Dam. September 27, 2007. http://www.mwr.gov.cn/english/20070927/86992.asp

Xinhua. 2007d. Investigation launched to protect Three Gorges' plants. September 17, 2007. http://www.mwr.gov.cn/english/20070917/86766.asp

Xu, K., Milliman, J.D., Yang, Z., and Wang, H. 2006. Yangtze sediment decline partly from Three Gorges Dam. *EOS* 87(19):185, 190, May 9, 2006.

Yan T., Hugo, G., Potter, L. 2005. Rural women, displacement and the Three Gorges Project. *Development and Change* 36(4):711–734.

Yang, Z., Wang, H., Saito, Y., Milliman, J.D., Xu, K., Qiao, S., and Shi, G. 2006. Dam impacts on the Changjiang (Yangtze) River sediment discharge to the sea: The past 55 years and after the Three Gorges Dam. *Water Resources Research* 42:Cite ID: W04407.

Yardley, J. 2007. Chinese dam projects criticized for their human costs. *The New York Times,* November 19, 2007.

Yue, P. and Chen, Y. 1998. *China Red Data Book of Endangered Animals: Pisces.* Beijing: Science Press.

Water Conflict Chronology

Peter H. Gleick

The World's Water "Water Conflict Chronology" appears here again, as it has since the first volume of *The World's Water* appeared in 1998. It continues to be one of the most popular and regular features of *The World's Water* and is available online at www.worldwater.org, where regular updates appear. New additions come to me from readers and researchers around the world, and the chronology is used regularly by the media and by academics interested in understanding more about both the history and character of disputes over water resources. We expand on the typology, which categorizes water conflicts as military targets, military tools, development disputes, and terrorism.

The history of violence over freshwater is long and distressing. The Pacific Institute has been evaluating and analyzing these connections for more than two decades, since our founding in 1987. A series of papers on the issue of water and conflict has been published. These range from historical reviews to regional case studies to theoretical analyses. We have organized workshops on lessons from regional conflicts in the Middle East, Central Asia, and Latin America, the connections between traditional and non-traditional arms control tools, and the role of science and religion in reducing the risks of water-related violence.

We regularly update, modify, and expand the Chronology. In 2004, we added a series of myths, legends, and history of water conflicts in the Middle East beginning 5,000 years before the present. The 2006 version added new connections between water and terrorism, as did the first chapter in that volume.[1] World events continue to expand the modern list, with examples in southern Asia, northern Africa, the Middle East, and elsewhere. Of particular note is the continuing trend toward conflicts related to terrorism, and especially to disputes over economic development, water allocations, and equity. More and more of the entries in the chronology are related to sub-national players and actors, and fewer are related to transnational conflicts. This supports the thesis identified a decade ago in the first volume of *The World's Water*:

> "Traditional political and ideological questions that have long dominated
> international discourse are now becoming more tightly woven with other
> variables that loomed less large in the past, including population growth,

---

[1] Gleick, P.H. 2006. "Water and terrorism." In P.H. Gleick, *The World's Water 2006–2007: The Biennial Report on Freshwater Resources*. Island Press, Washington, D.C. pp.1–28.

transnational pollution, resource scarcity and inequitable access to resources and their use."[2]

The updated chronology is presented here, with new entries and a range of corrections and modifications to the older ones. The current categories or types of conflicts include:

- Military Tool (state actors): where water resources, or water systems themselves, are used by a nation or state as a weapon during a military action.

- Military Target (state actors): where water resources or water systems are targets of military actions by nations or states.

- Terrorism, including cyberterrorism: (non-state actors): where water resources, or water systems, are the targets or tools of violence or coercion by non-state actors. A distinction is drawn between environmental terrorism and eco-terrorism (see Chapter 1).

- Development Disputes: (state and non-state actors): where water resources or water systems are a major source of contention and dispute in the context of economic and social development.

One factor remains constant: the importance of water to life means that providing for water needs and demands will never be free of politics. As social and political systems change and evolve, this chronology and the kinds of entries and categories will change and evolve. I look forward to the ongoing debate over water conflicts and to new contributions and comments from readers. Please email any contributions with full citations and supporting information to pgleick@pipeline.com.

---

[2] Gleick, P.H. 1998. "Conflict and cooperation over fresh water." In P.H. Gleick, *The World's Water 1998–1999: The Biennial Report on Freshwater Resources*. Island Press, Washington, D.C. pp.105.

Date	Parties Involved	Basis of Conflict	Violent Conflict or In the Context of Violence?	Description	Sources
3000 BC	Ea, Noah	Religious account	Yes	Ancient Sumerian legend recounts the deeds of the deity Ea, who punished humanity for its sins by inflicting the Earth with a 6-day storm. The Sumerian myth parallels the Biblical account of Noah and the deluge, although some details differ.	Hatami and Gleick 1994
2500 BC	Lagash, Umma	Military tool	Yes	Lagash-Umma Border Dispute: The dispute over the "Gu'edena" (edge of paradise) region begins. Urlama, King of Lagash from 2450 to 2400 B.C., diverts water from this region to boundary canals, drying up boundary ditches to deprive Umma of water. His son Il cuts off the water supply to Girsu, a city in Umma.	Hatami and Gleick 1994
1790 BC	Hammurabi	Development disputes	No	Code of Hammurabi for the State of Sumer: Hammurabi lists several laws pertaining to irrigation that address negligence of irrigation systems and water theft.	Hatami and Gleick 1994
1720–1684 BC	Abi-Eshuh, Iluma-Ilum	Military tool	Yes	Abi-Eshuh v. Iluma-Ilum: A grandson of Hammurabi, Abish or Abi-Eshuh, dams the Tigris to prevent the retreat of rebels led by Iluma-Ilum, who declared the independence of Babylon. This failed attempt marks the decline of the Sumerians who had reached their apex under Hammurabi	Hatami and Gleick 1994

continues

Date	Parties Involved	Basis of Conflict	Violent Conflict or In the Context of Violence?	Description	Sources
circa 1300BC	Sisera, Barak, God	Religious account, Military Tool	Yes	This is an Old Testament account of the defeat of Sisera and his "nine hundred chariots of iron" by the unmounted army of Barak on the fabled Plains of Esdraelon. God sends heavy rainfall in the mountains, and the Kishon River overflows the plain and immobilizes or destroys Sisera's technologically superior forces ("...the earth trembled, and the heavens dropped, and the clouds also dropped water," Judges 5:4; "...The river of Kishon swept them away, that ancient river, the river Kishon," Judges 5:21).	New Scofield Reference Bible, KJV; Judges 4:7–15 and Judges 5:4–22.
1200 BC	Moses, Egypt	Military tool, Religious account	Yes	Parting of the Red Sea: When Moses and the retreating Jews find themselves trapped between the Pharoah's army and the Red Sea, Moses miraculously parts the waters of the Red Sea, allowing his followers to escape. The waters close behind them and cut off the Egyptians.	Hatami and Gleick 1994
720–705 BC	Assyria, Armenia	Military tool	Yes	After a successful campaign against the Halidians of Armenia, Sargon II of Assyria destroys their intricate irrigation network and floods their land.	Hatami and Gleick 1994
705–682 BC	Sennacherib, Babylon	Military weapon /target	Yes	In quelling rebellious Assyrians in 695 B.C., Sennacherib razes Babylon and diverts one of the principal irrigation canals so that its waters wash over the ruins.	Hatami and Gleick 1994
6th Century BC	Assyria	Military target; Military tool	Yes	Assyrians poison the wells of their enemies with rye ergot.	Eitzen, E.M. and E.T. Takafuji. 1997

Date		Type	Parties	Description	Sources
Unknown	Yes	Military tool	Sennacherib, Jerusalem	As recounted in Chronicles 32.3, Hezekiah digs into a well outside the walls of Jerusalem and uses a conduit to bring in water. Preparing for a possible siege by Sennacherib, he cuts off water supplies outside of the city walls, and Jerusalem survives the attack.	Hatami and Gleick 1994
681–699 BC	Yes	Military tool, Religious account	Assyria, Tyre	Esarhaddon, an Assyrian, refers to an earlier period when gods, angered by insolent mortals, created destructive floods. According to inscriptions recorded during his reign, Esarhaddon besieges Tyre, cutting off food and water.	Hatami and Gleick 1994
669–626 BC	Yes	Military tool, Military target	Assyria, Arabia, Elam	Assurbanipal's inscriptions also refer to a siege against Tyre, although scholars attribute it to Esarhaddon. In campaigns against both Arabia and Elam in 645 B.C., Assurbanipal, son of Esarhaddon, dries up wells to deprive Elamite troops. He also guards wells from Arabian fugitives in an earlier Arabian war. On his return from victorious battle against Elam, Assurbanipal floods the city of Sapibel, and ally of Elam. According to inscriptions, he dams the Ulai River with the bodies of dead Elamite soldiers and deprives dead Elamite kings of their food and water offerings.	Hatami and Gleick 1994
612 BC	Yes	Military tool	Egypt, Persia, Babylon, Assyria	A coalition of Egyptian, Median (Persian), and Babylonian forces attacks and destroys Ninevah, the capital of Assyria. Nebuchadnezzar's father, Nebopolassar, leads the Babylonians. The converging armies divert the Khosr River to create a flood, which allows them to elevate their siege engines on rafts.	Hatami and Gleick 1994

continues

Date	Parties Involved	Basis of Conflict	Violent Conflict or In the Context of Violence?	Description	Sources
605–562 BC	Babylon	Military tool	No	Nebuchadnezzar builds immense walls around Babylon, using the Euphrates and canals as defensive moats surrounding the inner castle.	Hatami and Gleick 1994
590–600 BC	Cirrha, Delphi	Military tool	Yes	Athenian legislator Solon reportedly had roots of helleborus thrown into a small river or aqueduct leading from the Pleistrus River to Cirrha during a siege of this city. The enemy forces became violently ill and were defeated as a result. Some accounts have Solon building a dam across the Plesitus River cutting off the city's water supply. Such practices were widespread.	Absolute Astronomy 2006
558–528 BC	Babylon	Military tool	Yes	On his way from Sardis to defeat Nabonidus at Babylon, Cyrus faces a powerful tributary of the Tigris, probably the Diyalah. According to Herodotus' account, the river drowns his royal white horse and presents a formidable obstacle to his march. Cyrus, angered by the "insolence" of the river, halts his army and orders them to cut 360 canals to divert the river's flow. Other historians argue that Cyrus needed the water to maintain his troops on their southward journey, while another asserts that the construction was an attempt to win the confidence of the locals.	Hatami and Gleick 1994
539 BC	Babylon	Military tool	Yes	According to Herodotus, Cyrus invades Babylon by diverting the Euphrates above the city and marching troops along the dry riverbed. This popular account describes a midnight attack that coincided with a Babylonian feast.	Hatami and Gleick 1994

Date	Parties		Type	Description	Reference
430 BC	Athens	Yes	Military tool	During the second year of the Peloponnesian War in 430 BC when plague broke out in Athens, the Spartans were accused of poisoning the cisterns of the Piraeus, the source of most of Athens' water.	Strategy Page 2006.
355–323 BC	Babylon	Yes	Military tool	Returning from the razing of Persepolis, Alexander proceeds to India. After the Indian campaigns, he heads back to Babylon via the Persian Gulf and the Tigris, where he tears down defensive weirs that the Persians had constructed along the river. Arrian describes Alexander's disdain for the Persians' attempt to block navigation, which he saw as "unbecoming to men who are victorious in battle."	Hatami and Gleick 1994
210–209 BC	Rome and Carthage	Yes	Military tool	In 210 BC, Scipio crossed the Ebro to attack New Carthage. During a short siege, Scipio led a breaching column through a supposedly impregnable lagoon located on the landward side of the city; a strong northerly wind combined with the natural ebb of the tide left the lagoon shallow enough for the Roman infantry to wade through. New Carthage was soon taken.	Fonner 1996, Gowan 2004
537	Goths and Rome	Yes	Military tool and military target	In the 6th century AD, as the Roman Empire began to decline, the Goths besieged Rome and cut almost all of the aqueducts leading into the city. In 537 AD, this siege was successful. The only aqueduct that continued to function was that of the Aqua Virgo, which ran entirely underground.	Rome Guide 2004, InfoRoma 2004.
1187	Saladin and the Middle East	Yes	Military tool	Saladin was able to defeat the Crusaders at the Horns of Hattin in 1187 by denying them access to water. In some reports, Saladin had sanded up all the wells along the way and had destroyed the villages of the Maronite Christians who would have supplied the Christian army with water.	Lockwood 2006, Priscoli 1998

continues

157

Date	Parties Involved	Basis of Conflict	Violent Conflict or In the Context of Violence?	Description	Sources
1503	Florence and Pisa warring states.	Military tool	No: Plan only	Leonardo da Vinci and Machievelli plan to divert Arno River away from Pisa during conflict between Pisa and Florence.	Honan 1996
1573–74	Holland and Spain	Military tool	Yes	In 1573 at the beginning of the eighty years war against Spain, the Dutch flooded the land to break the siege of Spanish troops on the town Alkmaar. The same defense was used to protect Leiden in 1574. This strategy became known as the Dutch Water Line and was used frequently for defense in later years.	Dutch Water Line 2002
1642	China; Ming Dynasty	Military tool	Yes	The Huang He's dikes breached for military purposes. In 1642, "toward the end of the Ming dynasty (1368–1644), General Gao Mingheng used the tactic near Kaifeng in an attempt to suppress a peasant uprising."	Hillel 1991
1672	French, Dutch	Military tool	Yes	Louis XIV starts the third of the Dutch Wars in 1672, in which the French overran the Netherlands. In defense, the Dutch opened their dikes and flooded the country, creating a watery barrier that was virtually impenetrable.	Columbia 2000
1748	United States	Development dispute; terrorism	Yes	Ferry house on Brooklyn shore of East River burns down. New Yorkers accuse Brooklynites of having set the fire as revenge for unfair East River water rights.	Museum of the City of New York (MCNY n.d.)
1777	United States	Military tool	Yes	British and Hessians attacked the water system of New York. ". . . the enemy wantonly destroyed the New York water works" during the War for Independence.	Thatcher 1827

1841	Canada	Yes	Development dispute, terrorism	A reservoir in Ops Township, Upper Canada (now Ontario) was destroyed by neighbors who considered it a hazard to health.	Forkey 1998
1844	United States	Yes	Development dispute, terrorism	A reservoir in Mercer County, Ohio was destroyed by a mob that considered it a hazard to health.	Scheiber 1969
1850s	United States	Yes	Development dispute; terrorism	Attack on a New Hampshire dam that impounded water for factories downstream by local residents unhappy over its effect on water levels.	Steinberg 1990
1853–1861	United States	Yes	Development dispute, terrorism	Repeated destruction of the banks and reservoirs of the Wabash and Erie Canal in southern Indiana by mobs regarding it as a health hazard.	Fatout 1972, Fickle 1983
1860–1865	United States	Yes	Military tool; Military target	W.T. Sherman's memoirs contain an account of Confederate soldiers poisoning ponds by dumping the carcasses of dead animals into them. Other accounts suggest this tactic was used by both sides.	Eitzen and Takafuji 1997
1870s	China	No	Development dispute	Local construction and government removal (twice) of an unauthorized dam in Hubei, China.	Rowe 1988
1870s to 1881	United States	Yes	Development dispute	Recurrent friction and eventual violent conflict over water rights in the vicinity of Tularosa, New Mexico involving villagers, ranchers, and farmers.	Rasch 1968
1887	United States	Yes	Development dispute, Terrorism	Dynamiting of a canal reservoir in Paulding County, Ohio by a mob regarding it as a health hazard. State Militia called out to restore order.	Walters 1948
1890	Canada	Yes	Development dispute, terrorism	Partly successful attempt to destroy a lock on the Welland Canal in Ontario, Canada either by Fenians protesting English Policy in Ireland or by agents of Buffalo, NY grain handlers unhappy at the diversion of trade through the canal.	Styran and Taylor 2001

continues

Date	Parties Involved	Basis of Conflict	Violent Conflict or In the Context of Violence?	Description	Sources
1908–09	United States	Development dispute	Yes	Violence, including a murder, directed against agents of a land company that claimed title to Reelfoot Lake in northwestern Tennessee who attempted to levy charges for fish taken and threatened to drain the lake for agriculture.	Vanderwood 1969
1863	United States Civil War	Military tool	Yes	General U.S. Grant, during the Civil War campaign against Vicksburg, cut levees in the battle against the Confederates.	Grant 1885, Barry 1997
1898	Egypt; France; Britain	Military and political tool	Military maneuvers	Military conflict nearly ensues between Britain and France in 1898 when a French expedition attempted to gain control of the headwaters of the White Nile. While the parties ultimately negotiates a settlement of the dispute, the incident has been characterized as having "dramatized Egypt's vulnerable dependence on the Nile, and fixed the attitude of Egyptian policy-makers ever since."	Moorhead 1960
1907–1913	Owens Valley, Los Angeles, California	Terrorism, Development dispute	Yes	The Los Angeles Valley aqueduct/pipeline suffers repeated bombings in an effort to prevent diversions of water from the Owens Valley to Los Angeles.	Reisner 1986, 1993
1915	German Southwest Africa	Military tool	Yes	Union of South African troops capture Windhoek, capital of German Southwest Africa. (May.) Retreating German troops poison wells – "a violation of the Hague convention."	Daniel 1995
1935	California, Arizona	Development dispute	Military maneuvers	Arizona calls out the National Guard and militia units to the border with California to protest the	Reisner 1986, 1993

Date	Parties	Basis of Conflict	Description	Sources
1938	China and Japan	Military tool, Military target	construction of Parker Dam and diversions from the Colorado River; dispute ultimately is settled in court. Chiang Kai-shek orders the destruction of flood-control dikes of the Huayuankou section of the Huang He (Yellow) river to flood areas threatened by the Japanese army. West of Kaifeng dikes are destroyed with dynamite, spilling water across the flat plain. The flood destroyed part of the invading army and its heavy equipment was mired in thick mud, though Wuhan, the headquarters of the Nationalist government was taken in October. The waters flooded an area variously estimated as between 3,000 and 50,000 square kilometers, and killed Chinese estimated in numbers between "tens of thousands" and "one million."	Hillel 1991, Yang Lang 1989, 1994
1939–40	Netherlands, Germany	Military tool	During the mobilization of the Dutch at the beginning of World War II, 1939–40, the Dutch attempted to flood the Gelderse Vallei with the New Dutch Water Defense Line, which had been completed in 1885. During the German invasion in May 1940, large areas were inundated.	IDG 1996
1939–1942	Japan, China	Military target, Military tool	Japanese chemical and biological weapons activities reportedly include tests by "Unit 731" against military and civilian targets by lacing water wells and reservoirs with typhoid and other pathogens.	Harris 1994
1940–1945	Multiple parties	Military target	Hydroelectric dams routinely bombed as strategic targets during World War II.	Gleick 1993

continues

Date	Parties Involved	Basis of Conflict	Violent Conflict or In the Context of Violence?	Description	Sources
1943	Britain, Germany	Military target	Yes	British Royal Air Force bombed dams on the Möhne, Sorpe, and Eder Rivers, Germany (May 16, 17). Möhne Dam breech killed 1,200, destroyed all downstream dams for 50 km. The flood that occurred after breaking the Eder dam reached a peak discharge of 8,500 m³/s, which is nine times higher than the highest flood observed. Many houses and bridges were destroyed. 68 were killed.	Kirschner 1949, Semann 1950
1944	Germany, Italy, Britain, United States	Military tool	Yes	German forces used waters from the Isoletta Dam (Liri River) in January and February to successfully destroy British assault forces crossing the Garigliano River (downstream of Liri River). The German Army then dammed the Rapido River, flooding a valley occupied by the American Army.	Corps of Engineers 1953
1944	Germany, Italy, Britain, United States	Military tool	Yes	German Army flooded the Pontine Marches by destroying drainage pumps to contain the Anzio beachhead established by the Allied landings in 1944. Over 40 square miles of land were flooded; a 30-mile stretch of landing beaches was rendered unusable for amphibious support forces.	Corps of Engineers 1953
1944	Germany, Allied forces	Military tool	Yes	Germans flooded the Ay River, France (July) creating a lake two meters deep and several kilometers wide, slowing an advance on Saint Lo, a German communications center in Normandy.	Corps of Engineers 1953
1944	Germany, Allied forces	Military tool	Yes	Germans flooded the Ill River Valley during the Battle of the Bulge (winter 1944–45) creating a lake 16 kilometers long, 3–6 kilometers wide, and	Corps of Engineers 1953

Date	Parties	Violent conflict?	Type	Description	Sources
				1–2 meters deep, greatly delaying the American Army's advance toward the Rhine.	
1945	Romania, Germany	Yes	Military target	The only known German tactical use of biological warfare was the pollution of a large reservoir in northwestern Bohemia with sewage in May 1945.	SIPRI 1971
1947 onwards	Bangladesh, India	No	Development dispute	Partition divides the Ganges River between Bangladesh and India; construction of the Farakka barrage by India, beginning in 1962, increases tension; short-term agreements settle dispute in 1977–82, 1982–84, and 1985–88, and thirty-year treaty is signed in 1996.	Butts 1997, Samson & Charrier 1997
1947–1960s	India, Pakistan	No	Development dispute	Partition leaves Indus basin divided between India and Pakistan; disputes over irrigation water ensue, during which India stems flow of water into irrigation canals in Pakistan; Indus Waters Agreement reached in 1960 after 12 years of World Bank-led negotiations.	Bingham *et al.* 1994, Wolf 1997
1948	Arabs, Israelis	Yes	Military tool	Arab forces cut off West Jerusalem's water supply in first Arab-Israeli war.	Wolf 1995, 1997
1950s	Korea, United States, others	Yes	Military target	Centralized dams on the Yalu River serving North Korea and China are attacked during Korean War.	Gleick 1993
1951	Korea, United Nations	Yes	Military tool and Military target	North Korea released flood waves from the Hwachon Dam damaging floating bridges operated by UN troops in the Pukhan Valley. U.S. Navy plans were then sent to destroy spillway crest gates.	Corps of Engineers 1953
1951	Israel, Jordan, Syria	Yes	Military tool, Development disputes	Jordan makes public its plans to irrigate the Jordan Valley by tapping the Yarmouk River; Israel responds by commencing drainage of the Huleh swamps located in the demilitarized zone between Israel and Syria; border skirmishes ensue between Israel and Syria.	Wolf 1997, Samson & Charrier 1997

continues

Date	Parties Involved	Basis of Conflict	Violent Conflict or In the Context of Violence?	Description	Sources
1953	Israel, Jordan, Syria	Development dispute, Military target	Yes	Israel begins construction of its National Water Carrier to transfer water from the north of the Sea of Galilee out of the Jordan basin to the Negev Desert for irrigation. Syrian military actions along the border and international disapproval lead Israel to move its intake to the Sea of Galilee.	Naff and Matson 1984, Samson & Charrier 1997
1958	Egypt, Sudan	Military tool, Development dispute	Yes	Egypt sends an unsuccessful military expedition into disputed territory amidst pending negotiations over the Nile waters, Sudanese general elections, and an Egyptian vote on Sudan-Egypt unification; Nile Water Treaty signed when pro-Egyptian government elected in Sudan.	Wolf 1997
1960s	North Vietnam, United States	Military target	Yes	Irrigation water supply systems in North Vietnam are bombed during Vietnam War; 661 sections of dikes damaged or destroyed.	IWTC 1967, Gleick 1993, Zemmali 1995
1962	Israel, Syria	Development dispute, Military target	Yes	Israel destroys irrigation ditches in the lower Tarfiq in the demilitarized zone. Syria complains.	Naff and Matson 1984
1962 to 1967	Brazil; Paraguay	Military tool, Development dispute	Military maneuvers	Negotiations between Brazil and Paraguay over the development of the Paraná River are interrupted by a unilateral show of military force by Brazil in 1962, which invades the area and claims control over the Guaira Falls site. Military forces were withdrawn in 1967 following an agreement for a joint commission to examine development in the region.	Murphy and Sabadell 1986

Date	Parties	Basis of conflict	Violent conflict	Description	Sources
1963–1964	Ethiopia, Somalia	Development dispute, Military tool	Yes	Creation of boundaries in 1948 leaves Somali nomads under Ethiopian rule; border skirmishes occur over disputed territory in Ogaden desert where critical water and oil resources are located; cease-fire is negotiated only after several hundred are killed.	Wolf 1997
1964	Cuba, United States	Military tool	No	On February 6, 1964, the Cuban government ordered the water supply to the U.S. Naval Base at Guantanamo Bay cut off.	Guantanamo Bay Gazette. 1964.
1964	Israel, Syria	Military target	Yes	Headwaters of the Dan River on the Jordan River are bombed at Tell El-Qadi in a dispute about sovereignty over the source of the Dan.	Naff and Matson 1984
1965	Zambia, Rhodesia, Great Britain	Military target	No	President Kenneth Kaunda calls on British government to send troops to Kariba Dam to protect it from possible saboteurs from Rhodesian government.	Chenje 2001
1965	Israel, Palestinians	Terrorism	Yes	First attack ever by the Palestinian National Liberation Movement Al-Fatah is on the diversion pumps for the Israeli National Water Carrier. Attack fails.	Naff and Matson 1984, Dolatyar 1995
1965–1966	Israel, Syria	Military tool, Development dispute	Yes	Fire is exchanged over "all-Arab" plan to divert the Jordan River headwaters (Hasbani and Banias) and presumably preempt Israeli National Water Carrier; Syria halts construction of its diversion in July 1966.	Wolf 1995, 1997
1966–1972	Vietnam, US	Military tool	Yes	U.S. tries cloud-seeding in Indochina to stop flow of materiel along Ho Chi Minh trail.	Plant 1995
1967	Israel, Syria	Military target and tool	Yes	Israel destroys the Arab diversion works on the Jordan River headwaters. During Arab-Israeli War, Israel occupies Golan Heights, with Banias tributary to the Jordan; Israel occupies West Bank.	Gleick 1993, Wolf 1995, 1997, Wallenstein & Swain 1997

continues

165

Date	Parties Involved	Basis of Conflict	Violent Conflict or In the Context of Violence?	Description	Sources
1969	Israel, Jordan	Military target and tool	Yes	Israel, suspicious that Jordan is overdiverting the Yarmouk, leads two raids to destroy the newly built East Ghor Canal; secret negotiations, mediated by the US, lead to an agreement in 1970.	Samson & Charrier 1997
1970	United States	Terrorism	No: Threat	The Weathermen, a group opposed to American imperialism and the Vietnam war, allegedly attempted to obtain biological agents to contaminate the water supply systems of US urban centers.	Kupperman and Trent 1979, Eitzen and Takafuji 1997, Purver 1995
1970s	Argentina, Brazil, Paraguay	Development dispute	No	Brazil and Paraguay announce plans to construct a dam at Itaipu on the Paraná River, causing Argentina concern about downstream environmental repercussions and the efficacy of their own planned dam project downstream. Argentina demands to be consulted during the planning of Itaipu but Brazil refuses. An agreement is reached in 1979 that provides for the construction of both Brazil and Paraguay's dam at Itaipu and Argentina's Yacyreta dam.	Wallenstein & Swain 1997
1972	United States	Terrorism	No: Threat	Two members of the right-wing "Order of the Rising Sun" are arrested in Chicago with 30–40 kg of typhoid cultures that are allegedly to be used to poison the water supply in Chicago, St. Louis, and other cities. It was felt that the plan would have been unlikely to cause serious health problems due to chlorination of the water supplies.	Eitzen and Takafuji 1997
1972	United States	Terrorism	No: Threat	Reported threat to contaminate water supply of New York City with nerve gas.	Purver 1995

Year	Parties	Type	Violent conflict or in the context of violence	Description	Sources
1972	North Vietnam	Military target	Yes	United States bombs dikes in the Red River delta, rivers, and canals during massive bombing campaign.	Columbia Electronic Encyclopedia 2000
1973	Germany	Terrorism	No: Threat	Threat by a biologist in Germany to contaminate water supplies with bacilli of anthrax and botulinum toxin unless he was paid $8.5 million	Jenkins and Rubin 1978, Kupperman and Trent 1979
1974	Iraq, Syria	Military target, Military tool, Development dispute	Military maneuvers	Iraq threatens to bomb the al-Thawra dam in Syria and massed troops along the border, alleging that the dam had reduced the flow of Euphrates River water to Iraq.	Gleick 1994
1975	Iraq, Syria	Development dispute, Military tool	Military maneuvers	As upstream dams are filled during a low-flow year on the Euphrates, Iraqis claim that flow reaching its territory is "intolerable" and asks the Arab League to intervene. Syrians claim they are receiving less than half the river's normal flow and pull out of an Arab League technical committee formed to mediate the conflict. In May, Syria closes its airspace to Iraqi flights and both Syria and Iraq reportedly transfer troops to their mutual border. Saudi Arabia successfully mediates the conflict.	Gleick 1993, 1994, Wolf 1997
1975	Angola, South Africa	Military goal, military target	Yes	South African troops move into Angola to occupy and defend the Ruacana hydropower complex, including the Gové Dam on the Kunene River. Goal is to take possession of and defend the water resources of southwestern Africa and Namibia.	Meissner 2000
1977	United States	Terrorism	Yes	Contamination of a North Carolina reservoir with unknown materials. According to Clark: "Safety caps and valves were removed, and poison chemicals were sent into the reservoir....Water had to be brought in."	Clark 1980, Purver 1995

continues

Date	Parties Involved	Basis of Conflict	Violent Conflict or In the Context of Violence?	Description	Sources
1978–onwards	Egypt, Ethiopia	Development dispute, Political tool	No	Long standing tensions over the Nile, especially the Blue Nile, originating in Ethiopia. Ethiopia's proposed construction of dams on the headwaters of the Blue Nile leads Egypt to repeatedly declare the vital importance of water. "The only matter that could take Egypt to war again is water" (Anwar Sadat–1979). "The next war in our region will be over the waters of the Nile, not politics" (Boutrous Ghali–1988).	Gleick 1991, 1994
1978–1984	Sudan	Development dispute, Military target, Terrorism	Yes	Demonstrations in Juba, Sudan in 1978 opposing the construction of the Jonglei Canal led to the deaths of two students. Construction of the Jonglei Canal in the Sudan was forcibly suspended in 1984 following a series of attacks on the construction site.	Suliman 1998; Keluel-Jang 1997
1980s	Mozambique, Rhodesia/ Zimbabwe, South Africa	Military target, Terrorism	Yes	Regular destruction of power lines from Cahora Bassa Dam during fight for independence in the region. Dam targeted by RENAMO (the Mozambican National Resistance).	Chenje 2001
1981	Iran, Iraq	Military target and tool	Yes	Iran claims to have bombed a hydroelectric facility in Kurdistan, thereby blacking out large portions of Iraq, during the Iran-Iraq War.	Gleick 1993
1980–1988	Iran, Iraq	Military tool	Yes	Iran diverts water to flood Iraqi defense positions.	Plant 1995
1982	United States	Terrorism	No: Threat	Los Angeles police and the FBI arrest a man who was preparing to poison the city's water supply with a biological agent.	Livingston 1982, Eitzen and Takafuji 1997
1982	Israel, Lebanon, Syria	Military tool	Yes	Israel cuts off the water supply of Beirut during siege.	Wolf 1997

Date	Party	Type	Violent	Description	Sources
1981–1982	Angola	Military target, Military tool	Yes	Water infrastructure, including dams and the major Cunene-Cuvelai pipeline, was targeted during the conflicts in Namibia and Angola in the 1980s.	Turton 2005
1982	Guatemala	Development dispute	Yes	177 civilians killed in Rio Negro over opposition to Chixoy hydroelectric dam.	Levy 2000
1983	Lebanon	Terrorism	Yes	An explosives-laden truck disguised as a water delivery vehicle destroyed a barracks in a U.S. military compound, killing more than 300 people. The attack was blamed on Hezbollah with the support of the Iranian government.	BBC News 2007
1983	Israel	Terrorism	No	The Israeli government reports that it had uncovered a plot by Israeli Arabs to poison the water in Galilee with "an unidentified powder."	Douglass and Livingstone 1987
1984	United States	Terrorism	Yes	Members of the Rajneeshee religious cult contaminate a city water supply tank in The Dalles, Oregon, using Salmonella. A community outbreak of over 750 cases occurred in a county that normally reports fewer than five cases per year.	Clark and Deininger 2000
1985	United States	Terrorism	No	Law enforcement authorities discovered that a small survivalist group in the Ozark Mountains of Arkansas known as The Covenant, the Sword, and the Arm of the Lord (CSA) had acquired a drum containing 30 gallons of potassium cyanide, with the apparent intent to poison water supplies in New York, Chicago, and Washington, D.C. CSA members devised the scheme in the belief that such attacks would make the Messiah return more quickly by punishing unrepentant sinners. The objective appeared to be mass murder in the name of a divine mission rather than to change government policy. The amount of poison possessed by the group is believed to have been insufficient to contaminate the water supply of even one city.	Tucker 2000, NTI 2005

continues

169

Date	Parties Involved	Basis of Conflict	Violent Conflict or In the Context of Violence?	Description	Sources
1986	North Korea, South Korea	Military tool	No	North Korea's announcement of its plans to build the Kumgansan hydroelectric dam on a tributary of the Han River upstream of Seoul raises concerns in South Korea that the dam could be used as a tool for ecological destruction or war.	Gleick 1993
1986	Lesotho, South Africa	Military goal, Development dispute	Yes	South Africa supports coup in Lesotho over support for ANC and anti-apartheid, and water. New government in Lesotho then quickly signs Lesotho Highlands water agreement.	American University 2000b
1986	Lesotho, South Africa	Development dispute, Military goal	Yes	Bloodless coup by Lesotho's defense forces, with support from South Africa, lead to immediate agreement with South Africa for water from the Highlands of Lesotho, after 30 previous years of unsuccessful negotiations. There is disagreement over the degree to which water was a motivating factor for either party. Water pipeline to Owamboland cut and destroyed.	Mohamed 2001
1990	South Africa	Development dispute	No	Pro-apartheid council cuts off water to the Wesselton township of 50,000 blacks following protests over miserable sanitation and living conditions.	Gleick 1993
1990	Iraq, Syria, Turkey	Development dispute, Military tool	No	The flow of the Euphrates is interrupted for a month as Turkey finishes construction of the Ataturk Dam, part of the Grand Anatolia Project. Syria and Iraq protest that Turkey now has a weapon of war. In mid-1990 Turkish president Turgut Ozal threatens to restrict water flow to Syria to force it to withdraw support for Kurdish rebels operating in southern Turkey.	Gleick 1993 & 1995

Date	Parties	Type	Violent or in a Violent Context	Description	Sources
1991–present	Karnataka, Tamil Nadu (India)	Development dispute	Yes	Violence erupts when Karnataka rejects an Interim Order handed down by the Cauvery Waters Tribunal, set up by the Indian Supreme Court. The Tribunal was established in 1990 to settle two decades of dispute between Karnataka and Tamil Nadu over irrigation rights to the Cauvery River.	Gleick 1993, Butts 1997, American University 2000a
1991	Iraq, Kuwait, US	Military target	Yes	During the Gulf War, Iraq destroys much of Kuwait's desalination capacity during retreat.	Gleick 1993
1991	Canada	Terrorism	No: Threat	A threat is made via an anonymous letter to contaminate the water supply of the city of Kelowna, British Columbia, with "biological contaminates." The motive was apparently "associated with the Gulf War." The security of the water supply was increased in response and no group was identified as the perpetrator.	Purver 1995
1991	Iraq, Turkey, United Nations	Military tool	Yes	Discussions are held at the United Nations about using the Ataturk Dam in Turkey to cut off flows of the Euphrates to Iraq.	Gleick 1993
1991	Iraq, Kuwait, US	Military target	Yes	Baghdad's modern water supply and sanitation system are intentionally and unintentionally damaged by Allied coalition. "Four of seven major pumping stations were destroyed, as were 31 municipal water and sewerage facilities –20 in Baghdad, resulting in sewage pouring into the Tigris. Water purification plants were incapacitated throughout Iraq" (Arbuthnot 2000). In the first eight months of 1991, after Iraq's water infrastructure was damaged by the Persian Gulf War, the *New England Journal of Medicine* reported that nearly 47,000 more children than normal died in Iraq and the country's infant mortality rate doubled to 92.7 per 1,000 live births.	Gleick 1993, Arbuthnot 2000, Barrett 2003

continues

Date	Parties Involved	Basis of Conflict	Violent Conflict or In the Context of Violence?	Description	Sources
1992	Czechoslovakia, Hungary	Political tool, Development dispute	Military maneuvers	Hungary abrogates a 1977 treaty with Czechoslovakia concerning construction of the Gabcikovo/Nagymaros project based on environmental concerns. Slovakia continues construction unilaterally, completes the dam, and diverts the Danube into a canal inside the Slovakian republic. Massive public protest and movement of military to the border ensue; issue taken to the International Court of Justice.	Gleick 1993
1992	Turkey	Terrorism	Yes	Lethal concentrations of potassium cyanide are reported discovered in the water tanks of a Turkish Air Force compound in Istanbul. The Kurdish Workers' Party (PKK) claimed credit.	Chelyshev 1992
1992	Bosnia, Bosnian Serbs	Military tool	Yes	The Serbian siege of Sarajevo, Bosnia and Herzegovina, includes a cutoff of all electrical power and the water feeding the city from the surrounding mountains. The lack of power cuts the two main pumping stations inside the city despite pledges from Serbian nationalist leaders to United Nations officials that they would not use their control of Sarajevo's utilities as a weapon. Bosnian Serbs take control of water valves regulating flow from wells that provide more than 80 percent of water to Sarajevo; reduced water flow to city is used to 'smoke out' Bosnians.	Burns 1992, Husarska 1995
1993–present	Iraq	Military tool	No	To quell opposition to his government, Saddam Hussein reportedly poisons and drains the water supplies of southern Shiite Muslims, the Madan. The	Gleick 1993, American University 2000c, National Geographic

Date	Parties	Basis of Conflict	Violent Conflict or In the Context of Violence	Description	Sources
				marshes of southern Iraq are intentionally targeted. The European Parliament and UN Human Rights Commission deplore use of water as weapon in region.	News 2001
1993	Iran	Terrorism	No	A report suggests that proposals were made at a meeting of fundamentalist groups in Tehran, under the auspices of the Iranian Foreign Ministry, to poison water supplies of major cities in the West "as a possible response to Western offensives against Islamic organizations and states."	Haeri 1993
1993	Yugoslavia	Military target and tool	Yes	Peruca Dam intentionally destroyed during war.	Gleick 1993
1994	Moldavia	Terrorism	No: Threat	Reported threat by Moldavian General Nikolay Matveyev to contaminate the water supply of the Russian 14th Army in Tiraspol, Moldova, with mercury.	Purver 1995
1995	Ecuador, Peru	Military and political tool	Yes	Armed skirmishes arise in part because of disagreement over the control of the headwaters of Cenepa River. Wolf argues that this is primarily a border dispute simply coinciding with location of a water resource.	Samson & Charrier 1997, Wolf 1997
1997	Singapore, Malaysia	Political tool	No	Malaysia supplies about half of Singapore's water and in 1997 threatened to cut off that supply in retribution for criticisms by Singapore of policy in Malaysia.	Zachary 1997
1998	Tajikistan	Terrorism, Political tool	No: Threat	On November 6, a guerrilla commander threatened to blow up a dam on the Kairakkhum channel if political demands are not met. Col. Makhmud Khudoberdyev made the threat, reported by the ITAR-Tass News Agency.	WRR 1998
1998	Angola	Military and political tool	Yes	In September 1998, fierce fighting between UNITA and Angolan government forces broke out at Gove Dam on the Kunene River for control of the installation.	Meissner 2001

continues

Date	Parties Involved	Basis of Conflict	Violent Conflict or In the Context of Violence?	Description	Sources
1998/1994	United States	Cyber-terrorism	No	The Washington Post reports a 12-year old computer hacker broke into the SCADA computer system that runs Arizona's Roosevelt Dam, giving him complete control of the dam's massive floodgates. The cities of Mesa, Tempe, and Phoenix, Arizona are downstream of this dam. No damage was done. This report turns out to be incorrect. A hacker did break into the computers of an Arizona water facility, the Salt River Project in the Phoenix area. But he was 27, not 12, and the incident occurred in 1994, not 1998. And while clearly trespassing in critical areas, the hacker never could have had control of any dams, leading investigators to conclude that no lives or property were ever threatened.	Gellman 2002, Lemos 2002
1998	Democratic Republic of Congo	Military target, Terrorism	Yes	Attacks on Inga Dam during efforts to topple President Kabila. Disruption of electricity supplies from Inga Dam and water supplies to Kinshasa	Chenje 2001, Human Rights Watch 1998
1998 to 2000	Eritrea and Ethiopia	Military target	Yes	Water pumping plants and pipelines in the border town of Adi Quala were destroyed during the civil war between Eritrea and Ethiopia.	ICRC 2003
1999	Lusaka, Zambia	Terrorism, Political tool	Yes	Bomb blast destroyed the main water pipeline, cutting off water for the city of Lusaka, population 3 million.	FTGWR 1999
1999	Yugoslavia	Military target	Yes	Belgrade reported that NATO planes had targeted a hydroelectric plant during the Kosovo campaign.	Reuters 1999a
1999	Bangladesh	Development dispute, Political tool	Yes	50 hurt during strikes called to protest power and water shortages. Protest led by former Prime Minister Begum Khaleda Zia over deterioration of public services and in law and order.	Ahmed 1999

1999	Yugoslavia	Military target	Yes	NATO targets utilities and shuts down water supplies in Belgrade. NATO bombs bridges on Danube, disrupting navigation.	Reuters 1999b
1999	Yugoslavia	Political tool	Yes	Yugoslavia refuses to clear war debris on Danube (downed bridges) unless financial aid for reconstruction is provided; European countries on Danube fear flooding due to winter ice dams will result. Diplomats decry environmental blackmail.	Simons 1999
1999	Kosovo	Political tool	Yes	Serbian engineers shut down water system in Pristina prior to occupation by NATO.	Reuters 1999c
1999	South Africa	Terrorism	Yes	A home-made bomb was discovered at a water reservoir at Wallmansthal near Pretoria. It was thought to have been meant to sabotage water supplies to farmers.	Pretoria Dispatch 1999
1999	Angola	Terrorism, Political tool	Yes	100 bodies were found in four drinking water wells in central Angola.	International Herald Tribune 1999
1999	Puerto Rico, U.S.	Political tool	No	Protesters blocked water intake to Roosevelt Roads Navy Base in opposition to U.S. military presence and Navy's use of the Blanco River, following chronic water shortages in neighboring towns.	New York Times 1999
1999	China	Development dispute; terrorism	Yes	Around Chinese New Years, farmers from Hebei and Henan Provinces fought over limited water resources. Heavy weapons, including mortars and bombs, were used and nearly 100 villagers were injured. Houses and facilities were damaged and the total loss reached one million $US. Parties involved: Huanglongkou Village, Shexian County, Hebei Province and Gucheng Village, Linzhou City, Henan Province	China Water Resources Daily 2002
1999	East Timor	Military tool, Terrorism	Yes	Militia opposing East Timor independence kill pro-independence supporters and throw bodies in water well.	BBC 1999

continues

Date	Parties Involved	Basis of Conflict	Violent Conflict or In the Context of Violence?	Description	Sources
1999	Yemen	Development dispute	Yes	700 soldiers were sent to quell fighting that claimed six lives and injured 60 others in clashes that erupted between two villages fighting over a local spring near Ta'iz. The village of Al-Marzuh believed it was entitled to exclusive rights from a spring because it was located on their land; the neighboring village of Quradah believed their rights to the water was affirmed in a 50-year-old court verdict. The dispute erupted in violence. President Ali Abdullah Saleh intervened by summoning the sheikhs of the two villages to the capital, and sorted out the problem by dividing the water into halves.	Al-Qadhi 2006
1998–1999	Kosovo	Terrorism, Political tool	Yes	Contamination of water supplies/wells by Serbs disposing of bodies of Kosovar Albanians in local wells. Other reports of Yugoslav federal forces poisoning wells with carcasses and hazardous materials.	CNN 1999, Hickman 1999.
1999 to 2000	Namibia, Botswana, Zambia	Military goal: Development dispute	No	Sedudu/Kasikili Island, in the Zambezi/Chobe River. Dispute over border and access to water. Presented to the International Court of Justice	ICJ 1999
2000	Ethiopia	Development dispute	Yes	One man stabbed to death during fight over clean water during famine in Ethiopia	Sandrasagra 2000
2000	Central Asia: Kyrgyzstan, Kazakhstan, Uzbekistan	Development dispute	No	Kyrgyzstan cuts off water to Kazakhstan until coal is delivered; Uzbekistan cuts off water to Kazakhstan for non-payment of debt.	Pannier 2000

Year	Location	Type	Violent	Description	Source
2000	Belgium	Terrorism	Yes	In July, workers at the Cellatex chemical plant in northern France dumped 5,000 liters of sulfuric acid into a tributary of the Meuse River when they were denied workers' benefits. A French analyst pointed out that this was the first time "the environment and public health were made hostage in order to exert pressure, an unheard-of situation until now."	Christian Science Monitor. 2000
2000	Hazarajat, Afghanistan	Development dispute	Yes	Violent conflicts broke out over water resources in the villages Burna Legan and Taina Legan, and in other parts of the region, as drought depleted local resources.	Cooperation Center for Afghanistan 2000
2000	India: Gujarat	Development dispute	Yes	Water riots reported in some areas of Gujarat to protest against authority's failure to arrange adequate supply of tanker water. Police are reported to have shot into a crowd at Falla village near Jamnagar, resulting in the death of three and injuries to 20 following protests against the diversion of water from the Kankavati dam to Jamnagar town.	FTGWR 2000
2000	United States	Terrorism	No	A drill simulating a terrorist attack on the Nacimiento Dam in Monterey County, California got out of hand when two radio stations reported it as a real attack.	Gaura 2000
2000	Kenya	Development dispute	Yes	A clash between villagers and thirsty monkeys left eight apes dead and ten villagers wounded. The duel started after water tankers brought water to a drought-stricken area and monkeys desperate for water attacked the villagers.	BBC 2000, Okoko 2000
2000	Australia	Cyber-terrorism	Yes	In Queensland, Australia, on April 23rd, 2000, police arrested a man for using a computer and radio transmitter to take control of the Maroochy Shire wastewater system and release sewage into parks, rivers, and property.	Gellman 2002

continues

Date	Parties Involved	Basis of Conflict	Violent Conflict or In the Context of Violence?	Description	Sources
2000	China	Development dispute	Yes	Civil unrest erupted over use and allocation of water from Baiyangdian Lake – the largest natural lake in northern China. Several people died in riots by villagers in July 2000 in Shandong after officials cut off water supplies. In August 2000, six died when officials in the southern province of Guangdong blew up a water channel to prevent a neighboring county from diverting water.	Pottinger 2000
2001	Israel, Palestine	Terrorism, Military target	Yes	Palestinians destroy water supply pipelines to West Bank settlement of Yitzhar and to Kibbutz Kisufim. Agbat Jabar refugee camp near Jericho disconnected from its water supply after Palestinians looted and damaged local water pumps. Palestinians accuse Israel of destroying a water cistern, blocking water tanker deliveries, and attacking materials for a wastewater treatment project.	Israel Line 2001a,b; ENS 2001a.
2001	Pakistan	Development dispute, Terrorism	Yes	Civil unrest over severe water shortages caused by the long-term drought. Protests began in March and April and continued into summer. Riots, four bombs in Karachi (June 13), one death, 12 injuries, 30 arrests. Ethnic conflicts as some groups "accuse the government of favoring the populous Punjab province [over Sindh province] in water distribution."	Nadeem 2001, Soloman 2001
2001	Macedonia	Terrorism, Military target	Yes	Water flow to Kumanovo (population 100,000) cut off for 12 days in conflict between ethnic Albanians and Macedonian forces. Valves of Glaznja and Lipkovo Lakes damaged.	AFP 2001, Macedonia Information Agency 2001

Date	Parties Involved	Basis of Conflict	Violent Conflict or in the Context of Violence	Description	Sources
2001	China	Development dispute	Yes	In an act to protest destruction of fisheries from uncontrolled water pollution, fishermen in northern Jiaxing City, Zhejiang Province, dammed the canal that carries 90 million tons of industrial wastewater per year for 23 days. The wastewater discharge into the neighboring Shengze Town, Jiangsu Province, killed fish, and threatened people's health.	China Ministry of Water Resources 2001.
2001	Philippines	Terrorism, Political tool	No	Philippine authorities shut off water to six remote southern villages yesterday after residents complained of a foul smell from their taps, raising fears Muslim guerrillas had contaminated the supplies. Abu Sayyaf guerrillas, accused of links with Saudi-born militant Osami bin Laden, had threatened to poison the water supply in the mainly Christian town of Isabela on Basilan island if the military did not stop an offensive against them.	World Environment News 2001
2001	Afghanistan	Military target	Yes	U.S. forces bombed the hydroelectric facility at Kajaki Dam in Helmand province of Afghanistan, cutting off electricity for the city of Kandahar. The dam itself was apparently not targeted.	BBC 2001, Parry 2001
2002	Nepal	Terrorism, Political Tool	Yes	The Khumbuwan Liberation Front (KLF) blew up a hydroelectric powerhouse of 250 kilowatts in Bhojpur District January 26. The power supply to Bhojpur and adjoining areas was cut off. Estimated repair time was 6 months; repair costs were estimated at 10 million Rs. By June 2002, Maoist rebels had destroyed more than seven micro-hydro projects as well as an intake of a drinking water project and pipelines supplying water to Khalanga in western Nepal.	Kathmandu Post 2002; FTGWR 2002a

continues

Date	Parties Involved	Basis of Conflict	Violent Conflict or In the Context of Violence?	Description	Sources
2002	Rome, Italy	Terrorism	No: Threat	Italian police arrest four Moroccans allegedly planning to contaminate the water supply system in Rome with a cyanide-based chemical, targeting buildings that included the United States embassy. Ties to Al-Queda were suggested.	BBC 2002
2002	Kashmir, India	Development dispute	Yes	Two people were killed and 25 others injured in Kashmir when police fired at a group of villagers clashing over water sharing. The incident took place in Garend village in a dispute over sharing water from an irrigation stream.	The Japan Times 2002
2002	United States	Terrorism	No: Threat	Papers seized during the arrest of a Lebanese national who moved to the US and became an Imam at an Islamist mosque in Seattle included "instructions on poisoning water sources" from a London-based al-Qaida recruiter. The FBI issued a bulletin to computer security experts around the country indicating that al-Qaida terrorists may have been studying American dams and water-supply systems in preparation for new attacks. "U.S. law enforcement and intelligence agencies have received indications that al-Qaida members have sought information on Supervisory Control And Data Acquisition (SCADA) systems available on multiple SCADA-related Web sites," reads the bulletin, according to SecurityFocus. "They specifically sought information on water supply and wastewater management practices in the U.S. and abroad."	McDonnell and Meyer 2002, MSNBC 2002

Year	Location	Type	Outcome	Description	Sources
2002	Colombia	Terrorism	Yes	Colombian rebels in January damaged a gate valve in the dam that supplies most of Bogota's drinking water. Revolutionary Armed Forces of Colombia (FARC) detonated an explosive device planted on a German-made gate valve located inside a tunnel in the Chingaza Dam.	Waterweek 2002
2002	Karnataka, Tamil Nadu, India	Development dispute	Yes	Continuing violence over the allocation of the Cauvery River between Karnataka and Tamil Nadu. Riots, property destruction, more than 30 injuries, arrests through September and October.	The Hindu 2002a,b, The Times of India 2002a.
2002	United States	Terrorism	No: Threat	Earth Liberation Front threatens the water supply for the town of Winter Park. Previously, this group claimed responsibility for the destruction of a ski lodge in Vail, Colorado that threatened lynx habitat.	Crecente 2002, Associated Press 2002
2003	United States	Terrorism	No: Threat	Al-Qaida threatens US water systems via call to Saudi Arabian magazine. Al-Qaida does not "rule out. . .the poisoning of drinking water in American and Western cities."	Associated Press 2003a, Waterman 2003, NewsMax 2003, US Water News 2003
2003	United States	Terrorism	Yes	Four incendiary devices were found in the pumping station of a Michigan water-bottling plant. The Earth Liberation Front (ELF) claimed responsibility, accusing Ice Mountain Water Company of "stealing" water for profit. Ice Mountain is a subsidiary of Nestle Waters.	Associated Press 2003b
2003	Colombia	Terrorism, development dispute	Yes	A bomb blast at the Cali Drinking Water Treatment Plant killed 3 workers May 8th. The workers were members of a trade union involved in intense negotiations over privatization of the water system.	PSI 2003

continues

Date	Parties Involved	Basis of Conflict	Violent Conflict or In the Context of Violence?	Description	Sources
2003	Jordan	Terrorism	No: Threat	Jordanian authorities arrested Iraqi agents in connection with a botched plot to poison the water supply that serves American troops in the eastern Jordanian desert near the border with Iraq. The scheme involved poisoning a water tank that supplies American soldiers at a military base in Khao, which lies in an arid region of the eastern frontier near the industrial town of Zarqa.	MJS 2003
2003	Iraq, United States, Others	Military Target	Yes	During the U.S.-led invasion of Iraq, water systems were reportedly damaged or destroyed by different parties, and major dams were military objectives of the U.S. forces. Damage directly attributable to the war includes vast segments of the water distribution system and the Baghdad water system, damaged by a missile.	UNICEF 2003, ARC 2003
2003	Iraq	Terrorism	Yes	Sabotage/bombing of main water pipeline in Baghdad. The sabotage of the water pipeline was the first such strike against Baghdad's water system, city water engineers said. It happened around 7 in the morning, when a blue Volkswagen Passat stopped on an overpass near the Nidaa mosque, and an explosive was fired at the six-foot-wide water main in the northern part of Baghdad, said Hayder Muhammad, the chief engineer for the city's water treatment plants.	Tierney and Worth 2003
2003–2007	Sudan, Darfur	Military tool, Military target, Terrorism	Yes	The ongoing civil war in the Sudan has included violence against water resources. In 2003, villagers from around Tina said that bombings had destroyed	Toronto Daily 2004, Reuters Foundation 2004

				water wells. In Khasan Basao they alleged that water wells were poisoned. In 2004, wells in Darfur were intentionally contaminated as part of a strategy of harassment against displaced populations.	
2004	Mexico	Development dispute	Yes	Two Mexican farmers argued for years over water rights to a small spring used to irrigate a small corn plot near the town of Pihuamo. In March, these farmers shot each other dead.	The Guardian 2004
2004	Pakistan	Terrorism	Yes	In military action aimed at Islamic terrorists, including Al Qaida and the Islamic Movement of Uzbekistan, homes, schools, and water wells were damaged and destroyed.	Reuters 2004a
2004	India, Kashmir	Terrorism	Yes	Twelve Indian security forces were killed by an IED planted in an underground water pipe during "counter-insurgency operation in Khanabal area in Anantnag district."	TNN 2004
2004	China	Development Dispute	Yes	Tens of thousands of farmers staged a sit-in against the construction of the Pubugou dam on the Dadu River in Sichuan Province. Riot police were deployed to quell the unrest and one was killed. Witnesses also report the deaths of a number of residents. (See China 2006 for follow-up.)	BBC 2004b, VOA 2004.
2004	China, United States	Military target	No	A 2004 Pentagon report on China's military capacity raises the concept of Taipei adopting military systems capable of being used as a tool for deterring Chinese military coercion by "presenting credible threats to China's urban population or high-value targets, such as the Three Gorges Dam." China promptly denounces "a U.S. suggestion" that Taiwan's military target the Three Gorges dam, leading to a U.S. denial that it had so urged.	The China Daily 2004; Pentagon 2004.

continues

Date	Parties Involved	Basis of Conflict	Violent Conflict or In the Context of Violence?	Description	Sources
2004	South Africa	Development dispute	Yes	Poor delivery of water and sanitation services in Phumelela Township led to several months of protests, including some severe injuries and property damage. No one was killed during the protests, but a few people seriously injured, and municipal property was extensively damaged.	CDE 2007
2004	Gaza Strip	Terrorism, Development dispute	Yes	The United States halts two water development projects as punishment to the Palestinian Authority for their failure to find those responsible for a deadly attack on a U.S. diplomatic convoy in October 2003.	Associated Press 2004a
2004	India	Development dispute	Yes	Four people were killed in October and more than 30 injured in November in ongoing protests by farmers over allocations of water from the Indira Ghandi Irrigation Canal in Sriganganagar district, which borders Pakistan. A curfew was imposed in the towns of Gharsana, Raola, and Anoopgarh.	Indo-Asian News Service 2004
2004–2006	Somalia, Ethiopia	Development dispute	Yes	At least 250 people killed and many more injured in clashes over water wells and pastoral lands. Villagers call it the "War of the Well" and describe "well warlords, well widows, and well warriors." A three-year drought has led to extensive violence over limited water resources, worsened by the lack of effective government and central planning.	BBC 2004a, AP 2005, Wax 2006
2005	Kenya	Development dispute	Yes	Police were sent to the northwestern part of Kenya to control a major violent dispute between Kikuyu and Maasai groups over water. More than 20 people were killed in fighting in January. By July, the death toll	BBC 2005a, Ryu 2005, Lane 2005

Date	Parties	Basis of Conflict	Violent Conflict or in the Context of Violence	Description	Sources
2006				exceeded 90, principally in the rural center of Turbi. The tensions arose over grazing and water. Maasai herdsmen accused a local Kikuyu politician of diverting a river to irrigate his farm, depriving downstream livestock. Fighting displaced more than 2000 villagers and reflects tensions between nomadic and settled communities.	
2006	Yemen	Development dispute	Yes	Local media reported a struggle between Hajja and Amran tribes over a well located between the two governorates in Yemen. According to news reports, armed clashes between the two sides forced many families to leave their homes and migrate. News reports confirmed that authorities arrested 20 people in an attempt to stop the fighting.	Al-Ariqi 2006
2006	China	Development dispute	Yes	The Chinese authorities executed a man who took part in protests against the Pubugou dam in Sichuan province in 2004 (see China 2004 entry). Chen Tao had been convicted of killing a policeman, but was executed before legal appeals had been completed.	BBC 2006d, Coonan 2006
2006	Ethiopia	Development dispute, water scarcity	Yes	At least 12 people died and over 20 were wounded in clashes over competition for water and pasture in the Somali border region.	BBC 2006a
2006	Ethiopia and Kenya	Development dispute	Yes	At least 40 people died in Kenya and Ethiopia in continuing clashes over water, livestock, and grazing land. Fighting occurred in the southern Ethiopia in the region of Oromo and the northern Kenya Marsabit district.	Reuters 2006
2006	Sri Lanka	Military tool, military target, terrorism	Yes	Tamil Tiger rebels cut the water supply to government-held villages in northeastern Sri Lanka. Sri Lankan government forces then launched attacks on the reservoir, declaring the Tamil actions to be terrorism.	BBC 2006b, 2006c

continues

Date	Parties Involved	Basis of Conflict	Violent Conflict or In the Context of Violence?	Description	Sources
2006	Israel, Lebanon	Military target, terrorism	Yes	Hezbollah rockets damaged a wastewater treatment plant in Israel. The Lebanese government estimates that Israeli attacks damaged water systems throughout southern Lebanon, including tanks, pipes, pumping stations, and facilities along the Litani River.	Science 2006, Amnesty International 2006, Murphy 2006
2007	India	Development dispute	Yes	Thousands of farmers breached security and stormed the area of Hirakud dam to protest allocation of water to industry. Minor injuries were reported during the conflict between the farmers and police.	Statesman News Service 2007
2007	Canada	Terrorism	No	A Toronto man previously accused of attempted murder and illegal possession of explosives was charged with eight more counts of attempted murder after allegedly tampering with bottled water, which had been injected with an unspecified liquid.	The Star 2007
2007	Burkina Faso, Ghana, and Cote D'Ivoire	Development dispute	Yes	Declining rainfall has led to growing fights between animal herders and farmers with competing needs. In August 2000, people were forced to flee their homes by fighting in Zounweogo province.	UNOCHA 2007

Notes:

1. Conflicts may stem from the drive to possess or control another nation's water resources, thus making water systems and resources a *political or military goal*. Inequitable distribution and use of water resources, sometimes arising from a water development, may lead to *development disputes*, heighten the importance of water as a strategic goal or may lead to a degradation of another's source of water. Conflicts may also arise when water systems are used as instruments of war, either as *targets or tools*. These distinctions are described in detail in Gleick (1993, 1998). In 2001, the Institute began including incidents involving water and *terrorism*. We note, however, the difficulty in defining "terrorism" (as opposed to military target, tool, or goal or other category) and caution users to use care with apply these categories. We use this term when individuals or groups act against governments or official agencies.

2. Thanks to the many people who have contributed to this over time, including William Meyer who sent nine fascinating items from the 1800s, Patrick Marsh, Mike Lane, Arthur Westing, Avilash Roul, Tony Turton, Hans-Juergen.Liebscher, Robert Halliday, Ma Jun, Marcus Moench, and others I've no doubt forgotten.

Sources:

Absolute Astronomy webpage. Reviewed 2006. Incapacitating agent. http://www.absoluteastronomy.com/reference/incapacitating_agent.

Agence France Press (AFP). 2001. Macedonian troops fight for water supply as president moots amnesty. *AFP*, June 8, 2001. http://www.balkanpeace.org/hed/archive/june01/hed3454.shtml.

Ahmed, A. 1999. Fifty hurt in Bangladesh strike violence. *Reuters News Service*, Dhaka, April 18, 1999.

Al-Ariqi, A. 2006. Water war in Yemen. *Yemen Times* 14(932), April 24, 2006. http://yementimes.com/article.shtml?i=932&p=health&a=1.

Al-Qadhi, M. 2003. Thirst for water and development leads to conflict in Yemen. *Choices* 12(1): 13–14. United Nations Development Programme. See also: http://yementimes.com/article.shtml?i=642&p=health&a=1.

American Red Cross (ARC). 2003. Baghdad Hospitals Reopen But Health Care System Strained. Mason Booth, Staff Writer, RedCross.org . April 24, http://www.redcross.org/news/in/iraq/030424baghdad.html.

American University (Inventory of Conflict and the Environment ICE). 2000a. Cauvery River Dispute. http://www.american.edu/projects/mandala/TED/ice/CAUVERY.HTM.

American University (Inventory of Conflict and the Environment ICE). 2000b. Lesotho Water Coup. http://www.american.edu/projects/mandala/TED/ice/LESWATER.HTM

American University (Inventory of Conflict and the Environment ICE). 2000c. Marsh Arabs and Iraq. http://www.american.edu/projects/mandala/TED/ice/MARSH.HTM.

Amnesty International. 2006. Israel/Lebanon. Deliberate destruction or collateral damage? Israeli attacks on civilian infrastructure. http://web.amnesty.org/library/Index/ENGMDE180072006.

Arbuthnot, F. 2000. Allies deliberately poisoned Iraq public water supply in Gulf War. *Sunday Herald* (Scotland) September 17, 2000.

Associated Press. 2002. Earth Liberation Front members threaten Colorado town's water. *AP*, October 15, 2002.

Associated Press. 2003a. Water targeted, magazine reports. *AP* May 29, 2003.

Associated Press. 2003b. Incendiary devices placed at water plant. *AP*, September 25, 2003.

Associated Press. 2004a. US dumps water projects in Gaza over convoy bomb. *AP*, May 6, 2004.

Associated Press. 2005. At least 16 killed in Somalia over water, pasture battles. *AP* June 8, 2005.

Barrett, G. 2003. Iraq's bad water brings disease, alarms relief workers. *The Olympian*, Olympia Washington, Gannett News Service, June 29, http://www.theolympian.com/home/news/20030629/frontpage/39442.shtml.

Barry J.M. 1997. *Rising Tide: The Great Mississippi Flood of 1927 and How it Changed America.* Simon and Schuster, New York. p. 67.

Bingham, G., A. Wolf, and T. Wohlegenant. 1994. Resolving water disputes: Conflict and cooperation in the United States, the Near East, and Asia. US Agency for International Development (USAID). Bureau for Asia and the Near East. Washington DC.

BBC 1999. World: Asia-Pacific Timor atrocities unearthed. BBC News. September 22, 1999. http://news.bbc.co.uk/hi/english/world/asia-pacific/newsid_455000/455030.stm

BBC 2000. Kenyan monkeys fight humans for water. BBC News March 21, 2000. http://news.bbc.co.uk/1/hi/world/ south_asia/1632504.stm

BBC 2001. US bombed Afghan power plant. BBC News. http://news.bbc.co.uk/1/hi/world/ hi/english/world/europe/newsid_1831000/1831511.stm

BBC 2002. 'Cyanide attack' foiled in Italy. BBC News. February 20, 2002. http://news.bbc.co.uk/ hi/english/world/europe/newsid_1831000/1831511.stm

BBC 2004b (Louisa Lim). China tries to calm dam protests. BBC News. November 18, 2004. http://news.bbc.co.uk/go/pr/fr/-/2/hi/asia-pacific/4021901.stm.

BBC 2005a. Thousands flee Kenyan water clash. BBC News. January 24, 2005. http://news.bbc.co. uk/1/hi/world/africa/4201483.stm.

BBC 2006a. Somalis clash over scarce water. BBC News. February 17, 2006. http://news.bbc.co. uk/go/pr/fr/-1/hi/world/africa/4723008.stm.

BBC 2006b. Sri Lanka forces attack reservoir. BBC News. August 7, 2006. http://news.bbc.co.uk/ 2/hi/south_asia/5249884.stm?ls

BBC 2006c. Water and war in Sri Lanka. BBC News. August 3, 2006. http://news.bbc.co.uk/2/hi/ south_asia/5239570.stm.

BBC 2006d. China 'executes dam protester.' BBC News. December 7, 2006. http://www.bbc.co. uk/go/pr/fr/-/2/hi/asia-pacific/6217148.stm.

BBC 2007a. Iran faces $2.65 bn US bomb award. BBC News. September 7, 2007. http://news.bbc.co.uk/go/pr/fr/-/2/hi/middle_east/6984365.stm.

Burns, J.F. 1992. Tactics of the Sarajevo Siege: Cut Off the Power and Water. New York Times, September 25, 1992. p.A1.

Butts, K., ed. 1997. Environmental Change and Regional Security. Carlisle, PA: Asia-Pacific Center for Security Studies, Center for Strategic Leadership, US Army War College.

Cable News Network (CNN). 1999. U.S.: Serbs destroying bodies of Kosovo victims. May 5. www.cnn.com/WORLD/europe/9905/05/kosovo.bodies.

Centre for Development and Enterprise (CDE). 2007. Voices of anger: Phumelela and Khutsong: Protest and conflict in two municipalities. Number 10, April 2007. http://www.cde.org.za/article.php?a_id=246

Chenje, M. 2001. Hydro-politics and the quest of the Zambezi River Basin Organization. In Nakayama, M., editor. International Waters in Southern Africa, United Nations University, Tokyo, Japan.

Chelyshev, A. 1992. Terrorists Poison Water in Turkish Army Cantonment. Telegraph Agency of the Soviet Union (TASS), 29 March.

China Daily. 2004. PLA General: Attempt to destroy dam doomed. June 16, 2004. http://www.chinadaily.com.cn/english/doc/2004-06/16/content_339969.htm

China Ministry of Water Resources. 2001. http://shuizheng.chinawater.com.cn/ssjf/20021021/ 20021016087.htm (the website of the Policy and Regulatory Department).

China Water Resources Daily 2002. Villagers fight over water resources. 24 October 2002. Citation provided by Ma Jun, personal communication.

Christian Science Monitor. 2000. Ecoterrorism as negotiating tactic. July 21, 2000, p. 8.

Clark, R.C. 1980. Technological Terrorism. Devin-Adair, Old Greenwich, Connecticut.

Clark, R.M. and R.A. Deininger. 2000. Protecting the Nation's Critical Infrastructure: The Vulnerability of U.S. Water Supply Systems. Journal of Contingencies and Crisis Management 8(2):73–80.

Columbia Electronic Encyclopedia. 2000. Vietnam: History. Available at http://www.infoplease.com/ce6/world/A0861793.html.

Columbia Encyclopedia. 2000. Netherlands. 6th Edition. Columbia Encyclopedia available at http://www.bartleby.com/65/ne/Nethrlds.html

Coonan, C. 2006. China secretly executes anti-dam protester. The Independent. December 7, 2006.

Cooperation Center for Afghanistan. 2000. The Social Impact of Drought in Hazarajat. http://www.ccamata.com/impact.html

Corps of Engineers. 1953. Applications of Hydrology in Military Planning and Operations and Subject Classification Index for Military Hydrology Data. Military Hydrology R&D Branch, Engineering Division, Corps of Engineers, Department of the Army, Washington.

Crecente, B.D. 2002. ELF targets water: Group threatens eco-terror attack on Winter Park tanks. Rocky Mountain News, October 15, 2002. http://www.rockymountainnews.com/drmn/state/article/0,1299,DRMN_21_1479883,00.html

Daniel, C. (ed.). 1995. Chronicle of the 20th Century. Dorling Kindersley Publishing, Inc., New York.

Dolatyar, M. 1995. Water diplomacy in the Middle East. In E. Watson (editor) The Middle Eastern Environment. John Adamson Publishing, London. 256 pp.

Douglass, J.D., and Livingstone, N.C. 1987. America the Vulnerable: The Threat of Chemical and Biological Warfare. Lexington, Massachusetts: Lexington Books.

Drower, M.S. 1954. Water-supply, irrigation, and agriculture. In C. Singer, Holmyard, E.J., and Hall, A.R., eds. A History of Technology. Oxford University Press, New York.

Dutch Water Line. 2002. Information on the historical use of water in defense of Holland. http://www.xs4all.nl/;pho/Dutchwaterline/dutchwaterl.htm.

Eitzen, E.M. and E.T. Takafuji. 1997. Historical Overview of Biological Warfare. In Textbook of Military Medicine, Medical Aspects of Chemical and Biological Warfare. Published by the Office of The Surgeon General, Department of the Army, USA. Pages 415–424.

ENS: Environment News Service. 2001a. Environment a weapon in the Israeli-Palestinian conflict. February 5, 2001, http://www.ens-newswire.com/ens/feb2001/2001-02-05-01.asp.

188

Fatout, P. 1972. *Indiana Canals*. Purdue University Studies, West Lafayette, Indiana, pp. 158–162.

Ferguson, R. Brian. 2001. The birth of war. *Natural History* 122(6): 28–35 (July-August 2003).

Fickle, J.E. 1983. The 'people' versus 'progress': Local opposition to the construction of the Wabash and Erie Canal. *Old Northwest* 8(4): 309–328.

Financial Times Global Water Report. 1999. Zambia: Water Cutoff. *FTGWR* Issue 68, p. 15 (March 19, 1999).

Financial Times Global Water Report. 2000. Drought in India comes as no surprise. *FTGWR* Issue 94, p. 14 (April 28, 2000).

Financial Times Global Water Report. 2002a. Maoists destroy Nepal's infrastructure. *FTGWR*, Issue 146, pp. 4–5 (May 17, 2002).

Fonner, D.K. 1996. Scipio Africanus. Military History Magazine March 1996. Cited in http://historynet.com/mh/blscipioafricanus/index1.html

Forkey, N.S. 1998. Damning the dam: Ecology and community in Otps Township, Upper Canada. *Canadian Historical Review* 79(1):68–99.

Gaura, M.A. 2000. Disaster simulation too realistic. *San Francisco Chronicle*, page A1, October 27, 2000.

Gellman, B. 2002. Cyber-attacks by Al Qaida feared. *Washington Post*, June 27, 2002, page A1.

Gleick, P.H. 1991. Environment and security: The clear connections. *Bulletin of the Atomic Scientists* April:17–21.

Gleick, P.H. 1993. Water and conflict: Fresh water resources and international security. *International Security* 18(1):79–112.

Gleick, P.H. 1994. Water, war, and peace in the Middle East. *Environment* 36(3):6–on.

Gleick, P.H. 1995. Water and Conflict: Critical Issues. Presented to the 45th Pugwash Conference on Science and World Affairs. Hiroshima, Japan: 23–29 July.

Gleick, P.H. 1998. Water and conflict. In *The World's Water 1998–1999*. Washington: Island Press.

Gowan, H. 2004. Hannibal Barca and the Punic Wars. Website. http://www.barca.fsnet.co.uk/. Reviewed March 2005.

Grant, U.S. 1885. *Personal Memoirs of U.S. Grant*. New York: C.L. Webster. ["On the second of February, [1863] this dam, or levee, was cut. . . The river being high the rush of water through the cut was so great that in a very short time the entire obstruction was washed away. . . As a consequence the country was covered with water."]

Green Cross International. The Conflict Prevention Atlas: http://www.greencrossinternational. net/GreenCrossPrograms/waterres/gcwater/report.html

Guantanamo Bay Gazette. 1964. The History of Guantanamo Bay: An Online Edition. http://www.gtmo.net/gazz/hisidx.htm. Chapter XXI: The 1964 Water Crisis. http://www. gtmo.net/gazz/HISCHP21.HTM.

Guardian. 2004. Water duel kills elderly cousins. The Guardian Newspapers Limited. March 11, 2004.

Haeri, S. 1993. Iran: Vehement Reaction. *Middle East International* 19(3):8.

Harris, S.H. 1994. *Factories of Death: Japanese Biological Warfare 1932–1945 and the American Cover-up*. New York, N.Y.: Routledge.

Hatami, H. and Gleick, P. 1994. Chronology of conflict over water in the legends, myths, and history of the ancient Middle East. In: Water, War, and Peace in the Middle East. *Environment* 36(3):6–on.

Hickman, D.C. 1999. A Chemical and Biological Warfare Threat: USAF Water Systems at Risk. Couterproliferation Paper No. 3. uSAF Counterproliferation Center, Air War College, Maxwell Air Force Base, Alabama.

Hillel, D. 1991. Lash of the dragon. *Natural History* (August):28–37.

Hindu, The. 2002a. Ryots on the rampage in Mandya. *The Hindu, India's National Newspaper*. October 31, 2002. http://www.hinduonnet.com/thehindu/2002/10/31/stories/ 2002103106680100.htm

Hindu, The. 2002b. Farmers go berserk; MLA's house attacked. *The Hindu, India's National Newspaper*, October 30, 2002. http://www.hinduonnet.com/thehindu/2002/10/30/ stories/2002103004870400.htm

Honan, W.H. 1996. Scholar sees Leonardo's influence on Machiavelli. *The New York Times* (December 8), p. 18.

Human Rights Watch. 1998. Human Rights Watch Condemns Civilian Killings by Congo Rebels. http://www.hrw.org/press98/aug/congo827.htm

Husarska, A. 1995. Running dry in Sarajevo: Water fight. *The New Republic*. July 17, 24.

IDG. 1996. Information and Documentation Center for the Geography of the Netherlands. IDG-Bulletin 1995/96.

InfoRoma. 2004. Roman Aqueducts. http://www.inforoma.it/feature.php?lookup=aqueduct. Viewed March 2005.

Indo-Asian News Service. 2004. Curfew imposed in three Rajasthan towns. http://www.hindustantimes.com/news/181_1136315,000900010008.htm. *Hindustan Times* December 4, 2004. Also, see http://news.newkerala.com/india-news/?action=fullnews&id=46359. *India News* at newkerala.com.

International Committee of the Red Cross. 2003. Eritrea: ICRC repairs war-damaged health centre and water system. 15 Dec 2003. ICRC News No. 03/158. http://www.alertnet.org/thenews/fromthefield/107148342038.htm.

International Court of Justice. 1999. International Court of Justice Press Communiqué 99/53. Kasikili Island/Sedudu Island (Botswana/Namibia). The Hague, Holland 13 December 1999, p. 2 (http://www.icj-cij.org/icjwww/ipresscom/ipress1999/ipresscom9953_ibona_19991213.htm.)

International Herald Tribune. 1999. 100 bodies found in well. *International Herald Tribune*, August 14–15, p.4.

Israel Line. 2001a. Palestinians loot water pumping center, cutting off supply to refugee camp. *Israel Line* (http://www.israel.org/mfa/go.asp?MFAH0dmp0), downloaded January 5, 2001, http://www.mfa.gov.il/mfa/go.asp?MFAH0iy50.

Israel Line. 2001a. Palestinians vandalize Yitzhar water pipe. *Israel Line* January 9, 2001, http://www.mfa.gov.il/mfa/go.asp?MFAH0izu0.

IWCT. 1967. International War Crimes Tribunal. Some Facts on Bombing of Dikes. http://www.infotrad.clara.co.uk/antiwar/warcrimes/index.html.

Japan Times. 2002. Kashmir water clash. *The Japan Times* May 27, 2002, p. 3.

Jenkins, B.M., Rubin, A.P. 1978. New Vulnerabilities and the Acquisition of New Weapons by Nongovernment Groups. In: Evans, A.E. and Murphy, J.F., eds. Legal Aspects of International Terrorism. Lexington, Massachusetts: Lexington Books, pp. 221–276.

Kathmandu Post. 2002. KLF destroys micro hydro plant. *The Kathmandu Post*, January 28, 2002. http://www.nepalnews.com.np/contents/englishdaily/ktmpost/2002/jan/jan28/index.htm

Kirschner, O. 1949. Destruction and Protection of Dams and Levees. Military Hydrology, Research and Development Branch, U.S. Corps of Engineers, Department of the Army, Washington District. From Schweizerische Bauzeitung 14 March 1949, Translated by H.E. Schwarz, Washington.

Keluel-Jang, S.A. 1997. Alier and the Jonglei Canal. *Southern Sudan Bulletin* 2(3) (January) (from www.sufo.demon.co.uk/poli007.htm).

Kupperman, R.H., and Trent, D.M. 1979. *Terrorism: Threat, Reality, Response*. Stanford, California: Hoover Institution Press.

Lane, M. 2005. Personal communication to P. Gleick regarding conflicts in Northern Kenya, with reference to Sunday Nation newspaper reports of July 17, 2005.

Lemos, R. 2002. Safety: Assessing the infrastructure risk. CNET/new.com. http://news.com.com/ 2009–1001_3-954780.html August 26th.

Levy, K. 2000. Guatemalan dam massacre survivors seek reparations from financiers. *World Rivers Review*. Berkeley, California: International Rivers Network, December 2000, pp. 12–13.

Livingston, N.C. 1982. *The War Against Terrorism*. Lexington and Toronto, Canada: Lexington Books.

Lockwood, R.P. Reviewed April 2006. The battle over the Crusades. http://www.catholicleague. org/research/battle_over_the_crusades.htm.

Macedonia Information Agency. 2001. Humanitarian catastrophe averted in Kumanovo and Lipkovo. Republic of Macedonia Agency of Information Archive. June 18, 2001. http://www.reliefweb.int/w/rwb.nsf/0/dbd4ef105d93da4ac1256a6f005bc328?OpenDocument.

McDonnell, P.J. and Meyer, J. 2002. Links to Terrorism Probed in Northwest. *Los Angeles Times*, 13 July 2002.

Meissner, R. 2000. Hydropolitical hotspots in Southern Africa: Will there be a water war? The case of the Kunene river. In H. Solomon and Turton, A., editors. *Water Wars: Enduring Myth or Impending Reality?* Africa Dialogue Monograph Series No. 2. Accord, Creda Communications, KwaZulu-Natal, South Africa, pp. 103–131.

190

Meissner, R. 2001. Interaction and existing constraints in international river basins: The case of the Kunene River Basin. In: Nakayama, M., ed. *International Waters in Southern Africa.* Tokyo, Japan: United Nations University,

Milwaukee Journal Sentinel. 2003. Jordan foils Iraqi plot to poison U.S. troops' water, officials say. April 1, 2003. http://www.jsonline.com/news/gen/apr03/130338.asp.

Mohamed, A.E. 2001. Joint development and cooperation in international water resources: The case of the Limpopo and Orange River Basins in Southern Africa. In Nakayama, M., ed. *International Waters in Southern Africa.* Tokyo, Japan: United Nations University.

Moorehead, A. 1960. *The White Nile.* England: Penguin Books.

MSNBC 2002. FBI says al-Qaida after water supply. Numerous wire reports, see, for example, http://www.ionizers.org/water-terrorism.html.

Murphy, K. 2006. Old feud over Lebanese river takes new turn. Israel's airstrikes on canals renew enduring suspicions that it covets water from the Litani. August 10, 2006. http://www.latimes.com/news/nationworld/world/la-fg-litani10aug10,1 3447228.story?coll=1a-headlines-world

Murphy, I.L. and Sabadell, J.E. 1986. International river basins: A policy model for conflict resolution. *Resources Policy* 12(1):133–144. United Kingdom: Butterworth and Co. Ltd.

Museum of the City of New York (MCNY). No date. The Greater New York Consolidation Timeline. http://www.mcny.org/Exhibitions/GNY/timeline.htm

Nadeem, A. 2001. Bombs in Karachi Kill One. *Associated Press,* downloaded June 13, 2001. At http://dailynews.yahoo.com/h/ap/20010613/wl/pakistan_strike_3.html.

Naff, T. and Matson, R.C. (editors). 1984. *Water in the Middle East conflict or cooperation?* Boulder, Colorado: Westview Press.

National Geographic News. 2001. Ancient Fertile Crescent almost gone, satellite images show. May 18, 2001. http://news.nationalgeographic.com/news/2001/05/0518_crescent.html

New Scofield Reference Bible. 1967. C.I. Schofield, Editor. Oxford, United Kingdom: Oxford University Press.

New York Times. 1999. Puerto Ricans protest Navy's use of water. *The New York Times,* October 31, p. 30.

NewsMax. 2003. Al-Qaida Threat to U.S. Water Supply. NewsMax Wires, May 29, 2003. http://www.newsmax.com/archives/articles/2003/5/28/202658.shtml.

NTI Nuclear Threat Initiative. 2005. A Brief History of Chemical Warfare. http://www.nti.org/h_ learnmore/cwtutorial/chapter02_02.html

Okoko, T.O. 2000. Monkeys, Humans Fight over Drinking Water. Panafrican News Agency March 21, 2000.

Pannier, B. 2000. Central Asia: Water becomes a political issue. Racio Free Europe, www.rferl.org/ nca/features/2000/08/F.RU.000803122739.html.

Parry, R.L. 2001. UN fears 'disaster' over strikes near huge dam. *The Independent,* London, November 8.

Pentagon. 2004. FY04 Report to Congress on PRC Military Power; Pursuant to the FY2000 National Defense Authorization Act. Annual Report on the Military Power of the People's Republic of China 2004. http://www.defenselink.mil/pubs/d20040528FRC.pdf

Plant, G. 1995. Water as a weapon in war. Water and War, Symposium on Water in Armed Conflicts, Montreux 21–23 November 1994, Geneva, ICRC.

Pottinger, M. 2000. Major Chinese lake disappearing in water crisis. *Reuters Science News,* http://us.cnn.com/2000/NATURE/12/20/china.lake.reut/

Pretoria Dispatch Online. 1999. Dam bomb may be 'aimed at farmers'. http://www.dispatch.co. za/1999/07/21/southafrica/RESEVOIR.HTM. (July 21).

Priscoli, J.D. Water and Civilization: Conflict, Cooperation and the Roots of a New Eco Realism. A Keynote Address for the 8th Stockholm World Water Symposium, 10–13 August 1998. http://www.genevahumanitarianforum.org/docs/Priscoli.pdf

PSI. 2003. Urgent action: Bomb blast kills 3 workers at the Cali Water Treatment Plant. *Public Services International* www.world-psi.org. Also at http://209.238.219.111/Water.htm

Purver, R. 1995. Chemical and Biological Terrorism: The Threat According to the Open Literature. Canadian Security Intelligence Service, Ottawa, Canada. http://www.csis.gc.ca/en/publications/other/c_b_terrorism01.asp.

Rasch, P.J. 1968. The Tularosa Ditch War. *New Mexico Historical Review* 43(3): 229–235.

Reisner, M. 1986, 1993. *Cadillac Desert: The American West and its Disappearing Water.* New York: Penguin Books.

Reuters. 1999a. Serbs Say NATO Hit Refugee Convoys. April 14, 1999. http://dailynews.yahoo. com/headlines/ts/story.html?s=v/nm/19990414/ts/yugoslavia_192.html. http:// www.uia.ac.be/u/carpent/kosovo/messages/397.html

191

Reuters 1999b. NATO Keeps Up Strikes But Belgrade Quiet. June 5, 1999. Downloaded June 1999. http://dailynews.yahoo.com/headlines/wl/story.html?s=v/nm/19990605/wl/yugoslavia_strikes_129.html.

Reuters 1999c. NATO Builds Evidence Of Kosovo Atrocities. June 17, 1999. Downloaded June 1999. http://dailynews.yahoo.com/headlines/ts/story.html?s=v/nm/19990617/ts/yugoslavia_leadall_171.html.

Reuters. 2004a. Al Qaida spy chief killed in Pakistani raid. Reuters Yahoo.

Reuters. 2006. Clashes over water, pasture kill 40 in east Africa. Reuters/Asia News. June 7, 2006. http://asia.news.yahoo.com/060606/3/2lk9x.html.

Reuters Foundation. 2004. Darfur: 2.5 million people will require food aid in 2005. http://www.medair.org/en_portal/medair_news?news=258. R. Schofield. November 22, 2004.

Rome Guide. 2004. Fontana di Trevi: History. http://web.tiscali.it/romaonlineguide/Pages/eng/ rbarocca/sBMy5.htm. Viewed March 2005.

Rowe, W.T. 1988. Water control and the Qing political process: The Fankou Dam controversy, 1876–1883. *Modern China* 14(4):353–387.

Ryu, A. 2005. Water rights dispute sparks ethnic clashes in Kenya's Rift Valley. Voice of America, http://www.voanews.com/english/archive/2005-03/2005-03-21-voa28.cfm.

Samson, P. and Charrier, B. 1997. International freshwater conflict: Issues and prevention strategies. Green Cross International. http://www.greencrossinternational.net/GreenCross Programs/waterres/gcwater/report.html

Sandrasagra, M. J. 2000. Development Ethiopia: Relief agencies warn of major food crisis. *Inter Press Service* April 11, 2000.

Scheiber, H.N. 1969. *Ohio Canal Era.* Athens, Ohio: Ohio University Press, pp. 174-175.

Science. 2006. Tallying Mideast damage. *Science* 313(5793):1549.

Semann, D. 1950. Die Kriegsbeschädigungen der Edertalspermmauer, die Wiederherstellungsarbeiten und die angestellten Untersuchungen über die Standfestigkeit der Mauer. *Die Wasserwirtschaft* 41(1):2.

Shapiro, C. 2004. A search for flaws deep in the heart of the Surry reactor. *The Virginia-Pilot* December 6, 2004. http://home.hamptonroads.com/stories/print.cfm?story=78992&ran=226100.

Simons, M. 1999. Serbs refuse to clear bomb-littered river. *New York Times,* October 24, 1999.

Strategy Page. 2006. Reviewed April 2006. http://www.strategypage.com/articles/biotoxin_files/ BIOTOXINSINWARFARE.asp.

Stockholm International Peace Research Institute (SPIRI). 1971. The Rise of CB Weapons: The Problem of Chemical and Biological Warfare. Humanities Press, New York, NY

Soloman, A. 2001. Policeman dies as blasts rock strike-hit Karachi. *Reuters,* June 13, 2001. at http://dailynews.yahoo.com/h/nm/20010613/ts/pakistan_strike_dc_1.html. http://www.labline.de/indernet/partikel/karachi/bombse.htm

Star. 2007. Man face attempted murder charge over water. *The Star* November 6, 2007. http://www.thestar.com/printArticle/274128.

Statesman News Service. 2007. Clash takes place over water rights. *The Statesman* Sambalpur, India. http://www.thestatesman.net/page.arcview.php?clid=9&id=202958&usrsess=1

Steinberg, T.S. 1990. Dam-breaking in the nineteenth-century Merrimack Valley. *Journal of Social History* 24(1): 25–45.

Styran, R.M., and Taylor, R.R. 2001. *The Great Swivel Link': Canada's Welland Canal.* The Champlain Society, Toronto, Canada.

Suliman, M. 1998. Resource access: A major cause of armed conflict in the Sudan. The case of the Nuba Mountains. Institute for African Alternatives, London, UK (from http://srdis.ciesin.org/cases/Sudan-Paper.html.)

Thatcher, J. 1827. A Military Journal During the American Revolutionary War, From 1775 to 1783. Second Edition, Revised and Corrected. Cottons and Barnard. Boston, Massachusetts. (from http://www.fortklock.com/journal1777.htm.)

Tierney, J., and Worth, R.F. 2003. Attacks in Iraq May Be Signals of New Tactics. *The New York Times,* August 18, 2003. Page 1. Also at http://www.nytimes.com/2003/08/18/international/worldspecial/18IRAQ.html?hp

Times of India. 2002a. Cauvery row: Farmers renew stir. October 20, 2002. http://timesofindia. indiatimes.com/cms/html/uncomp/articleshow?art_id=26586125.

Times News Network (TNN). 2004. IED was planted in underground pipe. http://timesofindia. indiatimes.com/articleshow/947432.cms December 5, 2004.

Toronto Daily. 2004. Darfur: "Too many people killed for no reason. Amnesty International Index: AFR 54/008/2004, 3 February 2004.

Tucker, J.B. ed. 2000. *Toxic Terror: Assessing Terrorist Use of Chemical and Biological Weapons*. Cambridge, Massachusetts: MIT Press.

Turton, A. 2005. A Critical Assessment of the Basins at Risk in the Southern African Hydropolitical Complex. Workshop on the Management of International Rivers and Lakes Hosted by the Third World Centre for Water Management & Helsinki University of Technology, 17–19 August 2005 Helsinki, Finland. Council for Scientific and Industrial Research (CSIR). African Water Issues Research Unit (AWIRU) CSIR Report Number: ENV-P-CONF 2005-001, Pretoria, South Africa.

UNICEF 2003. Iraq: Cleaning up neglected, damaged water system, clearing away garbage. News Note Press Release, May 27. http://www.unicef.org/media/media_6998.html.

UNOCHR. 2007. Burkina Faso: Innovation and education needed to head off water war. UN Office for the Coordination of Humanitarian Affairs. http://www.irinnews.org/ PrintReport.aspx?ReportId=74308.

US Water News. 2003. Report suggests al-Qaida could poison U.S. water. *US Water News Online*. June. http://www.uswaternews.com/archives/arcquality/3repsug6.html.

Vanderwood, P.J. 1969. *Night riders of Reelfoot Lake*. Memphis, Tennessee: Memphis State University Press.

Voice of America News (VOA). 2004. China's Sichuan Province Tense in Aftermath of Violent Anti-Dam Protests. Luis Ramirez. November 24, 2004.

Wallenstein, P. and A. Swain. 1997. International freshwater resources - Conflict or cooperation? Comprehensive Assessment of the Freshwater Resources of the World: Stockholm: Stockholm Environment Institute.

Walters, E. 1948. *Joseph Benson Foraker: Uncompromising Republican*. Columbus, Ohio: Ohio History Press, pp. 44–45.

Waterman, S. 2003. Al-Qaida threat to U.S. water supply. United Press International (UPI), May 28, 2003.

Waterweek. 2002. Water facility attacked in Colombia. *Waterweek*, American Water Works Association. January 2002. http://www.awwa.org/advocacy/news/020602.cfm.

Wax, E. 2006. Dying for water in Somalia's drought: Amid anarchy, warlords hold precious resource. *Washington Post*. April 14, 2006. p. A1. http://www.washingtonpost.com/wp-dyn/content/article/2006/04/13/AR2006041302116.html.

Wolf, A.T. 1995. *Hydropolitics along the Jordan River: Scarce Water and its Impact on the Arab-Israeli Conflict*. United Nations University Press, Tokyo, Japan.

Wolf, A. T. 1997. 'Water wars' and water reality: Conflict and cooperation along international waterways. NATO Advanced Research Workshop on Environmental Change, Adaptation, and Human Security. Budapest, Hungary. 9–12 October.

World Environment News. 2001. Philippine rebels suspected of water 'poisoning.' http://www.planetark.org/avantgo/dailynewsstory.cfm?newsid=12807.

World Rivers Review (WRR). 1998. Dangerous Dams: Tajikistan. *World Rivers Review* 13(6):13 (December).

Yang Lang. 1989/1994. High Dam: The Sword of Damocles. In Dai Qing, ed. *Yangtze! Yangtze!* Probe International, Earthscan Publications, London, United Kingdom. pp. 229–240.

Zachary G.P. 1997. Water pressure: Nations scramble to defuse fights over supplies. *Wall Street Journal*, December 4, p. A17.

Zemmali, H. 1995. International humanitarian law and protection of water. Water and War, Symposium on Water in Armed Conflicts, Montreux 21–23 November 1994, Geneva, ICRC.

Total Renewable Freshwater Supply, by Country

Description

Average annual renewable freshwater resources are listed by country. This table is the same as the table in the previous version of *The World's Water.* Some newer data are reported for some countries in the latest UN FAO Aquastat database, but these data are typically produced by modeling or estimation, rather than measurement, and we have chosen not to update them in this version. We will be reviewing methods and data for the next volume.

The data in this table are typically comprised of both renewable surface water and groundwater supplies, including surface inflows from neighboring countries. The UN FAO refers to this as total natural renewable water resources. Flows to other countries are not subtracted from these numbers. All quantities are in cubic kilometers per year (km^3/yr). These data represent average freshwater resources in a country — actual annual renewable supply will vary from year to year.

Limitations

These detailed country data should be viewed, and used, with caution. The data come from different sources and were estimated over different periods. Many countries do not directly measure or report internal water resources data, so some of these entries were produced using indirect methods. For example, Margat compiles information from a wide variety of sources and notes that there is a wide variation in the reliability of the data. In the past few years, new assessments have begun to standardize definitions and assumptions.

Not all of the annual renewable water supply is available for use by the countries to which it is credited here; some flows are committed to downstream users. For example, the Sudan is listed as having 154 cubic kilometers per year, but treaty commitments require them to pass significant flows downstream to Egypt. Other countries, such as Turkey, Syria, and France, to name only a few, also pass significant amounts of water to other users. The annual average figures hide large seasonal, inter-annual, and long-term variations.

SOURCES

Compiled by P. H. Gleick and H. Cooley, Pacific Institute.
nd = no data

a. Total natural renewable surface and groundwater. Typically includes flows from other countries. (FAO: Natural total renewable water resources)

b. Estimates from Belyaev, Institute of Geography, USSR (1987).

c. Estimates from FAO (1995). Water resources of African countries. Food and Agriculture Organization, United Nations, Rome, Italy.

d. Estimates from WRI (1994). See this source for original data source.

e. Estimates from Goscomstat, USSR, 1989 as cited in Gleick 1993, Table A16.

f. Estimates from FAO (1997). Water resources of the Near East region: A review. Food and Agriculture Organization, United Nations, Rome, Italy.

g. Estimates from FAO (1997). Irrigation in the countries of the Former Soviet Union in figures. Food and Agriculture Organization, United Nations, Rome, Italy.

h. UNFAO. 1999. Irrigation in Asia in figures. Food and Agriculture Organization, United Nations, Rome, Italy.

i. Nix, H. 1995. Water/Land/Life: The eternal triangle. Water Research Foundation of Australia, Canberra, Australia.

j. UNFAO. 2000. Irrigation in Latin America and the Caribbean. Food and Agriculture Organization, United Nations, Rome, Italy.

k. AQUASTAT Web site as of February 2003.

l. Margat, J. OSS. 2001. Les ressources en eau des pays de l'OSS. Evaluation, utilisation et gestion. UNESCO/Observatoire du Sahara et du Sahel. (Updating of 1995).

m. Estimates from FAO (2003). Review of world water resources by country. Food and Agriculture Organization, United Nations, Rome, Italy (see specific references in this document for more information).

n. United States Geological Survey Revised: Conterminous US (2071); Alaska (980); Hawaii (18).

o. EUROSTAT, U. Wieland. 2003. Water resources in the EU and in the candidate countries. Statistics in Focus, Environment and Energy, European Communities.

p. Margat, J., and Vallée, D. 2000. Blue Plan—Mediterranean Vision on Water, Population and the Environment for the 21st Century. France: Sophia Antipolis. 62 pp.

q. Geres, D. 1998. Water resources in Croatia. International Symposium on Water Management and Hydraulic Engineering. Dubrovnic, Croatia (September 14–19).

r. AQUASTAT, website as of November 2005.

s. EUROSTAT. 2005.

t. Pearse, P. H., Bertrand, F., MacLaren, J. W. 1985. *Currents of Change, Final Report of Inquiry on Federal Water Policy*. Ottawa, Canada: Environment Canada.

DATA TABLE 1 Total Renewable Freshwater Supply, by Country (2006 Update)

Region and Country	Annual Renewable Water Resources[a] (km³/yr)	Year of Estimate	Source of Estimate
AFRICA			
Algeria	14.3	1997	c,f
Angola	184.0	1987	b
Benin	25.8	2001	l
Botswana	14.7	2001	l
Burkina Faso	17.5	2001	l
Burundi	3.6	1987	b
Cameroon	285.5	2003	m
Cape Verde	0.3	1990	c
Central African Republic	144.4	2003	m
Chad	43.0	1987	b
Comoros	1.2	2003	m
Congo	832.0	1987	b
Congo, Democratic Republic (formerly Zaire)	1283	2001	l
Cote D'Ivoire	81	2001	l
Djibouti	0.3	1997	f
Egypt	86.8	1997	f
Equatorial Guinea	26	2001	l
Eritrea	6.3	2001	l
Ethiopia	110.0	1987	b
Gabon	164.0	1987	b
Gambia	8.0	1982	c
Ghana	53.2	2001	l
Guinea	226.0	1987	b
Guinea-Bissau	31.0	2003	m
Kenya	30.2	1990	c
Lesotho	5.2	1987	b
Liberia	232.0	1987	b
Libya	0.6	1997	c,f
Madagascar	337.0	1984	c
Malawi	17.3	2001	l
Mali	100.0	2001	k
Mauritania	11.4	1997	c,f
Mauritius	2.2	2001	k
Morocco	29.0	2003	m
Mozambique	216.0	1992	c
Namibia	45.5	1991	c
Niger	33.7	2003	m
Nigeria	286.2	2003	m
Reunion	5.0	1988	m
Rwanda	5.2	2003	m
Senegal	39.4	1987	b
Sierra Leone	160.0	1987	b
Somalia	15.7	1997	f
South Africa	50.0	1990	c

continues

DATA TABLE 1 *continued*

Region and Country	Annual Renewable Water Resources[a] (km³/yr)	Year of Estimate	Source of Estimate
AFRICA (*continued*)			
Sudan	154.0	1997	c,f
Swaziland	4.5	1987	b
Tanzania	91	2001	l
Togo	14.7	2001	l
Tunisia	4.6	2003	m
Uganda	66.0	1970	c
Zambia	105.2	2001	l
Zimbabwe	20.0	1987	b
NORTH AND CENTRAL AMERICA			
Antigua and Barbuda	0.1	2000	j
Bahamas	nd	nd	
Barbados	0.1	2003	m
Belize	18.6	2000	j
Canada	3300.0	1985	t
Costa Rica	112.4	2000	j
Cuba	38.1	2000	j
Dominica	nd	nd	
Dominican Republic	21.0	2000	j
El Salvador	25.2	2001	l
Grenada	nd	nd	
Guatemala	111.3	2000	j
Haiti	14.0	2000	j
Honduras	95.9	2000	j
Jamaica	9.4	2000	j
Mexico	457.2	2000	j
Nicaragua	196.7	2000	j
Panama	148.0	2000	j
St. Kitts and Nevis	0.02	2000	j
Trinidad and Tobago	3.8	2000	j
United States of America	3069.0	1985	n
SOUTH AMERICA			
Argentina	814.0	2000	j
Bolivia	622.5	2000	j
Brazil	8233.0	2000	j
Chile	922.0	2000	j
Colombia	2132.0	2000	j
Ecuador	432.0	2000	j
Guyana	241.0	2000	j
Paraguay	336.0	2000	j
Peru	1913.0	2000	j
Suriname	122.0	2003	m
Uruguay	139.0	2000	j
Venezuela	1233.2	2000	j

Region and Country	Annual Renewable Water Resources[a] (km³/yr)	Year of Estimate	Source of Estimate
ASIA			
Afghanistan	65.0	1997	f
Bahrain	0.1	1997	f
Bangladesh	1210.6	1999	h
Bhutan	95.0	1987	b
Brunei	8.5	1999	h
Cambodia	476.1	1999	h
China	2829.6	1999	h
India	1907.8	1999	h
Indonesia	2838.0	1999	h
Iran	137.5	1997	f
Iraq	96.4	1997	f
Israel	1.7	2001	l,m
Japan	430.0	1999	h
Jordan	0.9	1997	f
Korea DPR	77.1	1999	h
Korea Rep	69.7	1999	h
Kuwait	0.02	1997	f
Laos	333.6	2003	m
Lebanon	4.8	1997	f
Malaysia	580.0	1999	h
Maldives	0.03	1999	h
Mongolia	34.8	1999	h
Myanmar	1045.6	1999	h
Nepal	210.2	1999	h
Oman	1.0	1997	f
Pakistan	233.8	2003	k
Philippines	479.0	1999	h
Qatar	0.1	1997	f
Saudi Arabia	2.4	1997	f
Singapore	0.6	1975	d
Sri Lanka	50.0	1999	h
Syria	46.1	1997	f
Taiwan	67.0	2000	r
Thailand	409.9	1999	h
Turkey	234.0	2003	k, l, m, o
United Arab Emirates	0.2	1997	f
Vietnam	891.2	1999	h
Yemen	4.1	1997	f
EUROPE			
Europe			
Albania	41.7	2001	p
Austria	84.0	2005	s
Belgium	20.8	2005	s
Bosnia and Herzegovina	37.5	2003	m
Bulgaria	19.4	2005	s

continues

DATA TABLE 1 *continued*

Region and Country	Annual Renewable Water Resources[a] (km³/yr)	Year of Estimate	Source of Estimate
EUROPE (*continued*)			
Croatia	105.5	1998	o, q
Cyprus	0.4	2005	s
Czech Republic	16.0	2005	s
Denmark	6.1	2003	o
Estonia	21.1	2005	s
Finland	110.0	2005	s
France	189.0	2005	s
Germany	188.0	2005	s
Greece	72.0	2005	s
Hungary	120.0	2005	s
Iceland	170.0	2005	s
Ireland	46.8	2003	o
Italy	175.0	2005	s
Luxembourg	1.6	2005	s
Macedonia	6.4	2001	p
Malta	0.07	2005	s
Netherlands	89.7	2005	s
Norway	381.4	2005	s
Poland	63.1	2005	s
Portugal	73.6	2005	s
Romania	42.3/211.90	2003	o, m
Slovakia	80.3/50.1	2003	o, m
Slovenia	32.1	2005	s
Spain	111.1	2005	s
Sweden	179.0	2005	s
Switzerland	53.3	2005	s
United Kingdom	160.6	2005	s
Serbia-Montenegro*	208.5	2003	m
Russia	4498.0	1997	e,g
Armenia	10.5	1997	g
Azerbaijan	30.3	1997	g
Belarus	58.0	1997	g
Estonia	12.8	1997	g
Georgia	63.3	1997	g
Kazakhstan	109.6	1997	g
Kyrgyzstan	46.5	1997	m
Latvia	49.9	2005	s
Lithuania	24.5	2005	s
Moldova	11.7	1997	g
Tajikistan	99.7	1997	m
Turkmenistan	60.9	1997	m
Ukraine	139.5	1997	g
Uzbekistan	72.2	2003	m

Region and Country	Annual Renewable Water Resources[a] (km³/yr)	Year of Estimate	Source of Estimate
OCEANIA			
Australia	398.0	1995	i
Fiji	28.6	1987	b
New Zealand	397.0	1995	i
Papua New Guinea	801.0	1987	b
Solomon Islands	44.7	1987	b

*Referred to as Yugoslavia in previous World's Water.

Freshwater Withdrawal by Country and Sector

Description

The use of water varies greatly from country to country and from region to region. Data on water use by regions and by different economic sectors are among the most sought after in the water resources area. Ironically, these data are often the least reliable and most inconsistent of all water-resources information. This table is the same as Data Table 2 in the previous version of *The World's Water*. Some different data are reported for some countries in the UN FAO Aquastat database, but these data are typically produced by modeling or estimation, rather than measurement, and we have chosen not to update them in this version. We will be reviewing methods and data for the next volume.

This table includes the data available on total freshwater withdrawals by country in cubic kilometers per year and cubic meters per person per year, using national population estimates from approximately the year of withdrawal. The table also gives the breakdown of that water use by the domestic, agricultural, and industrial sectors, in both percentage of total water use and cubic meters per person per year. The data sources are also explicitly identified.

"Withdrawal" typically refers to water taken from a water source for use. It does not refer to water "consumed" in that use. The domestic sector typically includes household and municipal uses as well as commercial and governmental water use. The industrial sector includes water used for power plant cooling and industrial production. The agricultural sector includes water for irrigation and livestock.

In 2003, the Food and Agriculture Organization of the United Nations published a comprehensive update of its water use estimates in the AQUASTAT dataset, and it has continued to provide new and updated numbers. As a result, new estimates of water use by country are available, as are new estimates of the sectoral breakdown. An advantage of the AQUASTAT dataset is that it provides what appears to be a consistent set of information, but users should be very careful to understand which numbers are measured and which are only calculated (see Limitations).

Limitations

Extreme care should be used when applying these data. They come from a variety of sources and are collected using a variety of approaches, with few formal standards. As a result, this table includes data that are actually measured, estimated, modeled using different assumptions, or derived from other data. The data also come from different years, making direct inter-comparisons difficult. For example, some water-use data are over twenty years old. Separate data are now provided for the independent states of the former Soviet Union, but not for the former states of Yugoslavia. Industrial withdrawals for Panama, St. Lucia, St. Vincent, and the Grenadines are included in the domestic category.

The revisions of the FAO AQUASTAT dataset is the most dramatic change in water use data in recent years, yet these data are marred by inadequate information on sources and assumptions. They should be used with great care, and with appropriate caveats about their quality. As an example, the latest actual measured data on water use in Canada comes from a detailed national assessment published in 1996 by Environment Canada (see later reference "o"). More recent data can be found on the AQUASTAT database, but these data simply take the 1996 measured value and modify it for the change in population in Canada — thus the 2000 AQUASTAT value is modeled, not measured. Although we often report the AQUASTAT values in this table, in the case of Canada, we have chosen to rely on the 1996 measured value, while awaiting a Canadian reassessment.

Another major limitation of these data is that they do not include the use of rainfall in agriculture. Many countries use a significant fraction of the rain falling on their territory for agricultural production, but this water use is neither accurately measured nor reported in this set. We repeat our regular call for a systematic reassessment of water-use data and for national and international commitments to collect and standardize this information. We note that budgetary constraints continue to delay any major new data initiatives.

SOURCES

a. AQUASTAT estimates from www.fao.org. (March 2004).
b. World Resources Institute. 1990. *World Resources 1990–1991*. New York: Oxford University Press.
c. World Resources Institute. 1994. *World Resources 1994–1995*, in collaboration with the United Nations Environment Programme and the United Nations Development Programme. New York: Oxford University Press.
d. Eurostat Yearbook. 1997. *Statistics of the European Union*, EC/C/6/Ser.26GT, Luxembourg.
e. UN FAO. 1999. *Irrigation in Asia in Figures*. Rome, Italy: Food and Agriculture Organization, United Nations.
f. Nix, H. 1995. *Water/Land/Life*. Water Research Foundation of Australia, Canberra.
g. UNFAO. 2000. *Irrigation in Latin America and the Caribbean*. Rome, Italy: Food and Agricultural Organization, United Nations.
h. AQUASTAT Web site January 2002. http://www.fao.org.
i. Ministry of Water Resources, China. 2001. *Water Resources Bulletin of China, 2000*. Beijing, China: People's Republic of China, (September).
j. Hutson, S. S., Barber, N. L., Kenny, J. F., Linsey, K. S., Lumia, D. S., Maupin, M. A. 2004. Estimated use of water in the United States in 2000. Reston, VA: U.S. Geological Survey, Circular 1268.
k. Eurostat. 2004. Statistics in focus. http://europa.eu.int/comm/eurostat.
l. New FAO AQUASTAT estimates. http://www.fao.org (November 2005): See text for details.
m. Eurostat. 2005. Updated July 2005.
n. National land and water resources Audit. 2001.
o. Environment Canada. *Water use in Canada in 1996*. http://www.ec.gc.ca/water/en/manage/use/e_wuse.htm.

Population data. Population Division of the Department of Economic and Social Affairs of the United Nations Secretariat. 2005. World population prospects: The 2004 revision. Highlights. New York: United Nations.

DATA TABLE 2 Freshwater Withdrawal, by Country and Sector (2006 Update)

Region and Country	Year	Total Freshwater Withdrawal (km³/yr)	Per Capita Withdrawal (m³/p/yr)	Domestic Use (%)	Industrial Use (%)	Agricultural Use (%)	Domestic Use m³/p/yr	Industrial Use m³/p/yr	Agricultural Use m³/p/yr	Source	2005 Population (millions)
AFRICA											
Algeria	2000	6.07	185	22	13	65	41	24	120	1	32.85
Angola	2000	0.35	22	23	17	60	5	4	13	1	15.94
Benin	2001	0.13	15	32	23	45	5	4	7	1	8.44
Botswana	2000	0.19	107	41	18	41	44	19	44	1	1.77
Burkina Faso	2000	0.80	60	13	1	86	8	1	52	1	13.23
Burundi	2000	0.29	38	17	6	77	6	2	30	1	7.55
Cameroon	2000	0.99	61	18	8	74	11	5	45	1	16.32
Cape Verde	2000	0.02	39	7	2	91	3	1	36	1	0.51
Central African Republic	2000	0.03	7	80	16	4	6	1	0	1	4.04
Chad	2000	0.23	24	17	0	83	4	0	20	1	9.75
Comoros	1999	0.01	13	48	5	47	6	1	6	1	0.80
Congo, Democratic Republic (formerly Zaire)	2000	0.36	6	53	17	31	3	1	2	1	57.55
Congo, Republic of	2000	0.03	8	59	29	12	4	2	1	1	4.00
Cote D'Ivoire	2000	0.93	51	24	12	65	12	6	33	1	18.15
Djibouti	2000	0.02	25	84	0	16	21	0	4	1	0.79
Egypt	2000	68.30	923	8	6	86	70	55	793	1	74.03
Equatorial Guinea	2000	0.11	220	83	16	1	183	35	2	1	0.50
Eritrea	2000	0.30	68	3	0	97	2	0	66	1	4.40
Ethiopia	2002	5.56	72	6	0	94	4	0	67	1	77.43
Gabon	2000	0.12	87	50	8	42	43	7	37	1	1.38
Gambia	2000	0.03	20	23	12	65	5	2	13	1	1.52
Ghana	2000	0.98	44	24	10	66	11	4	29	1	22.11
Guinea	2000	1.51	161	8	2	90	12	4	144	1	9.40
Guinea-Bissau	2000	0.18	113	13	5	82	15	6	93	1	1.59
Kenya	2000	1.58	46	30	6	64	14	3	29	1	34.26

Country	Year										
Lesotho	2000	0.05	28	40	40	20	11	11	6	1	1.80
Liberia	2000	0.11	34	27	18	55	9	6	18	1	3.28
Libya	2000	4.27	730	14	3	83	102	22	606	1	5.85
Madagascar	2000	14.96	804	3	2	96	23	13	769	1	18.61
Malawi	2000	1.01	78	15	5	80	12	4	63	1	12.88
Mali	2000	6.55	484	9	1	90	44	5	436	1	13.52
Mauritania	2000	1.70	554	9	3	88	49	16	489	1	3.07
Mauritius	2000	0.61	488	25	14	60	124	69	294	a	1.25
Morocco	2000	12.60	400	10	3	87	40	12	348	1	31.48
Mozambique	2000	0.63	32	11	2	87	4	1	28	1	19.79
Namibia	2000	0.3	148	24	5	71	35	7	105	1	2.03
Niger	2000	2.18	156	4	0	95	6	0	149	1	13.96
Nigeria	2000	8.01	61	21	10	69	13	6	42	1	131.53
Rwanda	2000	0.15	17	24	8	68	4	1	11	1	9.04
Senegal	2002	2.22	190	4	3	93	8	6	177	1	11.66
Sierra Leone	2000	0.38	69	5	3	92	4	2	63	1	5.53
Somalia	2000	3.29	400	0	0	100	2	0	398	1	8.23
South Africa	2000	12.50	264	31	6	63	82	16	166	1	47.43
Sudan	2000	37.32	1,030	3	1	97	27	7	996	1	36.23
Swaziland	2000	1.04	1,010	2	1	97	20	10	979	1	1.03
Tanzania	2000	5.18	135	10	0	89	14	0	120	1	38.33
Togo	2000	0.17	28	53	2	45	15	1	12	1	6.15
Tunisia	2000	2.64	261	14	4	82	37	10	214	1	10.10
Uganda	2002	0.30	10	43	17	40	4	2	4	1	28.82
Zambia	2000	1.74	149	17	7	76	25	10	113	1	11.67
Zimbabwe	2002	4.21	324	14	7	79	45	23	256	1	13.01
NORTH AND CENTRAL AMERICA											
Antigua and Barbuda	1990	0.005	63	60	20	20	38	13	13	g	0.08
Barbados	2000	0.09	333	33	44	22	111	147	73	1	0.27
Belize	2000	0.15	556	7	73	20	39	406	111	1	0.27
Canada	1996	44.72	1,386	20	69	12	271	952	163	o	32.27
Costa Rica	2000	2.68	619	29	17	53	182	106	331	1	4.33
Cuba	2000	8.20	728	19	12	69	138	89	500	1	11.27

continues

DATA TABLE 2 *continued*

Region and Country	Year	Total Freshwater Withdrawal (km³/yr)	Per Capita Withdrawal (m³/p/yr)	Domestic Use (%)	Industrial Use (%)	Agricultural Use (%)	Domestic Use m³/p/yr	Industrial Use m³/p/yr	Agricultural Use m³/p/yr	Source	2005 Population (millions
NORTH AND CENTRAL AMERICA (*continued*)											
Dominica	1996	0.02	213	-	-	-				h	0.08
Dominican Republic	2000	3.39	381	32	2	66	122	7	252	l	8.90
El Salvador	2000	1.28	186	25	16	59	46	29	111	l	6.88
Guatemala	2000	2.01	160	6	13	80	10	21	128	l	12.60
Haiti	2000	0.99	116	5	1	94	5	1	110	l	8.53
Honduras	2000	0.86	119	8	12	80	10	14	95	l	7.21
Jamaica	2000	0.41	155	34	17	49	53	26	76	l	2.65
Mexico	2000	78.22	731	17	5	77	127	40	564	l	107.03
Nicaragua	2000	1.30	237	15	2	83	36	5	197	l	5.49
Panama	2000	0.82	254	67	5	28	170	13	72	l	3.23
St. Lucia	1997	0.01	81	-	-	-				g	0.16
St. Vincent and the Grenadines	1995	0.01	83	-	-	-				g	0.12
Trinidad and Tobago	2000	0.31	237	68	26	6	161	62	13	l	1.31
United States of America	2000	477.00	1,600	13	46	41	203	736	660	j	298.21
SOUTH AMERICA											
Argentina	2000	29.19	753	17	9	74	128	71	558	l	38.75
Bolivia	2000	1.44	157	13	7	81	21	11	127	l	9.18
Brazil	2000	59.30	318	20	18	62	65	57	196	l	186.41
Chile	2000	12.55	770	11	25	64	87	194	489	l	16.30
Colombia	2000	10.71	235	50	4	46	118	9	108	l	45.60
Ecuador	2000	16.98	1,283	12	5	82	160	68	1055	l	13.23
Guyana	2000	1.64	2,187	2	1	98	37	19	2143	l	0.75
Paraguay	2000	0.49	80	20	8	71	16	6	56	l	6.16
Peru	2000	20.13	720	8	10	82	60	73	587	l	27.97
Suriname	2000	0.67	1,489	4	3	93	67	43	1379	l	0.45

Uruguay	2000	3.15	910	2	1	96	22	10	—	878	3.46
Venezuela	2000	8.37	313	6	7	47	19	22	—	149	26.75
ASIA											
Afghanistan	2000	23.26	779	2	0	98	14	0	—	765	29.86
Armenia	2000	2.95	977	30	4	66	293	43	—	642	3.02
Azerbaijan	2000	17.25	2,051	5	28	68	99	567	—	1385	8.41
Bahrain	2000	0.30	411	40	3	57	163	12	—	233	0.73
Bangladesh	2000	79.40	560	3	1	96	18	4	—	538	141.82
Bhutan	2000	0.43	199	5	1	94	10	2	—	187	2.16
Brunei	1994	0.09	243	nd	nd	nd			e		0.37
Cambodia	2000	4.08	290	1	0	98	3	0	—	284	14.07
China*	2000	549.76	415	7	26	68	27	107	i	281	1,323.35
Cyprus	2000	0.21	250	27	1	71	68	4	m	179	0.84
Georgia	2000	3.61	808	20	21	59	161	170	—	477	4.47
India	2000	645.84	585	8	5	86	47	32	—	506	1,103.37
Indonesia	2000	82.78	372	8	1	91	30	3	—	339	222.78
Iran	2000	72.88	1,048	7	2	91	71	24	—	953	69.52
Iraq	2000	42.70	1,482	3	5	92	47	68	—	1367	28.81
Israel	2000	2.05	305	31	7	62	94	20	—	189	6.73
Japan	2000	88.43	690	20	18	62	136	123	—	431	128.09
Jordan	2000	1.01	177	21	4	75	37	8	—	133	5.70
Kazakhstan	2000	35.00	2,360	2	17	82	40	390	—	1930	14.83
Korea Democratic People's Republic	2000	9.02	401	20	25	55	80	101	—	220	22.49
Korea Rep	2000	18.59	389	36	16	48	139	64	—	186	47.82
Kuwait	2000	0.44	164	45	2	52	73	3	—	86	2.69
Kyrgyz Republic	2000	10.08	1,916	3	3	94	60	59	—	1797	5.26
Laos	2000	3.00	507	4	6	90	22	29	—	457	5.92
Lebanon	2000	1.38	385	33	1	67	126	2	—	257	3.58
Malaysia	2000	9.02	356	17	21	62	60	75	—	221	25.35
Maldives	1987	0.003	9	98	2	0	9	0	e	0	0.33
Mongolia	2000	0.44	166	20	27	52	34	45	—	86	2.65
Myanmar	2000	33.23	658	1	1	98	8	4	—	646	50.52

continues

207

DATA TABLE 2 *continued*

Region and Country	Year	Total Freshwater Withdrawal (km³/yr)	Per Capita Withdrawal (m³/p/yr)	Domestic Use (%)	Industrial Use (%)	Agricultural Use (%)	Domestic Use m³/p/yr	Industrial Use m³/p/yr	Agricultural Use m³/p/yr	Source	2005 Population (millions
ASIA (*continued*)											
Nepal	2000	10.18	375	3	1	96	11	2	362	1	27.13
Oman	2000	1.36	529	7	2	90	38	11	476	1	2.57
Pakistan	2000	169.39	1,072	2	2	96	21	22	1030	1	157.94
Philippines	2000	28.52	343	17	9	74	57	32	254	1	83.05
Qatar	2000	0.29	358	24	3	72	86	10	257	1	0.81
Saudi Arabia	2000	17.32	705	10	1	89	69	8	628	1	24.57
Singapore	1975	0.19	44	45	51	4	20	22	2	c	4.33
Sri Lanka	2000	12.61	608	2	2	95	14	15	579	1	20.74
Syria	2000	19.95	1,048	3	2	95	34	19	994	1	19.04
Tajikistan	2000	11.96	1,837	4	5	92	68	86	1683	1	6.51
Thailand	2000	82.75	1,288	2	2	95	32	32	1225	1	64.23
Turkey	2001	39.78	544	15	11	74	80	59	404	m	73.19
Turkmenistan	2000	24.65	5,104	2	1	98	86	39	4978	1	4.83
United Arab Emirates	2000	2.30	511	23	9	68	118	44	349	1	4.50
Uzbekistan	2000	58.34	2,194	5	2	93	104	45	2045	1	26.59
Vietnam	2000	71.39	847	8	24	68	66	205	577	1	84.24
Yemen	2000	6.63	316	4	1	95	13	2	301	1	20.98
EUROPE											
Albania	2000	1.71	546	27	11	62	146	61	339	1	3.13
Austria	1999	3.67	448	35	64	1	157	286	4	m	8.19
Belarus	2000	2.79	286	23	47	30	67	134	86	1	9.76
Belgium	1998	7.44	714	13	85	1	95	610	9	k	10.42
Bosnia and Herzegovina											3.91
Bulgaria	2003	6.92	895	3	78	19	27	700	168	m	7.73
Croatia											4.55
Czech Republic	2002	1.91	187	41	57	2	76	107	4	m	10.22

	Year										
Denmark	2002	0.67	123	32	26	42	40	32	52	m	5.43
Estonia	2002	1.41	1,060	56	39	5	591	418	52	m	1.33
Finland	1999	2.33	444	14	84	3	61	371	12	m	5.25
France	2000	33.16	548	16	74	10	86	408	54	m	60.50
Germany	2001	38.01	460	12	68	20	57	312	91	m	82.69
Greece	1997	8.70	782	16	3	81	128	25	630	m	11.12
Hungary	2001	21.03	2,082	9	59	32	192	1222	668	m	10.10
Iceland	2003	0.17	567	34	66	0	193	373	1	m	0.30
Ireland	1994	1.18	284	23	77	0	64	220	0	m	4.15
Italy	1998	41.98	723	18	37	45	131	265	326	m	58.09
Latvia	2003	0.25	108	55	33	12	59	35	13	m	2.31
Lithuania	2003	3.33	971	78	15	7	758	149	64	m	3.43
Luxembourg	1999	0.06	121	42	45	13	51	55	16	m	0.47
Macedonia	2000	2.27	1,118							m	2.03
Malta	2000	0.02	50	74	1	25	37	0	13	m	0.40
Moldova	2000	2.31	549	10	58	33	55	316	181	l	4.21
Netherlands	2001	8.86	544	6	60	34	33	326	184	m	16.30
Norway	1996	2.40	519	23	67	10	118	347	54	k	4.62
Poland	2002	11.73	304	13	79	8	40	240	25	m	38.53
Portugal	1998	11.09	1,056	10	12	78	101	128	827	m	10.50
Romania	2003	6.50	299	9	34	57	26	103	171	m	21.71
Russian Federation	2000	76.68	535	19	63	18	100	340	95	l	143.20
Serbia and Montenegro											10.50
Slovakia	2003	1.04	193							m	5.40
Slovenia	2002	0.90	457							m	1.97
Spain	2002	37.22	864	13	19	68	116	160	588	m	43.06
Sweden	2002	2.68	296	37	54	9	109	161	26	m	9.04
Switzerland	2002	2.52	348	24	74	2	84	257	7	m	7.25
Ukraine	2000	37.53	807	12	35	52	98	286	424	l	46.48
United Kingdom	1994	11.75	197	22	75	3	43	148	6	d	59.67
OCEANIA											
Australia	2000	24.06	1,193	15	10	75	176	120	898	n	20.16
Fiji	2000	0.07	82	14	14	71	12	12	58	l	0.85

continues

DATA TABLE 2 *continued*

Region and Country	Year	Total Freshwater Withdrawal (km³/yr)	Per Capita Withdrawal (m³/p/yr)	Domestic Use (%)	Industrial Use (%)	Agricultural Use (%)	Domestic Use m³/p/yr	Industrial Use m³/p/yr	Agricultural Use m³/p/yr	Source	2005 Population (millions
OCEANIA (*continued*)											
New Zealand	2000	2.11	524	48	9	42	251	50	220	1	4.03
Papua New Guinea	1987	0.10	17	56	43	1	9	7	0	c	5.89
Solomon Islands	1987			40	20	40	0	0	0	c	0.48

210

Notes:

Figures may not add to totals due to independent rounding.

2005 Population numbers: medium UN Variant

* Population includes Hong Kong and Macao

Sources

a. New FAO Aquastat Estimates from www.fao.org (March 2004): See text for details

b. World Resources Institute, 1990, *World Resources 1990–1991*, Oxford University Press, New York.

c. World Resources Institute, 1994, *World Resources 1994–95*, in collaboration with the United Nations Environment Programme and the United Nations Development Programme, Oxford University Press, New York.

d. Eurostat Yearbook, 1997, *Statistics of the European Union*, EC/C/6/Ser.26GT, Luxembourg.

e. UN FAO 1999, *Irrigation in Asia in Figures*, Food and Agriculture Organization, United Nations, Rome, Italy.

f. Nix, H. 1995, *Water/Land/Life*, Water Research Foundation of Australia, Canberra.

g. UNFAO. 2000. *Irrigation in Latin America and the Caribbean*. Food and Agricultural Organization, United Nations, Rome, Italy.

h. AQUASTAT website January 2002, www.fao.org.

i. Ministry of Water Resources, China. 2001. *Water Resources Bulletin of China, 2000*. People's Republic of China, Beijing, China (September).

j. Hutson, S.S., N.L. Barber, J.F. Kenny, K.S. Linsey, D.S. Lumia, and M.A. Maupin. 2004. Estimated Use of Water in the United States in 2000. U.S. Geological Survey, Circular 1268. Reston, Virginia.

k. Eurostat. 2004. Statistics in Focus. At http://europa.eu.int/comm/eurostat. http://europa.eu.int/comm/eurostat/newcronos/queen/display.do?screen=detail&language=en&product=THEME8&root=THEME8_copy_151979619462/yearlies_copy_1067300085946/dd_copy_251110364103/dda_copy_64928961 0368/dda10512_copy_72937 9227605

l. New FAO Aquastat Estimates from www.fao.org (November 2005): See text for details.

m. Eurostat. 2005. Updated 7/2005

n. National Land and Water Resources Audit. 2001. (Note: the breakdown by sector is from old data; new data combines domestic and industry into one value (25%), thus I left the data as is)

o. Environment Canada. *Water Use in Canada in 1996*. http://www.ec.gc.ca/water/en/manage/use/e_wuse.htm.

Population data Population Division of the Department of Economic and Social Affairs of the United Nations Secretariat. 2005. World Population Prospects: The 2004 Revision. Highlights. New York: United Nations.

Access to Safe Drinking Water by Country, 1970 to 2004

Description

Safe drinking water is one of the most basic human requirements, and one of the Millennium Development Goals (MDGs) by 2015 is to reduce by half the proportion of people unable to reach or afford safe drinking water (see Chapter 4 in this volume). As a result, estimates of access to safe drinking water are a cornerstone of most international assessments of progress, or lack thereof, toward solving global and regional water problems.

Data are given here for the percent of urban, rural, and total populations, by country, with access to safe drinking water for 1970, 1975, 1980, 1985, 1990, 1994, 2000, 2002, and 2004—the most recent year for which data are available. The World Health Organization (WHO) collected the data presented here over various periods. Most of the data presented were drawn from responses by national governments to WHO questionnaires. Participants in data collection include the Joint Monitoring Programme (JMP) of WHO, the United Nations Children's Fund, and the Water Supply and Sanitation Collaborative Council, which has continued sector monitoring and aims to support and strengthen the monitoring efforts of individual countries. The forty largest countries in the developing world account for 90 percent of population in these regions. As a result, WHO spent extra effort to collect comprehensive data for these countries.

Data for 2000 and later reflect a significant change in definition. Data are now reported for populations without access to "improved" water supply. According to WHO, the following technologies were included in the assessment as representing "improved" water supply:

Household connection
Public standpipe
Borehole
Protected dug well
Protected spring
Rainwater collection

In comparison, "unimproved" drinking water sources refers to:

Unprotected well
Unprotected spring
Rivers or ponds

Vendor-provided water
Bottled water
Tanker truck water

Limitations

A review of water and sanitation coverage data from the 1980s and 1990s shows that the definition of safe, or improved, water supply and sanitation facilities differs from one country to another and for a given country over time. Indeed, some of the data from individual countries often showed rapid and implausible changes in the level of coverage from one assessment to the next. This indicates that some of the data are also unreliable, irrespective of the definition used. Countries used their own definitions of "rural" and "urban."

For the 1996 data, two-thirds of the countries reporting indicated how they defined "access." At the time, the definition most commonly centered on walking distance or time from household to water source, such as a public standpipe, which varied from 50 to 2,000 meters and 5 to 30 minutes. Definitions sometimes included considerations of quantity, with the acceptable limit ranging from 15 to 50 liters per capita per day. The WHO considers safe drinking water to be treated surface water or untreated water from protected springs, boreholes, and wells.

WHO Assessments since 2000 have attempted to shift from gathering information from water providers only to include consumer-based information. The current approach uses household surveys in an effort to assess the actual use of facilities. "Reasonable access" was broadly defined as the availability of at least 20 liters per person per day from a source within one kilometer of the user's dwelling. A drawback of this approach is that household surveys are not conducted regularly in many countries. Thus, direct comparisons between countries, and across time within the same country, are difficult. Direct comparisons are additionally complicated by the fact that these data hide disparities between regions and socioeconomic classes.

Access to water, as reported by WHO, does not imply that the level of service or quality of water is "adequate" or "safe." The assessment questionnaire did not include any methodology for discounting coverage figures to allow for intermittence or poor quality of the water supplies. However, the instructions stated that piped systems should not be considered "functioning" unless they were operating at over 50 percent capacity on a daily basis; and that hand pumps should not be considered functioning unless they were operating for at least 70 percent of the time with a lag between breakdown and repair not exceeding two weeks. These aspects were taken into consideration when estimating coverage for countries for which national surveys had not been conducted. More details of the methods used, and their limitations, can be found at http://www.who.int/docstore/water_sanitation_health/Globassessment/Global TOC.htm.

SOURCES

United Nations Environment Programme (UNEP). 1989. *Environmental Data Report*. GEMS Monitoring and Assessment Research Centre, Oxford: Basil Blackwell.

United Nations Environment Programme (UNEP). 1993–94. *Environmental Data Report.* GEMS Monitoring and Assessment Research Centre in cooperation with the World Resources Institute and the UK Department of the Environment, Oxford: Basil Blackwell.

World Health Organization (WHO). 1996. *Water Supply and Sanitation Sector Monitoring Report: 1996 (Sector status as of 1994).* In collaboration with the Water Supply and Sanitation Collaborative Council and the United Nations Children's Fund, UNICEF, New York.

World Health Organization (WHO). 2000. *Global Water Supply and Sanitation Assessment 2000 Report.* http://www.who.int/docstore/water_sanitation_health/Globassessment/Global TOC.htm.

World Health Organization (WHO) and United Nations Children's Fund (UNICEF). 2004. *Meeting the MDG Drinking Water and Sanitation Target: A Mid-term Assessment of Progress.* http://www.who.int/water_sanitation_health/monitoring/jmp2004/en/index.html

World Health Organization (WHO) and United Nations Children's Fund (UNICEF). 2006. *Meeting the MDG Drinking Water and Sanitation Target: The Urban and Rural Challenge of the Decade.* http://www.who.int/water_sanitation_health/monitoring/jmpfinal.pdf

World Resources Institute (WRI). 1988. World Health Organization data, cited by the World Resources Institute, *World Resources 1988–89*, World Resources Institute and the International Institute for Environment and Development in collaboration with the United Nations Environment Programme, New York: Basic Books.

DATA TABLE 3 Access to Safe Drinking Water by Country, 1970 to 2004

Fraction of Population with Access to Improved Drinking Water

Region and Country	URBAN 1970	1975	1980	1985	1990	1994	2000	2002	2004	RURAL 1970	1975	1980	1985	1990	1994	2000	2002	2004	TOTAL 1970	1975	1980	1985	1990	1994	2000	2002	2004
AFRICA	66	68	69	77									55							77		68					85
Algeria	84	100	85	85			88	92	88		61		55			94	80	80							94	87	85
Angola			85	87	73	69	34	70	75			10	15	20	15	40	40	40			26	33	35	32	38	50	53
Benin	83	100	26	80	73	41	74	79	78	20	20	15	34	43	53	55	60	57	29	34	18	50	54	50	63	68	67
Botswana	71	95		84	100		100	100	100		39		46	88			90	90	29	45		53	91		95		95
Burkina Faso	35	50	27	43			84	82	94	10	23	31	69	70		44	44	54	12	25	31	67	45	78		51	61
Burundi	77		90	98	92	92	96	90	92			20	21	43	49	78	78	77			23	25	45	52	62	79	79
Cameroon	77			43	42		82	84	86	21			24	45		42	41	44	32			32	44		62	63	66
Cape Verde			100	83		70	64	86	86			21	24	45	34	89	73	73			25	52		51	74	80	80
Central African Republic				13	19	18	80	93	93				50	26	18	43	61	61					23	18	60	75	75
Chad	47	43				48	31	40	41	24	23				17	26	32	43	27	26				24	27	34	42
Comoros							98	90	92							95	96	82							96	94	86
Congo	63	81	42				71	72	84		9	7				17	17	27	27	38	20				51	46	58
Congo, Democratic Rep.	33	38		52	68	37	89	83	82	4	12		21	24	23	26	29	29	11	19		32	36	27	45	46	46
Cote D'Ivoire	98				57	59	90	98	97	29				80	81	65	74	74	44				71	72	77	84	84
Djibouti			50	50	77	100	82		76			20	20			100	67	59			43	45	90	100	80	80	73
Egypt	94		88		95	82	96	100	99	93		64		86	50	94	97	97	93		84		90	64	95	98	98
Equatorial Guinea			47		65	88	45	45	45					18	100	42	42	42					32	95	43	44	43
Eritrea							63	72	74							42	54	57							46	57	60
Ethiopia	61	58		69			77	81	81		1		9			13	11	11	6	8		16			24	22	22
Gabon							73	95	95							55	47	47							70	87	88
Gambia	97		85	97	100		80	95	95	3				48		53	77	77	12			59	60	76	62	82	82
Ghana	86	86	72	93	63	70	87	93	88		14	33	39		49	49	68	64	35	35	45	56	21	56	64	79	75
Guinea	68	69	69	41	100	61	72	78	78			2	12	37	62	36	38	35			15	18	53	62	48	51	50
Guinea-Bissau			18	17		38	29	79	79			8	22		57	55	49	49			10	21		53	49	59	59
Kenya	100	100	85			67	87	89	83	2	4	15			49	31	46	46	15	17	26	36		53	49	62	61
Lesotho	100	65	37	65		14	98	88	92	1	14	11	30		64	88	74	76	3	17	15		52		91	76	79

214

Table continued — country immunization/coverage data (columns represent successive years; no column headers printed on this page).

Liberia	100	100		58	72	72	72	6		23	8	68	52	52	15	58	87	96	53	30	72	62	61
Libya	100	76	80	83	85	72	72	42	82	17	10	31	68	35	58	11	25	21	31	29	72	45	50
Madagascar	67	77	97	52	95	85	77	1	14	50	44	44	31	34				41	56	45	47	67	73
Malawi		37	46	36	74	95	98		37	10	38	61	44	62	17		41	56	16	37	65	48	50
Mali	29	46	36	84	34	74	78	10			69	40	61	35		84		16	13	76	37	56	53
Mauritania	98	80	73	95		63	59	69	85				40	44	17				37	76	37	56	53
Mauritius	100	100	100	98	100	100	100	100	22	25	100	100	100	100	61	60	99	100	100	98	100	100	100
Morocco	92	100	100	98	100	100	99	28	98	9	14	58	56	56	51			59		52	82	80	81
Mozambique				17	86	76	72				40	43	24	26				15		32	60	42	43
Namibia		38	90	87	100	98	98			37	42	67	72	81					52	57	77	80	87
Niger	37	35	98	46	70	80	80	19	26	32	55	56	36	36	20	27	33	47	56	53	59	46	46
Nigeria		100	100	63	81	72	67	28			26	39	49	31	51			38	49	39	57	60	48
Reunion																							
Rwanda	81	84	79	60		92	92	66	68	55	67	40	69	69	67	68	55	50	68		41	73	74
Sao Tome and Principe			84		89	89	89			45			73	73				45				79	79
Senegal	87	56	79	82	92	90	92	13	25	38	26	65	54	60	19	50	43	53	50	50	78	72	76
Seychelles	56	77			100	100	100			95	95		75	75		37	95	95	43		87	87	88
Sierra Leone	75	50	68	58	23	75	75	1	2	7	21	31	46	46	12	39	14	24	34	34	28	57	57
Somalia	17	58			32	32	32	14	22	22			27	27	15	38		34			29	29	29
South Africa	61	96		92	98	99	99	13	31		45	80	73	73	19	50			70	70	86	87	88
Sudan	61	96	100	66	86	78	78	13	43	7	45	69	64	64	19	50	51		50	50	75	69	70
Swaziland	83	100	41	87	87	87	44	29	7	44	44	42	54	54	37	31	31	43		52	62		
Tanzania	61	88	90	80	92	85	9	36	42	42	42	62	49	13	39	53	53	54	54	73	62		
Togo	100	49	70	100	74	85	80	80	5	10	58	38	36	36	17	16	38	54	63	54	54	51	52
Tunisia	92	93	100	100	94	99	99	17	17	31	89	31	60	82	49	35	60	70	99	99	82	82	93
Uganda	88	100	37	47	72	87	87	17	29	18	32	46	52	56	22	35	20	20	34	34	50	56	60
Zambia	70	86	76	64	88	90	90	22	16	41	27	48	36	40	37	42	58	58	43	43	64	55	58
Zimbabwe			95		100	100	98			32	80	77	74	72				84			85	83	81

NORTH & CENTRAL AMERICA & CARIBBEAN

Anguilla					60	60									60
Antigua and Barbuda				95	95						89	89		91	91

continues

DATA TABLE 3 *continued*

Fraction of Population with Access to Improved Drinking Water

NORTH & CENTRAL AMERICA & CARIBBEAN (*continued*)

Region and Country	URBAN									RURAL									TOTAL								
	1970	1975	1980	1985	1990	1994	2000	2002	2004	1970	1975	1980	1985	1990	1994	2000	2002	2004	1970	1975	1980	1985	1990	1994	2000	2002	2004
Aruba	100	100	100	100				100	100							100	100	100								100	100
Bahamas	95	100			98		99	98	100	12	13			75		86	86	86	65	65	100	100	90		96	97	97
Barbados	100	100	99	100	100		100	100	100	100	100	98	99	100		100	100	100	98	100	99	99	100		100	100	100
Belize			99	100	95	96	83	100	100			36	26	53	82	69	82	82			68	64	74	89	76	91	91
British Virgin Islands					100			98	98					100			98	98								98	100
Canada							100	100	100							99	99	99							100	100	100
Cayman Islands	100	100		98																							
Costa Rica	98	100	100	100	85		98	100	100	59	56	82	83		99	98	92	92	74	72	90	91		92	98	97	97
Cuba	82	96			100	96	99	95	95	15				91	85	82	78	78	56				98	93	95	91	91
Dominican Republic	72	88	85	85	82	74	83	98	97	14	27	34	33	45	67	70	85	91	37	55	60	62	67	71	79	93	95
Dominica							100	100	100							100	90	90							100	97	97
El Salvador	71	89	67	68	87	78	88	91	94	20	28	40	40	15	37	61	68	70	40	53	50	51	47	55	74	82	84
Grenada	100	100					97	97	97	47	77					93	93	93							94	95	95
Guadeloupe							94	98	98							94	93	93							94	98	98
Guatemala	88	85	90	72	92		97	99	99	12	14	18	14	43		88	92	92	38	39	46	37	62		92	95	95
Haiti		46	51	59	56	37	46	91	52		3	8	30	35	23	45	59	56		12	19	38	41	28	46	71	54
Honduras	99	99	93	56	85	81	97	99	95	10	13	40	45	48	53	82	82	81	34	41	59	49	64	65	90	90	87
Jamaica	100	100	55	99			81	98	98	48	79	46	93			59	87	88	62	86	51	96			71	93	93
Martinique																											
Mexico	71	70	90	99	94	91	94	97	100	29	49	40	47		62	63	72	87	54	62	73	83	69	83	86	91	97
Montserrat								100	100								100	100								100	100
Netherlands Antilles																											
Nicaragua	58	100	67	76		81	95	93	90	16	14	6	11		27	59	65	63	35	56	39	48		61	79	81	79
Panama	100	100	100				88	99	99	41	54	62	64			86	79	79	69	77	81	82		83	87	91	90
Puerto Rico																											
St Kitts							99	99	99								99	99								99	100
St Lucia							98	98	98								98	98								98	98
St Vincent									93									93									93
Trinidad/Tobago	100	79	100	100	100		92	92	92	95	100	93	95	88		88	88	88	96	93	97	98	96		86	91	91

	1	2	3	4	5	6	7	8	9	10	11	12	13	14	15	16	17	18
Turks/Caicos Islands						87			68					77		100	100	100
United States of America													100	100	100	100	100	100
United States Virgin Islands					100										100			100
SOUTH AMERICA																		
Argentina	69	76	61	63	85	97	98	12	26	17	17	30	56	66	54	56	79	96
Bolivia	92	81	69	75	93	95	95	2	6	10	13	55	33	34	36	43	79	85
Brazil	78	87	83	85	95	96	96	28	28	51	56	54	55	70	72	77	87	90
Chile	67	78	100	98	99	100	100	13	13	17	29	66	56	70	84	87	94	95
Colombia	88	86	93	87	98	99	99	28	33	73	76	73	63	64	86	86	91	93
Ecuador	76	67	79	63	81	92	97	7	8	20	31	51	34	36	50	57	71	94
Falkland Islands (Malvinas)					88	88	88											84
French Guiana	100	100	100	90	88	83	88	63	75	60	65	71	75	84	72	76	84	84
Guyana	100	100	53	61	98	83	83	5	5	9	8	91	11	13	21	28	94	83
Paraguay	22	25	39	68	95	100	99	8	15	18	17	58	35	47	50	55	79	86
Peru	58	72	68	73	87	87	89			24	24	66	51	47	55	60	77	83
Suriname		100	71		94	98	98		79	87	79	96	88		88	83	95	92
Uruguay	100	100	96	95	98	98	100	59	87	2	27	93	92	98	81	85	98	100
Venezuela	92	93	93	80	88	85	85	38		53	65	58	75	86	89	79	84	83
ASIA																		
Afghanistan	18	40	28	38	19	19	63	1	5	8	17	11	3	9	8	17	13	39
Armenia					99	99					80	80					92	92
Azerbaijan		50			95	95		95			59	59				77	77	77
Bahrain	100	100	100	100	100	100	100	94	100	100	100	100	99	100	99	100		
Bangladesh	13	22	26	24	82	82	82	47	61	89	97	72	45	56	39	46	81	74
Bhutan		50	60	75	86	86	86		5	30	54	60			7	32	62	62
Brunei Darus	100					100		95	95									
Cambodia					53	58	64			25	29	35					30	41
China		87		93	94	92	93	68	89	66	68	67		73		90	75	77
Cyprus	100	100	100	100	100	100	100	92	96	100	100	100	95	95	100	100	100	100
East Timor		94				73	77				51	56				58	52	58

(continued)

DATA TABLE 3 continued

Fraction of Population with Access to Improved Drinking Water

Region and Country	URBAN 1970	1975	1980	1985	1990	1994	2000	2002	2004	RURAL 1970	1975	1980	1985	1990	1994	2000	2002	2004	TOTAL 1970	1975	1980	1985	1990	1994	2000	2002	2004
ASIA (continued)																											
Gaza Strip								90	94									88									92
Georgia								90	96								61	67								76	82
Hong Kong					100							95		96									100				
India	60	80	77	76	86	85	92	96	95	6	18	31	50	69	79	86	82	83	17	31	42	56	73	81	88	86	86
Indonesia	10	41	35	43	35	78	91	89	87	1	4	19	36	33	54	65	69	69	3	11	23	38	34	62	76	78	77
Iran	68	76	82		100	89	99	98	99	11	30	50		75	77	89	83	84	35	51	66		89	83	95	93	94
Iraq	83	100		100	93		96	97	97	7	11	54	54	41		48	50	50	51			86	78	44	85	81	81
Israel							100	100	100							100	100	100							100	100	100
Japan							100	100	100							100	100	100							100	100	100
Jordan	98		100	100	100	100	100	91	99	59		65	88	97		84	91	91	77		86	96	99	89	96	91	97
Kazakhstan							98	96	97							82	72	73							91	86	86
Korea DPR							100	100	100							100	100	100							100	100	100
Korea Rep	84	95	86	90	100		97	97	97	38	33	61	48	76		71	71	71	58	66	75	75	93		92	92	92
Kuwait	60	100	86	97								100							51	89	87						
Kyrgyzstan							98	98	98							66	66	66							77	76	77
Laos	97	100	28		47	40	59	66	79	39	32	20		25	39		38	43	48	41	21		29	39		43	51
Lebanon						100	100	100	100						100	100	100	100						100	100	100	100
Macau																											
Malaysia	100	100	90	96	96		96	96	100	1	6	49	76	66	39	94	94	96	29	34	63	84	79	89	95		99
Maldives			11	58	77	98	100	99	98			3	12	68	86	100	78	76			2	21			100	84	83
Mongolia					100		77	87	87					58		30	30	30					82		60	62	62
Myanmar (Burma)	35	31	38	36	79	36	88	95	80	13	14	15	24	72	39	60	74	77	18	17	21	27	74	38	68	80	78
Nepal	53	85	83	70	66	66	85	93	96	2	5	7	25	34	41	80	82	89	2	8	11	28	38	44	81	84	90
Oman		100		90			41	81			48		49			30	72					53		63	39	79	
Pakistan	77	75	72	83	82	77	96	95	96	4	5	20	27	42	52	84	87	89	21	25	35	44	55	60	88	90	91
Philippines	67	82	49	49	93	93	92	90	87	20	31	43	54	72	77	80	77	82	36	50	45	52	81	85	87	85	85
Qatar	100	100	76		100		100	100	100	75	83	43				100	100	100	95	97	71				100	100	100
Saudi Arabia	100	97	92	100	100	100	100	97	97	37	56	87	88			64			49	64	90	94			95		
Singapore			100	100	100	100	100	100	100			100				100	100	100			100				100	100	100

Country	1	2	3	4	5	6	7	8	9	10	11	12	13	14	15	16	17	18	19	20	21	22	23	24	25	26
Sri Lanka	46	36	82	80	43	91	93	98	14	13	18	29	55	47	80	72	74	21	19	28	40	60	46	83	78	79
Syria	98	98			92	94	94	98	50		54		78	78	64	64	87	71		74		85		80	79	93
Tajikistan					93	93		92								47	48								58	59
Thailand	60	69		56		89	95	98	10	16	63	66	85		77	80	100	17	25	63	64			80	85	99
Turkey		95				82	96	98			62				84	87	93			76		83		83	93	96
Turkmenistan							93	93								54	54								71	72
United Arab Emirates		95			100			100			81						100			92						100
Uzbekistan					96	97									78	84					45	36	85	85	89	
Vietnam	45		70	47	53	81	93	99	32	32	39	33	33		50	67	80	4		31	40	36	36	56	73	85
Yemen A R		100	100			85	74	71	18	18	25	25			64	68	65	57		52				69	69	67
Yemen Dem	88	85				85			43	25	25				64									69		
OCEANIA																										
American Samoa				100	100	100	100	100						100	100	100	100						100	100		
Australia				100	100	100	100	100					88	100	100	100	100						100	100	100	100
Cook Islands	100		99	100	100	98	98	98					88	100	100	88	88		92				100	100	95	94
Fiji	78	94		96	43	43		43	15	56	66		69	100	51	51	51	37	69	77	80	100	47	47	47	47
French Polyneisa				100	100	100	100	100				18	100	100	100	100	100						100	100	100	100
Guam				100	100	100	100	100						100	100	100	100						100	100	100	100
Kiribati	93			91	82	77	77	77	25	25		63	63		25	53	53						47	47	64	65
Marshall Islands				100		80	82	82				45	45		95	95	96								85	87
Micronesia			100		100	95	95	95		100	100	38	38	100	94	94	94					100		94	94	94
Nauru																										
New Caledonia																										
New Zealand				100	100	100	100	100				100	100	100	100	100	100					100	100	100	100	100
Niue				0	100	100	100	100																		
Northern Mariana Islands	100			100	98	98		98					0			97	97								98	99
Palau	100			100	100	79	79	79				97	97		20	94	94	70	20	16	26	32	79	79	84	85
Papua New Guinea	44	30	55	95	94	88	88	88	72	19	10	15	20	17	32	32	32					28	42	42	39	39
Pitcairn	86	100																	43		62					
Samoa	97		100	100	95	91	90	90	23	94	94	77	77	100	88	88	87	17					99	99	88	88
Solomon Islands	96		82	82	94	94	94	94	45	45		58	58		65	65	65						71	71	70	70
Tokelau				100		89		88	100	23	25	100	100	100	100	89	88					100				88

continues

DATA TABLE 3 *continued*

Fraction of Population with Access to Improved Drinking Water

URBAN

Region and Country	1970	1975	1980	1985	1990	1994	2000	2002	2004
OCEANIA (*continued*)									
Tonga	100	100	86	99	92	100	100	100	100
Tuvalu				100		100	100	94	94
Vanuatu			65	95			63	85	86
Wallis and Futuna Islands									100
Western Samoa			97	75					
EUROPE									
Albania								99	99
Bosnia and Herzegovina									99
Bulgaria									100
Estonia									100
Hungary								100	100
Latvia									100
Netherlands								100	100
Republic of Moldova								97	97
Romania								91	91
Russian Federation								99	100
Serbia and Montenegro								99	99
Slovakia									100
Ukraine								100	99
Sources:	UNEP 1989, WRI 1988	UNEP 1989, WRI 1988	UNEP 1989, WRI 1988	UNEP 1989, WRI 1988	UNEP 1993	WHO 1996	WHO 2000	WHO/UNICEF 2004	WHO/UNICEF 2006

RURAL

Region and Country	1970	1975	1980	1985	1990	1994	2000	2002	2004
OCEANIA (*continued*)									
Tonga	53	71	70	99	98	100	100	100	100
Tuvalu				100		95	100	92	92
Vanuatu			53	54			94	52	52
Wallis and Futuna Islands									100
Western Samoa			94	67					
EUROPE									
Albania								95	94
Bosnia and Herzegovina									96
Bulgaria									97
Estonia									99
Hungary								98	98
Latvia									96
Netherlands								99	100
Republic of Moldova								88	88
Romania								16	16
Russian Federation								88	88
Serbia and Montenegro								86	86
Slovakia									99
Ukraine								94	91
Sources:	UNEP 1989, WRI 1988	UNEP 1989, WRI 1988	UNEP 1989, WRI 1988	UNEP 1989, WRI 1988	UNEP 1993	WHO 1996	WHO 2000	WHO/UNICEF 2004	WHO/UNICEF 2006

TOTAL

Region and Country	1970	1975	1980	1985	1990	1994	2000	2002	2004
OCEANIA (*continued*)									
Tonga	63	83	17	99	96	100	100	100	100
Tuvalu						98	100	93	100
Vanuatu				64			88	60	60
Wallis and Futuna Islands									100
Western Samoa				69					
EUROPE									
Albania								97	96
Bosnia and Herzegovina									97
Bulgaria									99
Estonia									100
Hungary								99	99
Latvia									99
Netherlands								100	100
Republic of Moldova								92	92
Romania								57	57
Russian Federation								96	97
Serbia and Montenegro								93	93
Slovakia									100
Ukraine								98	96
Sources:	UNEP 1989, WRI 1988	UNEP 1989, WRI 1988	UNEP 1989, WRI 1988	UNEP 1989, WRI 1988	Calculated from UNEP 1993	WHO 1996	WHO 2000	WHO/UNICEF 2004	WHO/UNICEF 2006

The UN considers all European countries, except those shown, to have 100 percent water supply and sanitation coverage.

Access to Sanitation by Country, 1970 to 2004

Description

Adequate sanitation is also a fundamental requirement for basic human well-being, and improving access is one of the Millennium Development Goals (MDGs). Data are given here for the percent of urban, rural, and total populations, by country, with access to sanitation services for 1970, 1975, 1980, 1985, 1990, 1994, 2000, 2002, and 2004—the most recent year for which data are available. The World Health Organization (WHO) collected these data over various periods. Most of the data presented were drawn from responses by national governments to WHO questionnaires. Participants in data collection include the JMP, the United Nations Children's Fund, and the Water Supply and Sanitation Collaborative Council, which has continued sector monitoring and aims to support and strengthen the monitoring efforts of individual countries. Countries used their own definitions of "rural" and "urban."

For all WHO Assessments since 2000, new definitions were provided for "improved" sanitation with allowance for acceptable local technologies. The forty largest countries in the developing world account for 90 percent of population. As a result, WHO spent extra effort to collect comprehensive data for these countries. The excreta disposal system was considered adequate if it was private or shared (but not public) and if it hygienically separated human excreta from human contact. The following technologies were included in the 2000 assessment as representing improved sanitation:

    Connection to a public sewer
    Connection to septic system
    Pour-flush latrine
    Simple pit latrine
    Ventilated improved pit latrine

In comparison, unimproved sanitation facilities refer to:

    Public or shared latrine
    Open pit latrine
    Bucket latrine

Limitations

As is the case with drinking water data, definitions for access to sanitation vary from country to country, and from year to year within the same country. Countries generally regard sanitation facilities that break the fecal-oral transmission route as adequate. In urban areas, adequate sanitation may be provided by connections to public sewers or by household systems such as pit privies, flush latrines, septic tanks, and communal toilets. In rural areas, pit privies, pour-flush latrines, septic tanks, and communal toilets are considered adequate. Direct comparisons between countries and across time within the same country are difficult and are additionally complicated by the fact that these data hide disparities between regions and socioeconomic classes.

WHO Assessments since 2000 have attempted to shift from gathering information from water providers only to include consumer-based information. The current approach uses household surveys to assess the actual use of facilities. Access to sanitation services, as reported by WHO, does not imply that the level of service is "adequate" or "safe." The assessment questionnaire did not include any methodology for discounting coverage figures to allow for intermittence or poor quality of the service provided. More details of the methods used, and their limitations, can be found at http://www.who.int/docstore/water_sanitation_health/Globassessment/GlobalTOC.htm.

SOURCES

United Nations Environment Programme (UNEP). 1989. *Environmental Data Report.* GEMS Monitoring and Assessment Research Centre, Oxford: Basil Blackwell.

United Nations Environment Programme (UNEP). 1993–94. *Environmental Data Report.* GEMS Monitoring and Assessment Research Centre in cooperation with the World Resources Institute and the UK Department of the Environment, Oxford: Basil Blackwell.

World Health Organization (WHO). 1996. *Water Supply and Sanitation Sector Monitoring Report: 1996 (Sector status as of 1994).* In collaboration with the Water Supply and Sanitation Collaborative Council and the United Nations Children's Fund, UNICEF, New York.

World Health Organization (WHO). 2000. *Global Water Supply and Sanitation Assessment 2000 Report.* http://www.who.int/docstore/water_sanitation_health/Globassessment/Global TOC.htm.

World Health Organization (WHO) and United Nations Children's Fund (UNICEF). 2004. *Meeting the MDG Drinking Water and Sanitation Target: A Mid-term Assessment of Progress.* http://www.who.int/water_sanitation_health/monitoring/jmp2004/en/index.html.

World Health Organization (WHO) and United Nations Children's Fund (UNICEF). 2006. *Meeting the MDG Drinking Water and Sanitation Target: The Urban and Rural Challenge of the Decade.* http://www.who.int/water_sanitation_health/monitoring/jmpfinal.pdf.

World Resources Institute (WRI). 1988. World Health Organization data, cited by the World Resources Institute, *World Resources 1988–89*, World Resources Institute and the International Institute for Environment and Development in collaboration with the United Nations Environment Programme, New York: Basic Books.

DATA TABLE 4 Access to Sanitation by Country, 1970 to 2004

Fraction of Population with Access to Improved Sanitation

Region and Country	URBAN									RURAL									TOTAL								
	1970	1975	1980	1985	1990	1994	2000	2002	2004	1970	1975	1980	1985	1990	1994	2000	2002	2004	1970	1975	1980	1985	1990	1994	2000	2002	2004
AFRICA	47	75	57	75			90	99	99	6	50		40			47	82	82	9	67		57			73	92	92
Algeria	13	100		80		34	70	56	56			15	16	20	8	30	16	16			20	19	21	16	44	30	31
Angola			40	29	25	34	46	56	56								16	16							23	32	33
Benin	83		48	58	60	54	46	58	59	1		4	20	35	6	6	12	11	14		16	33	45	20	23	32	33
Botswana				93	100		88	57	57				28	85			25	25				40	89		41	41	42
Burkina Faso	49	47	38	44		42	88	45	42			5	6		11	16	5	6			7	9		18	29	12	13
Burundi	96		40	84	64	60	79	47	47			35	56	16	50		35	35			35	58	18	51	36	36	36
Cameroon				100			99	63	58				1			85	33	43				43			92	48	51
Cape Verde			34	32		40	95	61	58			10	9		10	32	19	19			11	10		24	71	42	43
Central African Republic	64	100			45	73	43	47	47	96	100			46		23	12	12	72	100			46	31	31	27	27
Chad	7	9				73	81	30	24		1				7	13	0	4	1	1				21	29	8	9
Comoros							98	38	41							98	15	29							98	23	33
Congo	8	10					14	24	28	6	9					6	2	25	6	9					9	9	27
Congo, Democratic Republic	5	65			46	23	53	43	42	5	6		9	11	4	6	23	25	5	22		21		9	20	29	30
Cote D'Ivoire	23				81	59	99	55	46	8				100	51			29	5				92	54	40		37
Djibouti			43	78		77	99	55	88			20			100	50	27	50			39	64		90	91	50	82
Egypt					80	20	98	84	86			10		26	5	91	56	58					50	11	94	68	70
Equatorial Guinea					54	61	60	60	60					24	48	46	46	46					33	54	53	53	53
Eritrea							66	34	32							1	3	3							13	9	9
Ethiopia	67	56		96			58	19	44	8	8		96			6	4	7	14	14					15	6	13
Gabon							25	37	37							4	30	30							21	36	36
Gambia					100	83	41	72	72					27	23	35	46	46					44	37	37	53	53
Ghana	92	95	47	51	63	53	62	74	27	40	40	17	16	60	36	64	46	11	55	56	26	30	61	42	42	58	18
Guinea	70		54				94	25	31	2		1		0	17	41	6	11	13		11			70	58	13	18
Guinea-Bissau			21	29		32	88	57	57			13	18		17	34	23	23			15	21		20	47	34	35
Kenya	85	98	89			69	96	56	46	45	48	19			81	81	43	41	50	55	30			77	86	48	43
Lesotho	44	51	13	22		1	93	61	61	10	12	14	14	2	7	6	32	32	11	13	14	15		6	92	37	37
Liberia	100			6		38		49	49	9			2		2		7	7	19					18		26	27
Libya	100	100	100				97	97	97	54	69	72				30	96	96	67	79	88				97	97	97

continues

DATA TABLE 4 *continued*

Fraction of Population with Access to Improved Sanitation

Region and Country	URBAN									RURAL									TOTAL								
	1970	1975	1980	1985	1990	1994	2000	2002	2004	1970	1975	1980	1985	1990	1994	2000	2002	2004	1970	1975	1980	1985	1990	1994	2000	2002	2004
Madagascar	88		9	55		50	70	49	48		9				3	70	27	26						15	42	33	34
Malawi			100			70	96	66	62			81			51	98	42	61			83			53	77	46	61
Mali	63		79	90	81	58	93	59	59				3	10	21	100	38	39	8			19	27	31	69	45	46
Mauritania	100		5	8			44	64	49							19	9	8	7						33	42	34
Mauritius	51	63	100	100	100	100	100	100	95	99	100	90	86	100	100	99	99	94	77	82	94	92	100	100	99	99	94
Morocco	75			62	100	69	100	83	88	4			16		18	42	31	52	29			20		40	75	61	73
Mozambique				53		70	69	51	53				12		70	26	14	19							43	27	32
Namibia					24		96	66	50					11		17	14	13					15		41	30	25
Niger	10	30	36		71	71	79	43	43		1	3		4	4	5	4	4	1	3	7		17	15	20	12	13
Nigeria					80	61	85	48	53				5	11	21	45	30	36					35	36	63	38	44
Reunion																											
Rwanda	83	87	60	77	88		12	56	56	52	56	50	55	17		8	38	38	53	57	51	56	21		8	41	42
Sao Tome and Principe								32	32				15				20	20				15				24	25
Senegal			100	87	57	83	94	70	79			2		38	40	48	34	34			36		46	58	70	52	57
Seychelles								100	100								100	100									
Sierra Leone			31	60	55	17	23	53	53			6	10	31	8	31	30	30			12	24	39	11	28	39	39
Somalia		77		44				47	48		35		5				14	14		47		18				25	26
South Africa						79	99	86	79						12	73	44	46						46	86	67	65
Sudan	100	100	73	73		79	87	50	50	4	10				4	48	24	24	16	22				22	62	34	34
Swaziland		99	100	100		36		78	59		25		25		37		44	44		36	66	45		36	62	52	48
Tanzania		88		93			98	54	53		14		58			86	41	43		17					90	46	47
Togo	4	36	24	31		57	69	71	71	1	12	10	9		13	17	15	15	1	15	13	14		26	34	34	35
Tunisia	100		100	84		100		90	96	34			16		85		62	65	62			55		96	80	80	85
Uganda	84	82	32	32	32	75	96	53	54	76	95		30	60	55	72	39	41	76	94		30	57	57	75	41	43
Zambia	12	87	76	76		40	99	68	59	18	16		34		10	64	32	52	16	42		55		23	78	45	55
Zimbabwe					95		99	69	63				15	22		51	51	47					43		68	57	53
NORTH & CENTRAL AMERICA & CARIBBEAN																											
Anguilla																		99									99
Antigua and Barbuda								98	98								94	94								95	95

Aruba																											
Bahamas	100	100	88	100	98		93	100	100	13	13			2		94	100	100	66	65	88	100	63		93	100	100
Barbados	100	100		100	100		100	99	99	100	100			100		100	100	100	100	100			100		100	99	100
Belize			62	87	76	23	59	71	71			75	45	22	87	21	25	25			69	66	50	57	42	47	47
British Virgin Islands					100			100	100					100			100	100								100	100
Canada							100	100	100							99	99	99							100	100	100
Cayman Islands			94	96									94						96								
Costa Rica	66	94	99	99		85	98	89	89	43	93	84	89		99	95	97	97	52	93	91	95		92	96	92	92
Cuba	57	100			100	71	96	99	99					68	51	91	95	95					92	66	95	98	98
Dominican Republic	63	74	25	41	95	76	75	67	81	54	16	4	10	75	83	64	43	73	58	42	15	23	87	78	71	57	78
Dominica								86	86								75	75								83	84
El Salvador	66	71	48	82	85	78	88	78	77	18	17	26	43	38	59	78	40	39	37	39	35	58	59	68	83	63	62
Grenada							96	96	96							97	97	97							97	97	96
Guadeloupe							61		61							61		64							61		64
Guatemala			45	41	72		98	72	90	11	16	20	12	52		76	52	82			30	24	60		85	61	86
Haiti			42	42	44	42	50	52	57	43	1	10	13	17	16	16	23	14			19	21	25	24	28	34	30
Honduras	64	53	49	24	89	81	94	89	87	9	13	26	34	42	53	57	52	54	24	26	35	30	63	65	77	68	69
Jamaica	100	100	12	92			98	90	91	92	91	2	90			66	68	69	94	94	7	91			84	80	80
Martinique																											
Mexico			77	77	85	81	87	90	91	13	14	12	13		26	32	39	41			55	58		66	73	77	79
Montserrat								96	96								96	96								96	100
Netherlands Antilles																											
Nicaragua			34	35		34	96	78	56	8	24		16		27	68	51	34				27		31	84	66	47
Panama	87	78	83	99			87	89	89	69	76	59	61			94	51	51	78	77	71	81		86	99	72	73
Puerto Rico																											
St Kitts								95	96								96	96								96	95
St Lucia								89	89								89	89								89	89
St Vincent																	96	96									
Trinidad/Tobago					100			100	100					92			100	100					97		88	100	100
Turks/Caicos Islands	51	83	96	100				98	94	96	97	88	95				94	98	81	92	93	98				96	96
United States of America							100	100	100							100	100	100							100	100	100
United States Virgin Islands																											

continues

DATA TABLE 4 *continued*

Fraction of Population with Access to Improved Sanitation

Region and Country	URBAN									RURAL									TOTAL								
	1970	1975	1980	1985	1990	1994	2000	2002	2004	1970	1975	1980	1985	1990	1994	2000	2002	2004	1970	1975	1980	1985	1990	1994	2000	2002	2004
SOUTH AMERICA																											
Argentina	87	100	80	75			89		92	79	83	35	35			48		83	85	97		69			85		91
Bolivia	25		37	33	38	58	82	58	60	4	9	4	10	14	16	38	23	22	12		18	21	26	41	66	45	46
Brazil	85			86	84	55	85	83	83	24		1	1	32	3	40	35	37	58			63	71	44	77	75	75
Chile	33	36	100	100		82	98	96	95	10	11	10	4			93	64	62	29	32	83	84			97	92	91
Colombia	75	73	93	96	84	76	97	96	96	8	13	4	13	18	33	51	54	54	47	48	61		64	63	85	86	86
Ecuador			73	98	56	87	70	80	94		7	17	29	38	34	37	59	82			43	65	48	64	59	72	89
Falkland Islands																											
(Malvinas)																											
French Guiana							85		57							57		85							79		78
Guyana	95	99	73	100	97		97	86	86	92	94	80	79	81		81	60	60	93	96	78	86	86		87	70	70
Paraguay	16	28	95	89	31	62	95	94	94	6	10	80	83	60		95	58	61	6	10	86	85	46		95	78	80
Peru	52	57	57	67	76	62	90	72	74	16		0	12	20	10	40	33	32	36		36	49	59	44	76	62	63
Suriname			100	78			100	99	99			79	48			34	76	76			88	62			83	93	94
Uruguay	97	97	59	59	59		96	95	100	13	17	6	59			89	85	99	82	83	51	59			95	94	100
Venezuela			60	57		64	86	71	71	45		12	5	72	30	69	48	48			52	50		58	74	68	68
ASIA																											
Afghanistan	69	63		5	13	38	25	16	49	16	15			0	1	8	5	29	21	21				8	12	8	34
Armenia								96	96								61	61								84	83
Azerbaijan								73	73							36	36	36							55	54
Bahrain				100	100		100	100	100				100									100			100	100	100
Bangladesh	87	40	21	24	40	77	82	75	51	6	5	1	3	4	30	44	39	35	6	5	3	5	10	35	53	48	39
Bhutan					80	66	65	65	65					3	18	70	70	70					7	41	69	70	70
Brunei Darus																											
Cambodia			100				58	53	53			76				10	8	8							18	16	17
China					100	58	68	69	69					81	7	24	29	28					86	21	38	44	44
Cyprus	100	94	100	100	96		100	100	100	95	95	100	100	100		100	100	100	95	95	100	100	98		100	100	100
East Timor								65	66							30	30	33							33	33	36
Gaza Strip									78									61									73

Country	Data values (left to right)
Georgia	85, 90, 96, 96, 69, 91, 83, 94
Hong Kong	87, 60, 73, 58, 18, 22, 30, 33
India	50, 44, 79, 87, 71, 2, 3, 14, 38, 40, 18, 20, 15, 7, 9, 23, 29, 31, 51, 66, 52, 55
Indonesia	100, 79, 86, 86, 59, 30, 35, 37, 40, 74, 78, 78, 69, 37, 47, 67, 67, 51, 81, 67, 52
Iran	82, 96, 100, 89, 93, 95, 1, 35, 37, 31, 48, 47, 74, 47, 36, 72, 69, 79, 84, 79
Iraq	82, 75, 95, 11, 48, 48, 74, 80, 79
Israel	100, 100, 100, 100
Japan	100, 100, 100, 100, 100, 100
Jordan	94, 100, 94, 34, 100, 98, 98, 85, 87, 70, 95, 99, 99, 93, 93, 72
Kazakhstan	92, 100, 87, 98, 52, 52, 99, 99, 72, 72
Korea DPR	99, 58, 100, 60, 60, 99, 59, 59
Korea Rep	59, 100, 100, 67, 76, 50, 100, 100, 12, 4, 25, 64, 52, 63
Kuwait	100, 100, 100, 100, 100, 100
Kyrgyzstan	100, 75, 100, 51, 51, 100, 60, 59
Kyrgyzstan	
Laos	10, 13, 30, 84, 8, 2, 4, 3, 5, 12, 24, 24, 46, 30
Lebanon	100, 100, 100, 87, 61, 92, 14, 20, 100, 87, 99, 98, 98
Macau	100, 95, 94
Malaysia	100, 94, 94, 95, 43, 43, 55, 60, 94, 94, 41, 98, 93, 59, 60, 70, 75, 94, 56, 44, 94
Maldives	21, 60, 95, 100, 100, 1, 2, 4, 58, 42, 3, 13, 22, 56, 58, 59
Mongolia	100, 46, 75, 47, 2, 37, 37, 78, 30, 59, 59
Myanmar (Burma)	45, 38, 50, 42, 65, 96, 88, 13, 1, 13, 40, 39, 63, 72, 35, 33, 20, 22, 41, 46, 46, 73, 77
Nepal	14, 16, 34, 51, 75, 68, 62, 1, 3, 16, 20, 20, 30, 1, 1, 6, 20, 27, 27, 35
Oman	100, 88, 98, 97, 97, 5, 25, 61, 61, 12, 31, 76, 89, 92
Pakistan	12, 42, 51, 53, 94, 92, 2, 6, 12, 19, 42, 35, 41, 3, 6, 13, 19, 30, 61, 54, 59
Philippines	90, 76, 81, 83, 79, 92, 81, 80, 44, 67, 56, 63, 56, 67, 75, 61, 59, 57, 56, 75, 67, 83, 73, 72
Qatar	100, 100, 100, 100, 16, 100, 100, 85, 100, 100, 83, 100, 71, 82, 83, 100, 100
Saudi Arabia	67, 81, 100, 100, 100, 11, 35, 50, 33, 33, 21, 47, 70, 100, 100, 100
Singapore	80, 99, 99, 100, 100, 99, 80, 100, 100
Sri Lanka	76, 68, 65, 68, 33, 91, 80, 81, 61, 55, 63, 39, 45, 58, 80, 89, 64, 59, 67, 44, 52, 56, 83, 91, 90
Syria	74, 77, 98, 97, 28, 35, 81, 56, 50, 77, 90, 51
Tajikistan	71, 71, 47, 45, 53, 51
Thailand	65, 58, 64, 78, 97, 97, 98, 8, 36, 41, 46, 86, 96, 100, 99, 17, 40, 45, 52, 96, 99, 53, 99, 99
Turkey	56, 98, 94, 70, 62, 72, 91, 83, 88

227

continues

DATA TABLE 4 *continued*

Fraction of Population with Access to Improved Sanitation

Region and Country	URBAN									RURAL									TOTAL								
	1970	1975	1980	1985	1990	1994	2000	2002	2004	1970	1975	1980	1985	1990	1994	2000	2002	2004	1970	1975	1980	1985	1990	1994	2000	2002	2004
ASIA (*continued*)																											
Turkmenistan								77	77								50	50								62	62
United Arab Emirates			93					100	98			22					100	95			80					100	98
Uzbekistan							100	73	78							100	48	61							100	57	67
Vietnam	100				23	43	87	84	92		2	55		10	15	70	26	50	26				13	21	73	41	61
Yemen A R			60	83			99	76	86							31	14	28							45	30	43
Yemen Dem			70				99					15				31					35				45		
OCEANIA																											
American Samoa																											
Australia							100	100	100							100	100	100							100	100	100
Cook Islands			100	100	100		100	100	100			76	99	100		100	100	100				99			100	100	100
Fiji	100	100	85		91	100	75	99	87	87	93	60		65	85	12	98	55	91	96	70		75	92	43	98	72
French Polyneisa					98		99	99	97					95		97	97	99							98	98	98
Guam							99	99	98								98	99							99	98	99
Kiribati					91	100	54	59	59					49	100	44	22	22						100	48	39	40
Marshall Islands					100			93	93					45		59	59	58							82	82	82
Micronesia					99	100	61	61	61					46	100	14	14	14						100	28	28	28
Nauru																											
New Caledonia																											
New Zealand																											
Niue					0	100	100	100	100				100	100	100	100	100	100							100	100	100
Northern Mariana Islands					100			94	96					71		92	96	94								94	95
Palau					95		100	96	96					100		100	52	52							100	83	80
Papua New Guinea	100	96	96	99	57	82	92	67	67	5		3	35	11		80	41	41	14	18	15			22	82	45	44
Pitcairn																											
Samoa	100	100	86		100		95	100	100	80	99	83		92	17	100	100	100	84	99					99	100	100

Solomon Islands			80		73		98	98	98			21		2		18	18	18					13		34	31	31
Tokelau						100						41			100		74	78						100			78
Tonga	100	100	97	99	88	100		98	98	100	100	94	40	78	100		96	96	100	100	19	52	82	100		97	96
Tuvalu			100	81		90	100	92	93			80	73		85	100	83	84						87	100	88	90
Vanuatu			95	86			100	78	78			68	25			100	42	42				40			100	50	50
Wallis and Futuna Islands									80																		80
Western Samoa			86	88								83	83									84					
EUROPE																											
Albania								99	99								81	84								89	91
Belarus									93									61									84
Bosnia and Herzegovina									99									92									95
Bulgaria									100									96									99
Czech Republic									99									97									98
Estonia									97									96									97
Hungary								100	100								85	85								95	95
Latvia									82									71									78
Republic of Moldova								86	86								52	52								68	68
Romania								86	89								10									51	
Russian Federation								93	93								70	70								87	87
Serbia and Montenegro								97	97								77	77								87	87
Slovakia									100									98									99
Ukraine								100	98								97	93								99	96
Sources:	UNEP 1989, WRI 1988	UNEP 1989, WRI 1988	UNEP 1989, WRI 1988	UNEP 1989, WRI 1988	UNEP 1993	WHO 1996	WHO 2000	WHO/UNICEF 2004	WHO/UNICEF 2006	UNEP 1989, WRI 1988	UNEP 1989, WRI 1988	UNEP 1989, WRI 1988	UNEP 1989, WRI 1988	UNEP 1993	WHO 1996	WHO 2000	WHO/UNICEF 2004	WHO/UNICEF 2006	UNEP 1989, WRI 1988	UNEP 1989, WRI 1988	UNEP 1989, WRI 1988	UNEP 1989, WRI 1988	Calculated from UNEP 1993	WHO 1996	WHO 2000	WHO/UNICEF 2004	WHO/UNICEF 2006

MDG Progress on Access to Safe Drinking Water by Region

Description

The Millennium Development Goals (MDGs)—adopted by the United Nations in 2000—established a set of targets for improving the lives of the world's poor, ranging from eradicating extreme hunger to reducing child mortality and ensuring environmental sustainability. These targets, agreed to by all countries and leading development institutions throughout the world, consist of eight goals and 21 targets. See Chapter 4 for a fuller description of the MDGs.

Access to safe drink water is a basic human right, but this basic right is not being met universally. In many parts of the world, particularly sub-Saharan Africa and Oceania, a lack of clean water adversely affects human health and development. Using 1990 as a baseline, goal 7 of the MDGs seeks to reduce by half the proportion of people without sustainable access to safe drinking water by 2015.

At the global level, we are on track to meeting the target for improving access to safe drinking water. But some areas are performing better than others, highlighting a growth in regional disparities in access to safe drinking water. Europe, Latin America, the Caribbean, and much of Asia have met or are on track to meet the established targets. But in Sub-Saharan Africa and in many rural areas, there has been no progress or conditions have worsened. The global community must intensify efforts in these regions if they hope to achieve the established 2015 targets.

Limitations

These data give a good picture of the current lack of access to improved water and sanitation services, but comparison from different assessments should be done with extreme care, or not at all, because of changing definitions.

Country-reported data may reflect national definitions of "improved," unlike survey data, which were standardized as much as possible. For example, in many African countries the population "without access" to improved sanitation means people with no access to any sanitary facility. In Latin America and the Caribbean, however, it is more likely that those "without access" in fact have a sanitary facility, but the facility is deemed unsatisfactory by local or national authorities. Low coverage figures for Latin

America and the Caribbean may in part be a reflection of the comparatively narrow definitions used within that region.

Changes in the source of data also complicate comparisons over time. Prior to 2000, for example, data collected by WHO was provider-based and was collected from service providers, such as utilities, ministries, and water agencies. The data shown here, however, are sometimes user-based and was collected from household surveys and censuses. User-based data are more likely to include improvements installed by households or local communities and gives a more complete picture of water supply and sanitation coverage.

SOURCES

United Nations. 2007. *The Millennium Development Goals Report.* http://mdgs.un.org/unsd/mdg/Resources/Static/Data/2007%20Stat%20Annex%20current%20indicators.pdf.

United Nations. 2007. *MDG Progress Chart 2007.* http://mdgs.un.org/unsd/mdg/Resources/Static/Products/Progress2007/MDG_Report_2007_Progress_Chart_en.pdf.

DATA TABLE 5 MDG Progress on Access to Safe Drinking Water by Region

	1990			2004			2015 Target			On Target?
	Urban	Rural	Total	Urban	Rural	Total	Urban	Rural	Total	
Northern Africa	95	82	89	96	86	91	98	91	95	On Target
Sub-Saharan Africa	82	36	49	80	42	56	91	68	75	No progress or deteriorioration
Latin America and the Caribbean	93	60	83	96	73	91	97	80	92	Target Met or Close to Being Met
Eastern Asia	99	59	71	93	67	78	100	80	86	On Target
Southern Asia	90	66	72	94	81	85	95	83	86	Target Met or Close to Being Met
South-Eastern Asia	93	68	76	89	77	82	97	84	88	On Target
Western Asia	94	70	85	97	79	91	97	85	93	Target Met or Close to Being Met
Oceania	92	39	51	80	40	51	96	70	76	No progress or deteriorioration
Commonwealth of Independent States*	97	84	92	99	80	92	99	92	96	Target Nearly Met in Europe but No Progress or Deterioration in Asia

*Commonwealth of Independent States comprises Belarus, Republic of Moldova, Russian Federation and Ukraine in Europe, and Armenia, Azerbaijan, Georgia, Kazakhstan, Kyrgyzstan, Tajikistan, Turkmenistan, and Uzbekistan, in Asia

MDG Progress on Access to Sanitation by Region

Description

The Millennium Development Goals (MDGs)—adopted by the United Nations in 2000—established a set of targets for improving the lives of the world's poor, ranging from eradicating extreme hunger to reducing child mortality and ensuring environmental sustainability. These targets, agreed to by all countries and leading development institutions throughout the world, consist of eight goals and 21 targets. See Chapter 4 for a fuller description of the MDGs.

Adequate sanitation is also a fundamental requirement for basic human well-being, but like access to safe drinking water, this basic right is not universal. In many parts of the world, particularly the poor in rural and peri-urban areas, a lack of basin sanitation adversely affects human health and development. Using 1990 as a baseline, goal 7 of the MDGs seeks to reduce by half the proportion of people without sustainable access to sanitation by 2015.

Meeting the sanitation targets has proven to be more challenging than meeting the water target. While an estimated 2.6 billion people lacked access to basic sanitation in 1990, experts predict that 2.1 billion will still lack access to this basic right by 2015 (Ki Moon 2008). Some areas are performing better than others, highlighting a growing regional disparity in access to sanitation. Europe, Latin America, the Caribbean, and much of Asia are on track to meet the established targets. But in Sub-Saharan Africa, Southern Asia, Oceania, and many former Soviet Union countries, there has been no progress or conditions have worsened.

Because of a lack of progress, the United Nations has declared 2008 the International Year of Sanitation with a goal to raise awareness about the issue and improve progress toward achieving the 2015 target. One element of this effort is to secure sufficient financial resources. In March 2008, the Water Supply and Sanitation Collaborative Council, established by a United Nations mandate in 1990, launched the Global Sanitation Fund to provide an additional funding mechanism for sanitation and hygiene projects. Emphasis will be placed on existing, proven technologies rather than new research and approaches.

Limitations

These data give a good picture of the current lack of access to improved water and sanitation services, but comparison from different assessments should be done with extreme care, or not at all, because of changing definitions.

Country-reported data may reflect national definitions of "improved," unlike survey data, which were standardized as much as possible. For example, in many African countries the population "without access" to improved sanitation means people with no access to any sanitary facility. In Latin America and the Caribbean, however, it is more likely that those "without access" in fact have a sanitary facility, but the facility is deemed unsatisfactory by local or national authorities. Low coverage figures for Latin America and the Caribbean may in part be a reflection of the comparatively narrow definitions used within that region.

Changes in the source of data also complicate comparisons over time. Prior to 2000, for example, data collected by WHO was provider-based and was collected from service providers, such as utilities, ministries, and water agencies. The data shown here, however, are sometimes user-based and was collected from household surveys and censuses. User-based data are more likely to include improvements installed by households or local communities and gives a more complete picture of water supply and sanitation coverage.

SOURCES

Ki-Moon, B. 2008. Secretary-General, In Message for World Water Day. March 5.

United Nations. 2007. *The Millennium Development Goals Report.* http://mdgs.un.org/unsd/mdg/Resources/Static/Data/2007%20Stat%20Annex%20current%20indicators.pdf.

United Nations. 2007. *MDG Progress Chart 2007.* http://mdgs.un.org/unsd/mdg/Resources/Static/Products/Progress2007/MDG_Report_2007_Progress_Chart_en.pdf.

DATA TABLE 6 MDG Progress on Access to Sanitation by Region

	1990			2004			2015 Target			On Target?
	Urban	Rural	Total	Urban	Rural	Total	Urban	Rural	Total	Total
Northern Africa	84	47	65	91	62	77	92	74	83	On Target
Sub-Saharan Africa	52	24	32	53	28	37	76	62	66	No progress or deterioriation
Latin America and the Caribbean	81	36	68	86	49	77	91	68	84	On Target
Eastern Asia	64	7	24	69	28	45	82	54	62	On Target
Southern Asia	54	8	20	63	27	38	77	54	60	No progress or deterioriation
South-Eastern Asia	70	40	49	81	56	67	85	70	75	On Target
Western Asia	97	55	81	96	59	84	99	78	91	On Target
Oceania	80	46	54	80	43	53	90	73	77	No progress or deterioriation
Commonwealth of Independent States*	92	63	82	92	67	83	96	82	91	On Target in Europe, No Progress or Deterioration in Asia

*Commonwealth of Independent States comprises Belarus, Republic of Moldova, Russian Federation and Ukraine in Europe, and Armenia, Azerbaijan, Georgia, Kazakhstan, Kyrgyzstan, Tajikistan, Turkmenistan, and Uzbekistan, in Asia

United States Dams and Dam Safety Data, 2006

Description

Data are provided here on the number of dams in the United States, by State, as of 2006. Data are also provided on the number of dams in each state under state regulatory control, and the condition of those dams, categorized as "high hazard" or "significant hazard." State budgets for dam safety are also shown, with estimates of staff levels involved in monitoring or regulating dam safety. As these data show, many states have limited safety budgets and small staffs responsible for the oversight of the safety of thousands of dams. Many dams have already been identified as at high risk of failure, or as deficient in safety.

Limitations

Obtaining adequate data on dam safety remains a challenge and there are too few resources devoted to both ensuring the safety of dams and to evaluating dam conditions. These data should therefore be considered as offering only a rough glimpse of overall dam safety problems.

SOURCES

American Rivers and the Association of State Dam Safety Officials. 2006.
http://www.americanrivers.org/site/DocServer/State_by_State_Dam_Safety_Stats_2006.pdf?do
   cID=6581

DATA TABLE 7 United States Dams and Dam Safety Data, 2006

State	Total Dams in National Inventory	Dams Under State Regulation		State-Determined Deficient Dams			State Dam Safety Budget ($)	Staff Dedicated to Dam Safety Regulation	
		Total	HH	Total	HH	SH		Total FTEs	Dams Per FTE
Alabama	2,218	NA	NA	NA	NA	NA	0	0	NA
Alaska	100	81	17	29	6	6	100,500	1	81
Arizona	328	251	94	26	33	7	711,028	7	36
Arkansas	1,208	403	102	26	19	1	282,018	3	134
California	1,495	1,273	341	80	10	18	9,190,000	58	22
Colorado	1,808	1,928	345	27	0	2	1,692,300	14	138
Connecticut	723	1,187	226	13	0	6	490,000	7	170
Delaware	61	37	9	32	3	0	470,000	1	37
Florida	853	805	72	NR	7	28	20,878,995	77	10
Georgia	4,814	3,874	450	156	156	0	727,009	11	352
Hawaii	132	136	95	26	30	7	246,638	2	68
Idaho	407	569	107	26	4	6	249,294	8	71
Illinois	1,462	1,485	187	80	NR	NR	306,000	5	297
Indiana	1,047	993	241	27	76	154	425,000	5	199
Iowa	3,340	3,325	83	13	9	10	57,000	2	1,663
Kansas	5,707	6,031	194	32	11	6	557,104	10	603
Kentucky	1,057	1,060	177	NR	26	35	1,550,420	14	76
Louisiana	554	540	28	27	15	6	480,316	8	68
Maine[1]	337	831	25	26	3	10	36,914	1.5	554
Maryland	319	382	68	32	14	10	482,668	6	64
Massachusetts[2]	1,624	2,977	296	40	22	18	500,000	4.0	744
Michigan	985	1,034	84	24	5	5	255,400	3	345
Minnesota	1,030	1,151	23	77	6	19	305,000	3	384
Mississippi	3,433	3,698	258	32	28	0	62,079	4	925
Missouri	5,206	653	455	23	27	1	261,779	5	131
Montana	3,256	2,884	102	23	11	6	399,937	4	721
Nebraska	2,284	2,288	121	NR	NR	NR	326,145	6	381

continues

DATA TABLE 7 *continued*

State	Total Dams in National Inventory	Dams Under State Regulation		State-Determined Deficient Dams			State Dam Safety Budget ($)	Staff Dedicated to Dam Safety Regulation	
		Total	HH	Total	HH	SH		Total FTEs	Dams Per FTE
Nevada	461	672	157	26	4	2	197,304	2	336
New Hampshire	629	840	90	49	4	17	717,282	8	105
New Jersey	820	1,715	213	191	46	116	1,254,000	20	86
New Mexico	500	396	177	126	70	28	484,411	6	66
New York	1,971	5,060	386	NR	NR	NR	1,006,732	11	460
North Carolina	2,892	4,502	1,025	143	93	28	973,886	16	281
North Dakota	838	1,150	29	21	4	12	220,000	5	230
Ohio	1,587	1,698	442	825	170	285	1,483,944	14	121
Oklahoma	4,701	4,460	187	4	4	0	395,336	3	1,487
Oregon	896	1,204	122	5	4	1	212,400	2	602
Pennsylvania	1,517	3,177	789	369	215	30	2,211,046	25	127
Puerto Rico	35	35	35	NR	NR	NR	440,000	6	6
Rhode Island	181	671	17	1	0	0	113,976	1	671
South Carolina	2,419	2,317	153	4	2	1	NR	3	772
South Dakota	2,503	2,349	47	72	11	7	150,000	2	1,175
Tennessee	1,168	656	149	8	4	2	352,822	8	82
Texas	6,975	7,202	837	109	101	6	350,000	7	1,029
Utah	858	667	189	NR	NR	NR	666,200	6	111
Vermont	357	568	57	6	1	4	300,000	2	284
Virginia	1,640	1,604	146	112	34	34	1,247,124	5	321
Washington	745	950	145	30	15	13	938,952	8	119
West Virginia	558	341	245	33	30	3	465,773	6	57
Wisconsin	1,140	3,749	211	2	1	0	537,500	6	625
Wyoming	1,468	1,445	79	2	NR	NR	160,365	5	289
Total	82,647	87,304	10,127	3,040	1,334	950	55,922,597	447	354

NR = No Response
1: Data shown is for 2005.
2: Data shown is for 2004.
HH: High hazard
SH: Significant hazard
FTE: Full-time equivalent (1 FTE is the equivalent of one person, full time.)

Dams Removed or Decommissioned in the United States, 1912 to Present

Description

The list compiles information on dams removed or decommissioned from rivers in the United States since 1912. More than 680 dams are listed here, from all regions of the country. The name of the dam, the river, the state, information on dam height and length, and reasons for removal are provided, when available. Blank spaces mean no information was available for that item. The data in this table were collected by three non-governmental environmental organizations: American Rivers (www.american-rivers.org), Friends of the Earth (www.foe.org), and Trout Unlimited (www.tu.org). Original data sources included state dam safety offices, federal agencies, river conservation and fishing organizations, dam owners, media reports, and academic institutions.

When information was available about the reason for a dam's removal, it was included here in one of six broad categories. Many of these categories overlap and few dams are removed for a single reason. The categories are described as follows:

Ecology: dam was removed to restore fish and wildlife habitat; to provide fish passage; to improve water quality.

Economics: maintenance of dam was too costly; removal was cheaper than repair; dam was no longer used; dam was in deteriorating condition.

Failure: dam failed; dam was damaged in flooding.

NPS: deactivation of dams on or having an impact on National Park Service lands.

Recreation: dam was removed to increase recreational opportunities.

Safety: dam was deemed unsafe; owner no longer wanted liability associated with the dam.

Unauthorized dam: dam was built without a needed permit; dam was built improperly.

Limitations

Until relatively recently, no effort was made to record dam removals. Little information is available on the history of dam removal and no comprehensive review of dam removal experience exists. This list is among the first efforts to pull together records from widely divergent sources. As a result, this list should only be considered prelimi-nary, not comprehensive, and the actual number of dams removed is likely to be much higher. Many smaller dam removals (e.g., less than 2 meters high) are often not docu-mented at all. Many states and federal agencies have not kept thorough records of dams under their jurisdiction that have been removed.

Readers with information on other dams removed, or with corrections or additions to the list, are encouraged to contact one of the organizations listed above. Updates will regularly be made available at www.worldwater.org and www.americanrivers.org.

SOURCES

American Rivers, Friends of the Earth, and Trout Unlimited. 1999. Dam Removal Success Stories: Restoring Rivers through Selective Removal of Dams that Don't Make Sense. December. Special thanks to Margaret Bowman and Elizabeth Maclin.
American Rivers. 2007. Dams Slated for Removal in 2007 and Dams Removed From 1999–2006.

Data Table 8 Dams Removed or Decommissioned in the United States, 1912 to Present

State	River	Project Name	Removed	H (m)	L (m)	Reason
AK	Allison Creek	Unnamed Dam	2004	2	9	
AK	Chatanika River	Davidson Ditch Diversion Dam	2002			
AK	Switzer Creek (trib.)	Switzer One Dam	1988	5		S
AK	Switzer Creek (trib.)	Switzer Two Dam	1988	5		S
AL	Cahaba River	Marvel Slab Dam	2004	2	64	
AR		Hot Springs Park Ricks Lower #1 Dam	1986	3		NPS
AR		Winton Spring Dam		1		NPS
AR	Coop Creek	Marsfield Dam		6		
AR	Crow Creek	Lake St. Francis Dam	1989	14		U
AZ	Canada del Oro	Golder Dam	1980			S
AZ	Walsh Canyon	Concrete Dam	1982	12		S
AZ	Walsh Canyon	Perrin Dam	1980	10		S
CA		Arco Pond Dam		3		NPS
CA		Bear Valley Dam	1982	5		NPS
CA		C-Line Dam #1	1993	17		NPS
CA		Hagmaier North Dam		9		NPS
CA		Happy Isles Dam	1987	2		NPS
CA		John Muir #1 Dam				NPS
CA		Lower Murphy Dam		2		NPS
CA		Rogers Dam	1983	12		NPS
CA		Upper Murphy Dam		8		NPS
CA	Alameda Creek	Niles Dam	2006			
CA	Alameda Creek	Sunol Dam	2006			
CA	Alameda Creek	Swim Dam #1	2001			
CA	Alameda Creek	Swim Dam #2	2001			
CA	Beaver Creek	Three C. Picket Dam	1949			

continues

241

DATA TABLE 8 *(continued)*

State	River	Project Name	Removed	H (m)	L (m)	Reason
CA	Big Creek	Big Creek Mfg. Dam		4		
CA	Brandy Creek	A-Frame Dam	2003	9	30	E
CA	Butte Creek	McGowan Dam	1998	2		E
CA	Butte Creek	McPherrin Dam	1998	4		E
CA	Butte Creek	Point Four Dam	1993	2		
CA	Butte Creek	Western Canal East Channel Dam	1998	3		E
CA	Butte Creek	Western Canal Main Dam	1998	3		E
CA	Canyon Creek	Henry Danninbrink Dam	1927			
CA	Canyon Creek	Red Hill Mining Do. Dam	1951	9		
CA	Clear Creek	McCormick-Saeltzer Dam	2000	5	18	
CA	Cold Creek	Lake Christopher Dam	1994	3	122	
CA	Crocker Creek	Crocker Creek Dam	2002	9	24	
CA	East Panther Creek	East Panther Creek Dam	2003			
CA	Ferrari Creek	Unnamed Dam	2002			
CA	Guadalupe River	unnamed small dam #1	1998			
CA	Guadalupe River	unnamed small dam #2	1998			
CA	Hayfork Creek	Hessellwood Dam	1925	3		E
CA	Hayfork Creek	Russell (Hinkley) Dam	1922	3		E
CA	Haypress Pond	Haypress Pond Dam	2003	6		
CA	Horse Creek	Big Nugget Mine Dam	1949	4	12	
CA	Horse Creek	Horse Creek Dam	2006	4		E
CA	Indian Creek	D.B. Fields Dam	1947	2		
CA	Indian Creek	D.B. Fields/Johnson Dam	1946			
CA	Indian Creek	Minnie Reeves Dam		6		
CA	Kidder Creek	Altoona Dam	1947	4	18	
CA	Los Angeles River (trib.)	North Debris Dam	2002	6		
CA	Lost Man Creek	Upper Dam	1989	2	17	
CA	Mad River	Sweasey Dam	1970	17		

State	Waterway	Dam/Structure	Year			
CA	Merced River	Cascade Diversion Dam	2003	5		
CA	Monkey Creek	Trout Haven Dam			56	E
CA	Murphy Creek	Unnamed Dam	2003	4		
CA	Redding Creek	Clarissa V. Mining Dam	1950	6		
CA	Rock Creek	Rock Creek Dam	1985	4	19	
CA	Rush Creek	Anderline Dam	1936	6		
CA	Russian River	Mumford Dam	2003	0		
CA	Salmon River	Bennett-Smith Dam	1950	3		
CA	Salmon River	Bonally Mining Co. Dam	1946	3	54	
CA	Salt Creek	Salt Creek Dam		3		
CA	Scott River	Barton Dam	1950	4	8	
CA	Solstice Creek	Unnamed Arizona Crossing	2002	2		
CA	Soquel Creek	Tucker Road Ford	2006			
CA	Swillup Creek	Moser Dam	1949			
CA	Trancas Canyon (trib.)	Trancas Debris Dam	2002	5		
CA	Trinity River	Lone Jack Dam		7		
CA	Trinity River	North Fork Placers Dam	1950	5		
CA	Trinity River	Quinn Dam	1951	4		
CA	Trinity River	Todd Dam	1949	4		
CA	Trinity River	Trinity Cty. Water & Power Co. Dam	1946	3		
CA	Ward Creek	Unnamed	2006			$
CA	West Panther Creek	West Panther Dam	2003	5		
CA	White's Gulch	Smith Dam	1949	2	8	
CA	Wildcat Creek	Unnamed dam #1	1992	2		E
CA	Wildcat Creek	Unnamed dam #2	1992	2		E
CA	York Creek	York Creek Diversion Structure	2004			
CO		Glacier #1 Dam	1985	3		NPS
CO		No Name #15 Dam		5		NPS
CO		No Name #17 Dam		5		NPS
CO		No Name #21 Dam	1990			NPS

continues

DATA TABLE 8 *(continued)*

State	River	Project Name	Removed	H (m)	L (m)	Reason
CO		No Name #22 Dam		5		NPS
CO		No Name #8 Dam	1990	4		NPS
CO	Cony Creek	Pear Lake Dam	1988	9		NPS
CO	Ouzel Creek	Blue Bird Dam	1990	17	61	NPS;S
CO	Platt River	Unnamed Dam #1	2002			
CO	Platt River	Unnamed Dam #2	2002			
CO	Sand Beach Creek	Sand Beach Dam	1988	8		NPS
CT	Bigelow Creek (trib.)	Little Pond Dam	1994	3		U
CT	Blackwelll Brook (trib.)	Paradise Lake Dam	1991	2		$
CT	Bradley Brook	unnamed dam	1993	3		S
CT	Cedar Swamp Brook	Lower Pond Dam	1991	4		$
CT	Eight Mile River	Pizzini Dam	2005	1		E
CT	Indian River	Indian Lake Dam	1994	4		$
CT	Mad River	John Dee's Dam		5	14	
CT	Mad River (trib.)	Frost Road Pond Dam	1983	2		S
CT	Mill Brook	Sprucedale Water Dam	1980	3		
CT	Muddy Brook	Muddy Pond Dam	1992	2		S
CT	Naugatuck River	Anaconda Dam	1999	3	101	E
CT	Naugatuck River	Chase Brass Dam	2004	1	30	
CT	Naugatuck River	Freight Street Dam	1999	1	48	E
CT	Naugatuck River	Platts Mill Dam	1999	3	70	E
CT	Naugatuck River	Union City Dam	1999	5	61	E
CT	Qunnipiac River (trib.)	Woodings Pond Dam	1971	5		F
CT	Shetucket River	Baltic Mills Dam	1938	8		
CT	Wharton Brook	Simpson's Pond Dam	1995	2		$
DC	Rock Creek	Ford Dam #3	1991			
DC	Rock Creek	Millrace Dam		5		NPS

	Stream	Dam	Year			
DC	Rock Creek	Unnamed Ford	2004	1		
DC	Rock Creek	Unnamed Ford	2003	1		
FL	Chipola River	Pace's Dike Dam	1991	2		NPS
FL		Dead Lakes Dam	1987	5	240	E
FL	Kissimmee River	Dam and Lock	2000			
FL	Withlacoochee River	Wysong Dam	1988	1		$
GA	Wahoo Creek	Harrilton Mill Lake Dam				
IA		Hopkinton Dam	2004			
ID	Bear River	Cove Dam	2006			
ID	Clearwater River	Grangeville Dam	1963	17	134	E
ID	Clearwater River	Lewiston Dam	1973	14	323	E
ID	Colburn Creek	Colburn Mill Pond Dam	1999	4	11	E
ID	Dip Creek	Dip Creek Dam				
ID	Elkhorn Gulch	Lane Dam				
ID	Garden Creek	Buster Lake Dam				
ID	John Day Creek (trib.)	Kshmitter Dam	1988			U
ID	Lake Fork Creek	Malony Lake Dam	1986			E
ID	Little Timber Creek	Timber Creek Dam	1970			
ID	Packsaddle Creek	Packsaddle Dam				
ID	Salmon River	Sunbeam Dam	1931			
ID	Skein Lake	Skein Lake Dam	1980			
ID	Soldier Creek	Kunkel Dam	1994			E
IL		Woodhaven North Impoundment Dam		4		
IL		Woodhaven South Impoundment Dam		3		
IL	Brewster Creek	YWCA Dam	2003			
IL	Brush Creek (trib.)	Amax Delta Basin 31 Dam		3		
IL	Cypress Ditch (trib.)	Peabody #1A Dam		7		
IL	Cypress Ditch (trib.)	Peabocy #5 Dam		13		
IL	Delta Creek	Lake Marion Dam				

continues

DATA TABLE 8 *(continued)*

State	River	Project Name	Removed	H (m)	L (m)	Reason
IL	Ewing Creek (trib.)	Old Ben Dam		9		
IL	Fox River	South Batavia Dam	2005	2	213	
IL	Little Muddy River (trib.)	Consol/Burning Star 5/20 Dam		5		
IL	Mississippi River	Mississippi River Lock & Dam #26		30		
IL	Mississippi River (trib.)	Turkey Bluff Dam		13		
IL	Negro Creek (trib.)	Lake Adelpha Dam		5		
IL	Sangamon River (trib.)	Faries Park Dredge Disposal Dam		9		
IL	Sevenmile Branch	Olsens Lake Dam		5		
IL	Tributary to Sugar Creek	Springfield Dam		8		
IL	Waubonsee Creek	Stone Gate Dam	1999	1	30	E; $; F
IL	Wolf Branch (trib.)	Garden Forest Pond Dam				
IL	Wood River (trib.)	Paradise Lake Dam		6		
IN		Pinhook Dam		5		NPS
KS		Chapman Lake Dam		12		
KS		City of Wellington Dam		11		
KS		Edwin K. Simpson Dam		8		
KS		Kansas Gas & Electric Dam				
KS		Lake Bluestem Dam		21		
KS		Moline Middle City Lake Dam		6		
KS		Mott Dam		6		
KS		Robert Yonally Dam				
KS		Soldier Lake Dam		4		
KS		Wyandotte County Dam		42		
KY	Great Onyx Pond	No Name #1 Dam	1982	2		NPS
KY	Great Onyx Pond	No Name #2 Dam	1982	2		NPS
KY	Little Flat Creek	Sharpsburgh Reservoir Dam	1985	11		
KY	Pond Creek (trib.)	Ebenezer Lake Dam	1985	5		

State	River	Dam		Year		
KY	Pond River	West Fork Pond River #2 Dam	5			
LA	Bayou Dorcheat	Shirley Willis Pond Dam	3			
LA	Bayou Dupont (trib.)	Bayou Dupont #13 Dam	7			
LA	Dry Pong Creek	Kisathie Lake Dam	8			
LA	Pond Branch	Castor Lake Dam	3			
MA	Galloway Brook	Upper Cook's Canyon Dam	3	2006	26	E,$
MA	Housatonic River	Old Berkshire Mill Dam		2000		E
MA	Red Brook	Robbins Dike Dam	2	2006	30	
MA	Town Brook	Billington Street Dam		2002		
MA	Yokum Brook	Ballou Dam	3	2006	15	E,$
MA	Yokum Brook	Silk Mill Dam	5	2003		E
MD	Bacon Ridge Branch	Bacon Ridge Branch Weir		1991		
MD	Deep Run	Deep Run Dam		1989		
MD	Dorsey Run	Railroad Trestle Dam		1994		
MD	Horsepen Branch	Horsepen Branch Dam		1995		
MD	Little Elk Creek	Railroad Bridge at Elkton Dam		1992		
MD	Octoraro Creek	Octoraro Rubble Dam		2005		E
MD	Potomoc River (trib.)	Polly Pond Dam	8	2002		
MD	Stony Run	Stony Run Dam		1990		
MD	Western Branch	Route 214 Dam		1998		
ME	East Machias River	East Machias Dam		1983;2000		E
ME	Kennebec River	Edwards Dam	7	1999	279	E
ME	Machias River	Canaan Lake Outlet Dam		1999		
ME	Penobscot River	Bangor Dam		1995		
ME	Pleasant River	Brownville Dam	4	1999	91	
ME	Pleasant River	Columbia Falls Dam	3	1990;1998	107	$
ME	Presumpscot River	Smelt Hill		2002		
ME	St. George River	Sennebec Dam	5	2002	73	
ME	Sandy River	Madison Electric Works Dam		2006		

continues

DATA TABLE 8 *(continued)*

State	River	Project Name	Removed	H (m)	L (m)	Reason
ME	Sebasticook River	Main Street Dam	2002			
ME	Souadabscook Stream	Grist Mill Dam	1998	4	23	E
ME	Souadabscook Stream	Hampden Recreation Area Dam	1999	1		
ME	Souadabscook Stream	Souadabscook Falls Dam	1999		46	
ME	Stetson Stream	Archer's Mill Dam	1999	4	15	
MI		Foster Trout Pond Dam	1983	1		NPS
MI		Three River City Dam	1992	4		
MI	Anguilm Creek	Fibron Trout Pond Dam	2000			
MI	Au Sable River	Grayling Dam	2005	3		E
MI	Au Sable River	Salling Dam	1991	5	76	E; R
MI	Battle Creek River	Charlotte City Dam	2004	2		
MI	Battle Creek River	City of Charlotte Dam	2006	2		E
MI	Battle Creek River	Elm Street Dam	2004	1	30	
MI	Bear Creek	Copemish Dam	2003	2		E
MI	Chippewa River	Mill Pond Dam	2002	5	34	
MI	Coldwater River	Randall Dam	2002			$
MI	Dead River	Marquette Dam	1912			
MI	Dead River	Marquette City Dam #1	2004	3	61	E
MI	Grand River	Dimondale Dam	2006	2	91	E,R
MI	Grand River	Wager Dam	1985	3		E
MI	Hersey River	Hersey Dam	2006			E
MI	Lake Hudson Recreation Area	Haley Dam	2003			R,$
MI	Looking Glass River	Wacousta Dam	1966	1		
MI	Muskegon River	Big Rapids Dam	2000			
MI	Muskegon River	Newago Dam	1969	5		
MI	North Branch Spars Creek	Kimberly-Clark Dam	2004	1	61	F
MI	Pawpaw River	Wasman Dam	1999			

MI	Pine River	Stronach Dam	2002	5	107	
MI	Potagannissing River	Potagannissing Dam	2006	2	23	E
MI	Silver Lead Creek	Air Force Dam	1998			
MI	Sturgeon River	Sturgeon River Dam	2003	14		
MI	Tannery Creek	Tannery Creek Dam	2004			
MN	Cannon River	Stockton Dam	1994	9		F
MN	Cannon River	Welch Dam	1994	3	37	E;S
MN	Cottonwood River	Flandrau Dam	1995	4		E
MN	Crow River	Berning Mill Dam	1986	3		F
MN	Crow River	Hanover Dam	1984	4		F
MN	Garvin Brook	Stockton Dam				
MN	Kettle River	Sandstone Dam	1995	6	46	E; R
MN	Otter Tail River	Frazee Dam	1999			
MN	Pomme de Terre River	Appleton Mill Pond Dam	1999			
MN	Pomme de Terre River	Pomme de Terre River Dam				
MN	Red Lake River	Otter Tail Power Dam	2005	3		S
MN	Root River	Lake Florence Dam		4		
MO		Alkire Lake Dam	1990	9		
MO		Goose Creek Lake Dam	1987	16		
MO		Indian Rock Lake Dam	1986	17		S
MT	Bear Creek	Three Bears Lake-East Dam		3		NPS
MT		Three Bears Lake-West Dam		6		NPS
MT	Blackfoot River	Stimson Dam	2006			F
MT	Lone Tree Creek	Vaux #1 Dam	1995	10		S
MT	Lone Tree Creek	Vaux #2 Dam	1995	17		S
MT	Peet Creek	Peet Creek Dam	1994	13		S
MT	Rock Creek	small dam				
MT	Wallace Creek	Wallace Creek Dam	1997	9	219	

continues

249

DATA TABLE 8 *(continued)*

State	River	Project Name	Removed	H (m)	L (m)	Reason
NC		Ash Bear Pen Dam	1990	3		NPS; $
NC		Forny Ridge Dam	1988	1		NPS
NC	Little River	Cherry Hospital Dam	1998	2	41	E
NC	Little River	Rains Mill Dam	1999			
NC	Little Sugar Creek	Freedom Park Dam	2002	3	18	
NC	Marks Creek (trib.)	Unnamed Dam	2002	8	122	
NC	Neuse River (trib.)	Carbonpin Dam	2005			
NC	Neuse River (trib.)	Lowell Dam	2005	3		
NC	Neuse River	Quaker Neck Dam	1998	2	79	E
ND	Knife River	Antelope Creek Dam	1979	7		S
ND	Little Missouri River	Kunick Dam		7		U
ND	Stony Creek	Epping Dam	1979	14		
NE	Bozle Creek	Lake Crawford Dam	1987	8		S
NE	Camp Creek	Diehl Dam	1981	10		$
NE	Cedar River	Fullerton Power Plant Dam		5		F
NE	Lodgepole Creek	Bennet Dam	1982	6		$
NE	Timber Creek (trib.)	Helen Fehrs Trust Dam	1995	11		U
NE		Golf Course Dam		8		NPS
NH	Ashuelot River	McGoldrick Dam	2001			E
NH	Ashuelot River	Winchester Dam	2002	1	32	S,E
NH	Bellamy River	Bellamy River Dam	2004	1	27	
NH	Bearcamp River	Bearcamp River Dam	2003	6	70	S,E
NH	Clark Brook	Champlin Pond Dams	2005			
NH	Contoocook River	West Henniker Dam	2004	3	40	
NH	Tioga River	Badger Pond Dam	2004	5		
NJ	Cold Brook	Pottersville Dam	1985	6	55	S
NJ	Crooked Brook	Patex Pond Dam	1990	6	104	S

State	River	Dam	Year			
NJ	Delaware River (trib.)	Lake Success Dam	1995	6	91	S
NJ	Lopatcong Creek	Harry Pursel Dam	2006	5		F
NJ	Raritan River	Fieldsville Dam	1990	3	122	E
NJ	S.B. Timber Creek	Glenside Dam	1997	4	40	S
NJ	Van Camptens Brook (trib.)	Pool Colony Dam	1999			NPS
NJ	Van Camptens Brook	Upper Blue Mountain Dam	1995	8	64	NPS; S
NJ	Whippany River (trib.)	Knox Hill Dam	1996	5	46	S
NM	Pecos River	McMillan Dam	1989	20		S
NM	Sante Fe River	Two Mile Dam	1994	26	219	S
NV		Katherine Borrow Pit Embankment	1992	5		NPS
NY		Curry Pond Dam		1		NPS
NY		Luxton Lake Dam		5		NPS
NY	Black Creek	Gray Reservoir Dam	2002	10	117	S,$
NY	Hudson River	Fort Edward Dam	1973	9	179	S
NY	Neversink River	Cuddebackville Dam	2004	2		
OH		Armington Dam #2	1991	5		NPS
OH		Foxtail Dam		9		NPS
OH		Slippery Run (Stahl) Dam	1990	4		NPS
OH	Black Fork (trib.)	Altier Pond Dam	1989	10		
OH	Brannon Fork	Ohio Power Company Pond Dam	1987	5		
OH	Brush Creek (trib.)	Williams Dam		12		
OH	Cuyahoga River	Kent Dam	2004			
OH	Cuyahoga River	Munroe Falls Dam	2005	3		
OH	Collins Fork	Ohio Power Company Pond Dam		4		
OH	East Reservoir (trib.)	Wonder Lake Dam	1986	5		
OH	Hamley Run (trib.)	Poston Fresh Water Pond Dam	1988	13		
OH	Hocking River (trib.)	Cottingham Lake Dam	1991	5		
OH	Huron River	Milan Wildlife Area Dam	2002	2	30	
OH	Ice Creek (trib.)	Fair Haven Lake Dam	1980	9		

continues

State	River	Project Name	Removed	H (m)	L (m)	Reason
OH	Jackson Run (trib.)	Howard's Lake Dam				
OH	Johnny Woods River (trib.)	Carr Lake Dam	1985	3		
OH	Licking River (trib.)	Dutiel Pond Dam	1986	4		
OH	Little Auglaize River (trib.)	Burt Lake Dam	1992	5		
OH	Little Darby Creek	Little Darby Dam	1989	6		
OH	Little Darby Creek	Okie Rice Dam	1990	4		S
OH	Little Miami River	Foster Dam	1984			
OH	Little Miami River	Jacoby Road Dam	1997	2	30	E
OH	Little Pine Creek (trib.)	Mastrine Pond Dam	1978	5		
OH	Little Yellow Creek (trib.)	Old Jenkins Lake Dam		7		
OH	Olentangy River	Dennison Dam	2002			
OH	McLuney Creek (trib.)	Strip Mine Pond Dam		8		
OH	Modoc Run	Modoc Reservoir Dam	1981	7		
OH	Ogg Creek	Jones Lake Dam		6		
OH	Ottawa River	Unnamed Dam	2003	2	15	
OH	Porter Creek	Marshfield Lake Dam	1973	5		
OH	Robinson Run (trib.)	Lake Hill #2 Dam		9		
OH	Robinson Run (trib.)	Lake Hill Dam #1		9		
OH	Rocky Fork (trib.)	Village at Rocky Fork Lake Dam		2		rebuilt
OH	Sandusky River	St. John's Dam	2003	2	46	$
OH	Seven Mile Creek (trib.)	Ashworth Lake Dam		8		
OH	Silver Creek	Silver Creek Dam				
OH	Silver Creek (trib.)	Chapel Church Lake Dam	1989			
OH	South Fork (trib.)	Georgetoen Freshwater Dam	1988	4		
OH	Spencer Creek (trib.)	State Route 800 Dam	1989	8		
OH	Stillwater Creek	Consol Pond Dam		4		
OH	Sugartree Creek (trib.)	Brashear Lake Dam	1991	5		
OH	Timber Run (trib.)	Derby Petroleum Lake Dam	1984	9		

State	River/Basin	Dam Name	Year			
OH	Town Fork	Toronto Band Father's Lake Dam	1991	5		
OH	Wills Creek (trib.)	Killiany Lake Dam	1980	2		rebuilt
OH	Yankee Run	Yankee Lake Dam		8		
OR	Applegate River	Irrigation Push-Up Dam	2002	1		
OR	Ashland Creek	Unnamed Dam #1	2000			
OR	Ashland Creek	Unnamed Dam #2	2000			
OR	Ashland Creek	Unnamed Dam #3	2000			
OR	Bear Creek	Jackson Street Dam	1998	3	37	E
OR	Beaver Creek	Byrne Diversion Dam	2002	1		
OR	Dinner Creek	Dinner Creek Dam	2003	3	11	
OR	Evans Creek	Alphonso Dam	1999	3	17	E
OR	Evans Creek	Maple Gulch Diversion Dam	2002			
OR	Little Applegate River	Buck & Jones Dam	2006			
OR	Little Applegate River	Buck & Jones Diversion Dam	2003	2	30	F
OR	Poorman Creek	Unnamed Dam	1999			
OR	Powder River (trib.)	Rock Creek Dam	2002			
OR	Wagner Creek	Unnamed Dam	2003	1		
OR	Walla Walla River	Marie Dorian Dam	1997	2	30	E
OR	Willamette River	Catching Dam	1994	9	69	
OR	Yamhill Basin	Lafayette Locks Dam	1963			
PA		Butterfield Pond Dam	1992	4		NPS
PA		Carpenters Pond Dam		5		NPS
PA		Fire Pond Dam at Incline #10		5		NPS
PA		Lake Lettini Dam		2		NPS
PA		Lemon House Pond Dam	1984	5		NPS
PA		Lower Friendship Dam	1982	9		NPS
PA		unnamed dam, Peace Light Inn	1991	2		NPS
PA		Upper Friendship Dam	1982	4		NPS
PA		Van Horn #2 Dam		3		NPS
PA		Van Horn Dam #1	1991	2		NPS

continues

DATA TABLE 8 *(continued)*

State	River	Project Name	Removed	H (m)	L (m)	Reason
PA		Van Horn Dam #5	1991	4		NPS
PA	Bennett Run (trib.)	Shissler Dam	2006			F
PA	Ben's Creek	Benscreek Intake	2005	2	18	S,$
PA	Birch Run	Birch Run Dam	2005			S
PA	Black Log Creek	Old Furnace Dam	2006	2	46	S,E
PA	Clear Shade Creek	Clear Shade Creek Reservoir Dam	1998	4	58	
PA	Coal Creek	Coal Creek Dam #2	1995	7	35	
PA	Coal Creek	Coal Creek Dam #3	1995	7		
PA	Coal Creek	Coal Creek Dam #4	1995	4	109	
PA	Coal Creek	Diverting Dam		2	17	
PA	Cocalico Creek	Martins Dam	2000			S
PA	Codorus Creek	Muren's (Seitzville Mill) Dam	2000	4	30	S,$
PA	Codorus River (trib.)	Yorkane Dam	1997			
PA	Conestoga River	American Paper Products Dam	1998	1	40	E
PA	Conestoga River	Hinkletown Mill Dam	2000			
PA	Conestoga River	Rock Hill Dam	1997	4	91	E
PA	Conewago Creek	Detter's Mill Dam	2004	2	76	
PA	Conewago Creek	Sharrer's Mill Dam	2005	2	79	S,E
PA	Conococheague Creek	Siloam Dam	2005			S,E
PA	Conodoguinet Creek	Black Dam	2003	3	107	
PA	Conodoguinet Creek	Good Hope Dam	2001	2		
PA	Cooks Creek	Durham Dam	2004	3		S
PA	Fishing Creek	Goldsboro Dam	2005	1		E
PA	Fishing Creek	Snavely's Mill Dam	1997	1	32	
PA	Gillians Run	Maple Hollow Reservoir Dam	1995	7	59	
PA	Hammer Creek	Hammer Creek Dam	2001	2		S
PA	Hess Run	Kohut Pond Dam	2005	4		S
PA	Hunting Run (trib.)	Chancellorsville Brygadier A Dam	2000			

continues

PA	Hunting Run (trib.)	Chancellorsville Brygadier B Dam	2000			
PA	Huston Run	Unnamed Dam #1	2001			
PA	Huston Run	Unnamed Dam #2	2001			
PA	Juniata River	Williamsburg Station Dam	1996	4	79	E
PA	Kettle Creek	Rose Hill Intake Dam	1998	4	46	
PA	Kishacoquillas Creek	unnamed dam	1998	3	53	
PA	Laural Run	unnamed dam	1998	2	15	
PA	Lehigh River	Palmerton Dam	2006	1	91	E
PA	Lititz Run	Mill Port Conservancy Dam	1998	3	3	E
PA	Lititz Run	Unnamed dam	1998	1	3	E
PA	Lititz Run	Young's Dam	2002	1		
PA	Little Conemaugh River	Lower Lloydell Dam	2005	2	21	S,$
PA	Little Conestoga River	East Petersburg Authority Dam	1998	1	6	E
PA	Little Conestoga River	Maple Grove Dam	1997	2	18	E
PA	Little Leheigh Creek	Wild Lands Conservancy Dam	2000	2	23	
PA	Manantango Creek	Meisers Mill Dam	2001	2	23	$,S
PA	Manatawny Creek	Unnamed Dam	2000			
PA	Middle Creek	Franklin Mill Dam	2000			$,S
PA	Middle Creek	Mussers Dam	1992	9	117	
PA	Milesburn Run	Cleversburg Water Supply Dam	2004	1		
PA	Mill Creek	Daniel Esh Dam	2003	1		
PA	Mill Creek	Niederriter Farm Pond Dam	1995	6	107	
PA	Mill Creek	Yorktowne Paper Dam	1997	2	18	
PA	Muddy Creek	Amish Dam		1	12	E
PA	Muddy Creek	Castle Fin Dam	1997	2	117	
PA	Muddy Run	Amish Dam #1	2000			
PA	Muddy Run	Amish Dam #2	2000			
PA	Muddy Run	Amish Dam #3	2000			
PA	Muddy Run	Amish Dam #4	2000			
PA	Muddy Run	Amish Dam #5	2001			

DATA TABLE 8 *(continued)*

State	River	Project Name	Removed	H (m)	L (m)	Reason
PA	Muddy Run	Amish Dam #6	2001			
PA	Muddy Run	Amish Dam #7	2001			
PA	Muddy Run	Amish Dam #8	2001			S
PA	Neshannock Creek (trib.)	Graceland Dam	2006	5	9	
PA	Penns Creek	Millmont Dam	2006			E
PA	Pennypack Creek	Rhawn Street Dam	2006			E
PA	Pequea Creek	James Ford's Dam	2006	2	91	S
PA	Perkiomen Creek	Collegeville Mill Dam	2003	2	76	
PA	Perkiomen Creek	Goodrick Dam	2005	4		E
PA	Pickering Creek (trib.)	Binky Lee Preserve	2005	2		$
PA	Poplar Run	Unnamed Dam #1	2004	3		$
PA	Poplar Run	Unnamed Dam #2	2004	3		
PA	Red Run	Red Run Dam	1996	2	12	S
PA	Ridley Creek	Irving Mill Dam	2004	4	30	S,$
PA	Ridley Creek	Sharpless Dam	2005	4		
PA	Rife Run	Intake Dam	2001	2	15	
PA	Sandy Run (trib.)	Twining Valley Golf Course Dam	2004	5		
PA	Spring Creek	Cabin Hill Dam	1998			E
PA	Sugar Creek	Pomeroy Memorial Dam	1996	7	135	
PA	Tea Creek	Reedsville Milling Company Dam	2004	4	14	E,S
PA	Tinicum Creek (trib.)	unnamed dam	1998	2	12	
PA	Trindle Spring Run	Silver Spring Dam	2006	3		
PA	Tulpehocken Creek	Charming Forge Dam	2004	2	40	
PA	Smithtown Creek	Lochner Dam	2006	2		E,S
PA	Smithtown Creek	Ward Dam	2006	2	15	E,S
PA	Strodes Run	Hackenberg Dam	2006	1	11	
PA	Swatara Creek	Iron Stone Mine Dam	2006	1	152	
PA	Wallace Run	Unnamed Dam	2005	1		

State	Watercourse	Dam	Year			
PA	Wolf Creek	Main Street Dam	2005	4		S,E
PA	Wolf Creek	Upper Grove City Dam	2004	2	32	
PA	Wyomissing Creek	Mohnton Dam	2006			
PA	Wyomissing Creek	Reading Public Museum Dam #1	2004	1	14	$
PA	Wyomissing Creek	Reading Public Museum Dam #2	2004	2	18	$
PA	Yellow Breaches Creek	Barnitz Mill Dam	2000			S,$
PA	Yellow Breaches Creek	Hoffman Dam	2005	2	40	
RI	Pawtuxet River	Jackson Pond Dam	1979	6		$
SC	Burgess Creek	Gallagher Pond Dam	1989	13		S
SC	Cowpens National Battlefield	unnamed dam, State Road 11–58	1979	2		NPS
SC	Pole Branch River	Pole Branch Dam	1990	8		F
SC	Tools Fork (trib.)	Miller Trust Pond Dam	1993	12		S
SC	Turkey Quarter Creek	Old City Reservoir Dam	1988	8		S
SD		Arikara Dam	1978	12		
SD		Farmingdale Dam	1986	7		
SD		Lake Farley Dam	1980	8		rebuilt
SD		Menno Lake Dam	1984	12		
SD		Mission Dam	1987	8		
SD		Norbeck Dam & SD Highway 87		12		NPS
SD		P6L-Lower Bigger Dam		3		NPS
SD		unnamed dam #26	1987	3		NPS
SD		unnamed dam #30	1987	3		NPS
SD		unnamed dam #32		3		NPS
SD		unnamed dam #35	1987	3		NPS
TN		L. Thompson Dam #1	1990	3		NPS; $
TN		L.C. Hancock #1	1990	2		NPS
TN		L.C. Hancock #3		2		NPS
TN	Adkinson Creek	Gin House Lake Dam	1994	10		$
TN	Burra-Burra Creek	Cities Service Company Dam	1995	9		

continues

DATA TABLE 8 *(continued)*

State	River	Project Name	Removed	H (m)	L (m)	Reason
TN	Decant Pipes	Monsanto Dam #3	1988	12		
TN	Duck Creek	Occidental Chem Pond Dam A	1995	37		
TN	Duck Creek	Occidental Chem Pond Dam D	1995	49		
TN	Duck River	Monsanto Dam #7	1990	24		
TN	Flat Creek	Sandy Stand Dam	1987	12		$
TN	Flat Creek	Shangri-la Lake Dam	1985	10		S
TN	Fork Creek (trib.)	Ballard Mill Mine Dam	1992	9		
TN	Greenlick Creek	Monsanto Dam #4	1990	16		
TN	Greenlick Creek	Monsanto Dam #5A	1990	16		
TN	Helms Branch	Monsanto Dam #9	1990	10		
TN	Hurricane Creek	Cumberland Springs Dam	1989	9		$
TN	Johnson Creek	Lake Deforest Dam	1991	11		$
TN	Ollis Creek	Eblen-Powell Dam #1		10		$
TN	Quality Creek	Rhone Poulenc Dam #17	1995	10		
TN	Quality Creek	Rhone Poulenc Dam #19	1995	18		
TN	Quality Creek	Rhone Poulenc Dam #20	1995	10		
TN	Rocky Branch	Monsanto Dam #12	1990	38		
TN	Rutherford Creek (trib.)	Occidental Chem Dam #6	1991	16		
TN	Snake Creek (trib.)	Spence Farm Pond Dam #5	1983	11		S
TN	Tipton Branch	Laurel Lake Dam	1990	13		S
TN	Walker Stream	Walkers Dam	1992	10		
TX		Alamo Arroyo Dam	1979	15		
TX		Boot Spring Dam		5		NPS
TX		Duke Dam		2		NPS
TX		H and H Feedlot Dam	1980	11		
TX		Harris Back Lake Dam		5		
TX	Big Sandy Creek (trib.)	Lake Downs Dam		8		

State	River	Dam				
TX	Daves White Branch	Millsap Reservoir Dam			8	
TX	Mill Creek	Barefoot Lake Dam			8	
TX	Mustang Creek (trib.)	Bland Lake Dam	1989		6	
TX	Pecan River (trib.)	Hilsboro Lake Park Dam			6	
TX	Tributary to Willis Creek	Railroad Reservoir Dam			3	
TX	Wasson Branch	Nix Lake Dam			7	
UT		Atlas Mineral Dam	1994		28	
UT		Bell Canyon Dam	1979		9	$
UT	Box Elder Creek	Box Elder Creek Dam	1995		15	S
UT	Muddy Creek	Brush Dam	1983		15	$
VA		Adney Gap Pond Dam	1984		4	NPS
VA		Berryville Reservoir			5	NPS
VA		Fredricksburgh & Spotsylvania Dam #2			2	NPS
VA		Fredricksburgh & Spotsylvania Dam #3			2	NPS
VA		Fredricksburgh & Spotsylvania Dam #5			2	NPS
VA		Fredricksburgh & Spotsylvania Dam #6			1	NPS
VA		Osborne Dam			4	NPS
VA		Sykes Dam	1992		7	NPS
VA	Manassas NP Battlefield	Picnic Area Dam	1984		2	NPS
VA	Rappahannock River	Embrey Dam	2004			
VA	Shenandoah River	Knightly Dam	2004			
VA	Shenandoah River	Rockland Dam	2005		5	E,S
VA	Shenandoah River	McGaheysville Dam	2004	107		
VT	Ball Mountain Brook (trib.)	Dalewood Dam	2006	58	5	
VT	Batten Kill River	Red Mill Dam				
VT	Charles Brown Brook	Norwich Reservoir Dam			6	S
VT	Clyde River	Newport No. 11 Dam	1996	27	6	E
VT	Cold River	Cold River Dam	2003	27	2	E
VT	LaMoille River (trib.)	Johnson State College Dam	2003		9	E,$

continues

DATA TABLE 8 *(continued)*

State	River	Project Name	Removed	H (m)	L (m)	Reason
VT	Mussey Brook	Lower Eddy Pond Dam	1981	6		S
VT	Ompompanoosuc River (trib.)	Hillside Farm Dam	2003	5		F
VT	Passumpsic River (trib.)	Lyndon State College Lower Dam				
VT	Wells River	Groton Dam	1998	2		
VT	Winooski River (trib.)	Winooski Water Supply Upper Dam	1983	6		S
VT	Youngs Brook	Youngs Brook Dam	1995	14		S
WA		Black Mud Waste Pond A Dam		5		
WA		Black Mud Waste Pond B Dam		5		
WA		Black Mud Waste Pond C Dam		5		
WA		Bow Lake Reservoir		2		
WA		City Lakes Dam		5		
WA		North End Reservoir		9		
WA		Pomeroy Gulch Dam		12		
WA	Boise Creek	White River Mill Pond Dam		1		
WA	Coffee Creek	Coffee Creek Dam		3		
WA	Columbia River (trib.)	Stromer Lake Dam		2		
WA	Goldsborough Creek	Goldsborough Creek Dam	2001			
WA	Hanford Creek (trib.)	PEO Dam #32A		4		
WA	Hanford Creek (trib.)	PEO Dam #48		1		
WA	Headquarters Creek	Unnamed Dam	2000	2		
WA	Hunters Creek	Hunters Dam		20		
WA	Icicle Creek	Unnamed Barrier #1	2003	3		E
WA	Icicle Creek	Unnamed Barrier #2	2003	3		E
WA	Icicle Creek	Unnamed Barrier #3	2003	3		E
WA	Mill Creek	Mill Creek Settling Basin Dam		5		
WA	Sauk River (trib.)	Darrington Water Works Dam		6		
WA	Touchet River	Maiden Dam	1998			

State	River/Creek	Dam	Year			S
WA	Wagleys Creek	Sultan Mill Pond Dam		5		
WA	Whitestone Creek	Rat Lake Dam	1989	10	73	
WA	Wind River	Wind River Dam		6		
WI	Iron Run Creek	Iron Run Dam	2006			NPS
WI		McNally Trout Pond Dam	1983	2		NPS
WI		Poppe Dam	1982	1		NPS
WI		Rassussen #1 Dam	1982	1		NPS
WI		Rassussen #2 Dam	1982	1		NPS
WI		Rassussen #3 Dam	1982	1		NPS
WI		Schaaf #_ Dam	1982	1		NPS
WI		Schaaf #2 Dam	1982	1		NPS
WI		unnamed dam #1 (Larrabee Tract)	1990			NPS
WI		Weingarten Dam	1982	1		NPS
WI	Apple River	Huntington Dam	1968			
WI	Apple River	McClure Dam	1968			E
WI	Apple River	Somerset Dam	1965			
WI	Bad River	Mellen Dam	1967			E
WI	Baraboo River	Island Woolen Co. Dam	1972			
WI	Baraboo River	LaValle Dam	2001			
WI	Baraboo River	Linen Mill Dam	2001			$
WI	Baraboo River	Oak Street Dam	2000	4	63	
WI	Baraboo River	Reedsburg Dam	1973	3		
WI	Baraboo River	Waterworks Dam	1998	3	67	$
WI	Baraboo River	Wonewoc Dam	1996	9		
WI	Bark River	Hebron Dam	1996	5	52	
WI	Bark River	Slabtown Dam	1992	3	18	
WI	Bass Creek	Afton Dam	2002			
WI	Beaver Creek	Ettrick Dam	1976			
WI	Black Earth Creek	Black Earth Dam	1957	3		
WI	Black Earth Creek	Cross Plains Dam	1955	3		

continues

DATA TABLE 8 *(continued)*

State	River	Project Name	Removed	H (m)	L (m)	Reason
WI	Black River	Greenwood Dam	1994	5		
WI	Boulder Creek	Boulder Creek Dam #1	2003			E,S
WI	Boulder Creek	Boulder Creek Dam #2	2003			E,S
WI	Branch River	Unnamed Dam	2003	2	12	E
WI	Carpenter Creek	Carpenter Creek Dam	1995			
WI	Cedar Creek	Hamilton Mill Dam	1996	2	30	
WI	Cedar Creek	Schweitzer Dam	2002	2	9	
WI	Centerville Creek	Centerville Dam	1996	4		
WI	City Creek	Mellen Waterworks Dam	1995	4		
WI	Deerskin River	Deerskin Dam	2001			
WI	Dunlop Creek	Dunlop Creek Dam	1955			
WI	Eau Galle River	Spring Valley Dam	1997	1		
WI	Eighteen Mile Creek	Colfax Dam	1998	6	107	
WI	Embarrass River	Hayman Falls Dam	1995	5	61	
WI	Embarrass River	Upper Tigerton Dam	1997	3		
WI	Flambeau River	Port Arthur Dam	1968	3		E
WI	Flume Creek	Northland Dam	1992	3		
WI	Fox River	White River Dam	2003	4	76	S
WI	Fox River	Wilmot Dam	1992	2	61	
WI	Genesee Creek	Genesee Roller Mill Dam	2005			E,S
WI	Genesee Creek	Unnamed Dam	2005			E,S
WI	Grand River	Grand River Dam	2002			
WI	Grand River	Manchester Dam	2005	5		S
WI	Grand River	Manchester Mill Dam	2006			S
WI	Handsaw Creek	Huigen Dam	1970	2		
WI	Handsaw Creek	Schiek Dam	1970	2		
WI	Iron River	Orienta Falls Dam	2001	13		
WI	Kickapoo River	Ontario Dam	1992			E

State	River	Dam	Year			
WI	Kickapoo River	Realstown Dam	1985			
WI	Koshkonong Creek	Rocksdale Dam	2000			
WI	Lemonweir River	Lemonweir Dam	1992	4		
WI	Lowe Creek	Lowe Creek 1 Dam				
WI	Lowe Creek	Lowe Creek 2 Dam				
WI	Madden Branch (trib.)	Beardsley Dam	1990	4		
WI	Magdantz Creek	Clark's Mill Dam	2003	2	51	$
WI	Manitowoc River	Manitowoc Rapids Dam	1984	5	122	E
WI	Manitowoc River	Oslo Dam	1991	2		E
WI	Marengo River	Marengo Dam	1993	5		E
WI	Maunesha River	Ball Park Dam	2004	3		
WI	Maunesha River	Upper Waterloo Dam	1995	5	35	
WI	Milhome Creek	Spitzer Dams	2005			
WI	Milwaukee River	Chair Factory Dam	2001			
WI	Milwaukee River	New Fane Dam	2001			$
WI	Milwaukee River	North Avenue Dam	1997	6	132	
WI	Milwaukee River	Wauteka Dam	2003	3	67	S
WI	Milwaukee River	Woolen Mills Dam	1988	5		$
WI	Milwaukee River	Young America Dam	1994	3		
WI	Mullet River	Meyer Dam	2005			
WI	Oconomowoc River	Funks Dam	1993	2		
WI	Oconto River	Hemlock Dam	2004	2		
WI	Oconto River	Knowles Dam	2004	2		
WI	Oconto River	Pulcifer Dam	1994	2		
WI	Onion River (trib.)	Kamrath Dam	2001	2		
WI	Onion River (trib.)	Silver Springs multi-dam complex	2002			
WI	Osceola Creek	Millpond Dam	2005	2		F
WI	Otter Creek	Klondike Dam	1978	9		
WI	Peshtigo River	Crivitz Dam	1993			
WI	Pike River	Kenosha Country Club Dam	2004	1		

continues

DATA TABLE 8 *(continued)*

State	River	Project Name	Removed	H (m)	L (m)	Reason
WI	Pine River	Bowen Mill Dam	1996	4		
WI	Pine River	Parfrey Dam	1996	6	137	
WI	Potato Creek	Athens Dam	2004	3		
WI	Prairie River	Prairie Dells Dam	1991	18		
WI	Prairie River	Prairie Dells Dam	1991	18		E
WI	Prairie River	Ward Paper Mill Dam	1999	5	24	
WI	Rathbone Creek	Evans Pond Dam	1998	3		
WI	Red Cedar River	Colfax Light Power Dam	1969	6		
WI	Sheboygan River	Franklin Dam	2001	4	41	
WI	Shell Creek	Cartwright Dam	1995	2		
WI	Six-Mile Creek	Six-Mile Creek Dam	2004			
WI	Sugar River	Mount Vernon Dam	1950	3		
WI	Token Creek	Token Creek Dam	*	4		
WI	Tomorrow/Waupaca River	Nelsonville Dam	1988			
WI	Trempealeau River	Whitehall Dam	1988			
WI	Turtle Creek	Shopiere Dam	2000	4	42	S,$
WI	Waupaca River	Planing Mill Dam	2005			S
WI	Willow River	Mounds Dam	1998	18	131	$
WI	Willow River	Willow Falls Dam	1992	18	49	$
WI	Woods Creek	Woods Creek Dam	2002	5	61	
WI	Yahara River	Fulton Dam	1993	5		E
WV		Ladoucer Pond Dam	1993			NPS
WY		East Dam		2		NPS
WY		No Name Dam #1		2		NPS
WY		North Dam		5		NPS
WY		South Dam		2		NPS
WY		West Dam		2		NPS
WY		White Grass Dude Ranch Dam	1988			NPS
WY	City of Sheridan (trib.)	Sheridan Heights Reservoir				
WY	Laramie River	unnamed dam	1997			$

Dams Removed or Decommissioned in the United States, 1912 to Present, by Year and State

Description

The list compiles information on dams removed or decommissioned from rivers in the United States, summarized by state and by the number each year. Figure DT9.1 shows the number of dam removals by state for those states with more than 5 removals. Nearly 700 dams have been removed. Wisconsin and Pennsylvania lead the nation in the number of dams removed, with over 100 each. As Figure DT9.2 shows, the pace of dam removals has greatly accelerated in recent years, as unsafe, ecologically damaging, or economically useless dams have been identified and removed.

Limitations

Until relatively recently, no effort was made to record dam removals, and it is likely that many smaller dams have been removed without being recorded. As a result, this list should be considered preliminary rather than comprehensive.

SOURCE

Army Corp of Engineers. 2005. National Inventory of Dams. http://crunch.tec.army.mil/nidpublic/webpages/nid.cfm. These data are no longer completely open to public viewing.

DATA TABLE 9   Dam Removed or Decommissioned in the United States, 1912 to Present, by Year and State

State	Number of Dam Removals	Year	Number of Dam Removals
Wisconsin	106	1912	1
Pennsylvania	103	1913	0
California	68	1914	0
Ohio	45	1915	0
Michigan	27	1916	0
Tennessee	26	1917	0
Washington	24	1918	0
Illinois	19	1919	0
Connecticut	18	1920	0
Oregon	17	1921	0
Maine	14	1922	1
Idaho	14	1923	0
Virginia	13	1924	0
Vermont	12	1925	1
Texas	12	1926	0
Minnesota	12	1927	1
South Dakota	11	1928	0
Colorado	11	1929	0
Kansas	10	1930	0
North Carolina	9	1931	1
New Jersey	9	1932	0
Maryland	9	1933	0
Wyoming	8	1934	0
Montana	8	1935	0
New Hampshire	7	1936	1
Nebraska	6	1937	0
Massachusetts	6	1938	1
South Carolina	5	1939	0
New York	5	1940	0
Kentucky	5	1941	0
Utah	4	1942	0
Louisiana	4	1943	0
Florida	4	1944	0
District of Columbia	4	1945	0
Arkansas	4	1946	3
Alaska	4	1947	2
North Dakota	3	1948	0
Missouri	3	1949	5
Arizona	3	1950	5
New Mexico	2	1951	2
West Virginia	1	1952	0
Rhode Island	1	1953	0
Nevada	1	1954	0
Iowa	1	1955	2
Indiana	1	1956	0
Georgia	1	1957	1
Alabama	1	1958	0
Oklahoma	0	1959	0
Mississippi	0	1960	0

State	Number of Dam Removals		Year	Number of Dam Removals
Hawaii	0		1961	0
Delaware	0		1962	0
	681		1963	2
			1964	0
			1965	1
			1966	1
			1967	1
			1968	3
			1969	2
			1970	4
			1971	1
			1972	1
			1973	4
			1974	0
			1975	0
			1976	1
			1977	0
			1978	3
			1979	6
			1980	8
			1981	3
			1982	13
			1983	7
			1984	8
			1985	9
			1986	7
			1987	10
			1988	15
			1989	12
			1990	21
			1991	18
			1992	18
			1993	9
			1994	15
			1995	28
			1996	11
			1997	14
			1998	28
			1999	20
			2000	25
			2001	22
			2002	32
			2003	34
			2004	37
			2005	30
			2006	33
			Unknown	137
			Total	681

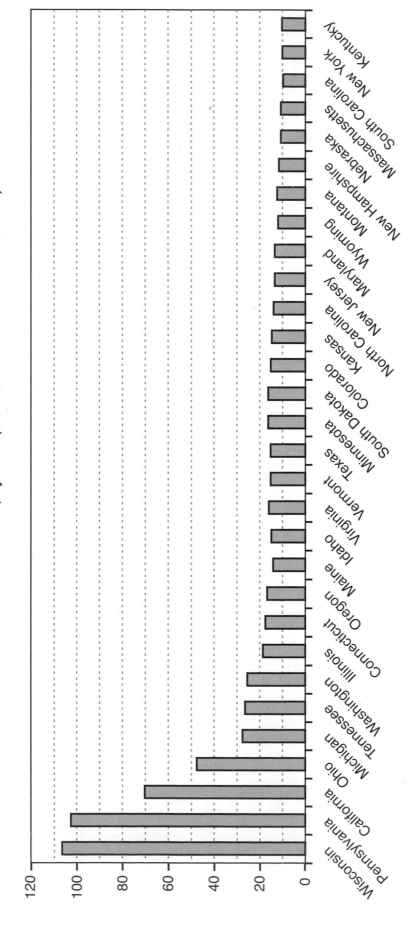

FIGURE DT9.1 NUMBER OF DAM REMOVALS, BY STATE.

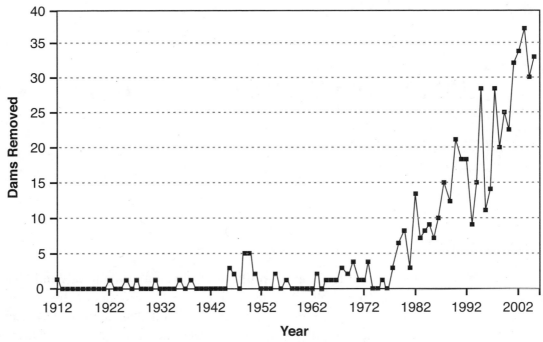

FIGURE DT9.2 DAM REMOVALS HAVE GREATLY ACCELERATED OVER THE YEARS.

United States Dams by Primary Purposes

Description

Dams are constructed for a wide variety of purposes, from controlling floods and fires to generating electricity. This Table summarizes the number of dams in the United States by their primary purposes. In the United States, nearly 34,000 dams, or 1/3 of all dams, were built for recreational purposes. Flood control and ponds for fires or farms accounted for about 16,000 dams each. Water supply and irrigation account for less than 17,000 dams in total.

Data for this table were compiled from the U.S. Army Corps of Engineers' National Inventory of Dams (NID). The NID contains data on about 80,000 dams throughout the United States and Puerto Rico that are more than 25 feet high, hold more than 50 acre-feet of water, or are considered a significant hazard if they fail. The U.S. Army Corps has made it extremely difficult to access the full National Inventory on Dams now, closing off an important source of basic data on water infrastructure, presumably because of worries about security.

Limitations

The NID only includes dams that:

1. Have a high or significant hazard potential

2. Have a low hazard potential dam but exceed 25 feet in height and 15 acre-feet of storage

3. Have a low hazard potential dam but exceed 50 acre-feet storage and 6 feet in height

Thus a potentially large number of dams are excluded from the database. In addition, because many dams are built for multiple purposes, the sum exceeds the total number of dams.

SOURCE

Army Corp of Engineers. 2005. National Inventory of Dams. http://crunch.tec.army.mil/nidpublic/webpages/nid.cfm. These data are no longer completely open to public viewing.

DATA TABLE 10 United States Dams by Primary Purposes

Purpose	Number
Recreation	33,944
Flood Control	15,769
Fire and Farm Ponds	15,952
Irrigation	9,405
Other	7,704
Water Supply	7,430
Tailings	1,448
Fish and Wildlife Pond	3,370
Hydroelectric	2,551
Debris Control	1,573
Navigation	693
Total	99,839

The NID contains data on 79,000 dams in the United States. Because dams are often built for multiple purposes, the total shown in the table above exceed the total number of dams included in the inventory.

United States Dams by Owner

Description

Data from this table were compiled from the U.S. Army Corps of Engineers' National Inventory of Dams (NID). The NID contains data on about 80,000 dams throughout the United States and Puerto Rico that are more than 25 feet high, hold more than 50 acre-feet of water, or are considered a significant hazard if they fail. Listed here is the category of primary owner. The vast majority of dams are privately owned, but these dams are typically the smallest – often on farms. The smaller number of federally and state owned dams include most of the largest dams in the country. The U.S. Army Corps has made it extremely difficult to access the full National Inventory on Dams now, closing off an important source of basic data on water infrastructure, presumably because of worries about security.

Limitations

The NID only includes dams that:

1. Have a high or significant hazard potential

2. Have a low hazard potential dam but exceed 25 feet in height and 15 acre-feet of storage

3. Have a low hazard potential dam but exceed 50 acre-feet storage and 6 feet in height

Thus a potentially large number of dams are excluded from the database.

SOURCES

Army Corp of Engineers. 2005. National Inventory of Dams. http://crunch.tec.army.mil/nidpublic/webpages/nid.cfm. These data are no longer completely open to public viewing.

DATA TABLE 11   United States Dams by Owner

Dam Owner	Number	Percent
Private	44,980	56%
Local	16,013	20%
Undetermined	9,212	12%
State	3,861	5%
Federal	3,771	5%
Public Utility	1,940	2%
	79,777	

African Dams: Number and Total Reservoir Capacity by Country

Description

Africa has far fewer major dams than more developed parts of the world. This table lists the number of dams in countries of Africa, along with the total reservoir capacity, in thousand cubic meters.

Limitations

The Aquastat database is self-described as "neither complete nor . . .error-free. It corresponds to the best available information at the time of the study."

SOURCES

AQUASTAT (http://www.fao.org/nr/water/aquastat/damsafrica/index.stm According to Aquastat, the references used for the database were: i) International Commission on Large Dams (ICOLD). 1985. The World Register of Dams; ii) National Reports; iii) Information obtained from national experts through AQUASTAT national surveys; iv) the Internet.

DATA TABLE 12A African Dams: Number and Total Reservoir Capacity by Country

Country	Number of Dams	Total Reservoir Capacity (×1000 m³)
Algeria	17	4,470,210
Angola	2	23,500
Benin	9	331,205
Botswana	78	2,001,692
Burkina Faso	10	15,725,050
Cameroon	2	9,000
Congo	22	37,219,200
Côte d'Ivoire	15	52,868
Democratic Republic of the Congo	5	167,000,000
Egypt	2	22,000
Eritrea	10	3,457,900
Ethiopia	1	220,000
Gabon	19	148,304,367
Ghana	2	237,000
Guinea	19	4,758,108
Kenya	6	2,820,395
Lesotho	1	0
Liberia	16	—
Libyan Arab Jamahiriya	12	493,460
Madagascar	8	41,755
Malawi	4	13,615,000
Mali	1	500,000
Mauritania	11	92,870
Mauritius	105	16,551,730
Morocco	27	64,473,570
Mozambique	21	708,521
Namibia	81	44,165,839
Nigeria	1	250,000
Senegal	1	220,000
Sierra Leone	517	28,350,208
South Africa	4	8,730,000
Sudan	9	584,990
Swaziland	4	1,710,900
Togo	52	2,512,376
Tunisia	2	980
Uganda	4	4,196,000
United Republic of Tanzania	5	99,751,400
Zambia	123	99,008,120
Zimbabwe	0	0
Africa Total	**1228**	**772,610,214**

Null or missing values excluded.

African Dams: The 30 Highest

Description

This table lists the 30 highest dams in Africa, with information on the location by country, river basin, and river, the area of the reservoir behind the dam in thousand of square meters, and the volume of the reservoir in thousands of cubic meters. The table is sorted by the height of the dams in meters and includes the date the dam was completed (or "u.c." for under construction). Africa has far fewer major dams than more developed parts of the world. The tallest dam in Africa is the Katse Dam in Lesotho; the dam that creates the largest reservoir is Kariba Dam on the Zambezi River, shared by Zambia and Zimbabwe.

Limitations

The Aquastat database is self-described as "neither complete nor . . .error-free. It corresponds to the best available information at the time of the study."

SOURCE

AQUASTAT (http://www.fao.org/nr/water/aquastat/damsafrica/index.stm According to Aquastat, the references used for the database were: i) International Commission on Large Dams (ICOLD). 1985. The World Register of Dams; ii) National Reports; iii) Information obtained from national experts through AQUASTAT national surveys; iv) the Internet.

DATA TABLE 12B African Dams: The 30 Highest

Name of Dam	Country	River	Sub-basin	Completed and/or Operational Since	Height of Dam (m)	Capacity of the Reservoir (×1000 cubic meters)	Area of the Reservoir (×1000 square meters)
Katse	Lesotho	Malibamatso	Orange	1997	185.00	1 950 000	35 800
Chahora Bassa	Mozambique	Zambeze	Zambezi	1974	171.00	39 000 000	2 660 000
Turkwel	Kenya	Turkwel	Lake Turkana	1991	155.00	1 645 000	66 100
Mohale	Lesotho	Senqunyane	Orange	u.c.	145.00	857 100	21 200
Hassan 1°	Morocco	Lakhdar	Oum er Rbia	1986	145.00	273 000	670
Akosombo (main)	Ghana	Volta	Volta	1965	134.00	147 960 000	8 482 250
Bin El Ouidane	Morocco	El Abid	Oum er Rbia	1953	133.00	1 484 000	3 820
Kariba	Border Zambia-Zimbabwe	Zambezi	Zambezi	1959	128.00	188 000 000	NA
Shiroro	Nigeria	Kaduna/Dinya	Kaduna	1984	125.00	7 000	312 000
Sidi Said	Morocco	Moulouya		2003	124.00	400 000	1 270
Maguga	Swaziland	Komati	Incomati	2001	115.00	332 000	10 420
Kiambere	Kenya	Tana	Tana	1987	112.00	585 000	25 000
Asfalou	Morocco	Asfalou		1999	112.00	317 000	NA
High Aswan dam	Egypt	Nile	Nile	1970	111.00	162 000 000	6 500 000
Keddara	Algeria	Boudouaou	Algerian east coast	1985	108.00	145 600	5 200
Ghrib	Algeria	Cheliff	Chelif	1938	105.00	280 000	11 985
Oued Fodda	Algeria	Fodda	Chelif	1932	101.00	228 000	7 000
Ahmed Al Hansali	Morocco	Oum Er Rbia		2001	101.00	740 000	NA
Bou Roumi	Algeria	Bou Roumi	Algerian east coast	1985	100.00	188 000	6 300
Moulay Youssef	Morocco	Tessaout	Oum er Rbia	1969	100.00	197 000	495
Pavua	Mozambique	Pungoè		u.c.	100.00	636 000	NA

continues

DATA TABLE 12B *continued*

Name of Dam	Country	River	Sub-basin	Completed and/or Operational Since	Height of Dam (m)	Capacity of the Reservoir (×1000 cubic meters)	Area of the Reservoir (×1000 square meters)
Bou Hanifia	Algeria	El Hamkam	Algerian west coast	1948	99.00	73 000	5 369
Sidi Mohamed Ben Abdellah	Morocco	Bouregreg	Bou Regreg	1974	99.00	509 000	2 800
Abdelmoumen	Morocco	Issen	Souss-Massa	1981	94.00	216 000	750
H. Meskoutine	Algeria	B. Hamdane	Algerian east coast	1987	93.00	220 000	11 000
Al Wahda	Morocco	Ouergha		1996	88.00	3 730 000	12 300
Sidi Yakoub	Algeria	Sly	Chelif	1983	87.00	280 000	8 850
Hassan Addakhil	Morocco	Ziz	Ziz	1971	85.00	369 000	1 900
Youssf Ben Tachfine	Morocco	Massa	Souss - Massa	1972	85.00	320 000	1 375
Turkwel	Kenya	Turkwel		1991	84.00	680 000	14 000

u.c.: Under construction.
NA: Not available.

Under-5 Mortality Rate by Cause and Country, 2000

Description

This table presents data from the World Health Organization (WHO) on the under-5 mortality rate and cause of death. Under-5 mortality rate provides an indication of child health and overall development; reducing under-5 mortality is one objective of the Millennium Development Goals (see Chapter 4). Information on the cause of death provides information needed to prioritize and evaluate various intervention methods and assess progress toward meeting national and international goals. In all regions, as noted below, a leading cause of death in children under the age of five is diarrheal disease.

Under-5 mortality is defined as the probability of a child born in a specific year or period dying before reaching the age of five. The measure is derived from a life table and expressed as a rate per 1,000 live births. The rates are calculated based on data from civil registration, census, and/or household surveys. Data on the cause of death are based on information provided on the medical certificate and recorded by the civil registration system. Epidemiological studies and modeling are used in countries with incomplete or no data.

There is tremendous regional variation in the probability of death. Under-5 mortality rates are highest in African countries, at an estimated 167 deaths per 1,000 live births, and lowest in Europe, at about 22 deaths per 1,000 live births. Of all countries, Sierra Leone has the highest mortality rate, 283 deaths per 1,000 live births, while Iceland and Singapore have the lowest mortality, at 3 deaths per 1,000 live births each.

The cause of death is also subject to significant variation. In nearly all countries, death during the first 0–27 days of life is the leading cause of death. Notable exceptions include Lesotho, Namibia, Zimbabwe, Swaziland, and South Africa, where HIV/AIDS is the leading cause of death among children under 5 years of age. Death from water-related diseases, e.g., diarrheal diseases and malaria, is highest in the African region, accounting for 34% of all under-5 mortalities. In Southeast Asia, water-related diseases are also high, accounted for 21% of all under-5 mortalities. Deaths from water-related diseases are lowest in Europe, accounting for almost 11% of all under-5 mortalities. Within Europe, rates are highest among the former Soviet Union countries.

Limitations

Data are collected from a variety of sources. In some cases, the age and cause of death is collected from surveys and is thus subject to some degree of error. As noted by the World Health Organization, "there clearly is substantial variation in data quality and consistency across countries."

SOURCE

World Health Organization. 2007. World Health Statistics 2007. http://www.who.int/whosis/en/index.html

DATA TABLE 13 Under-5 Mortality Rate by Cause and Country, 2000

WHO Regions/Country	Probability of Dying per 1,000 Live Births Under 5 Years (under 5 mortality)	Neonatal Causes	HIV/AIDS	Diarrheal Diseases	Measles	Malaria	Pneumonia	Injuries	Other
	2004	2000	2000	2000	2000	2000	2000	2000	2000
African Region	**167**	**26.2**	**6.8**	**16.6**	**4.3**	**17.5**	**21.1**	**1.9**	**5.6**
Algeria	40	48.0	0.0	11.9	0.9	0.5	13.7	5.0	20.0
Angola	260	22.2	2.2	19.1	4.8	8.3	24.8	1.4	17.2
Benin	152	25.0	2.2	17.1	5.3	27.2	21.1	2.1	0.0
Botswana	116	40.3	53.8	1.1	0.1	0.0	1.4	3.3	0.0
Burkina Faso	192	18.3	4.0	18.8	3.4	20.3	23.3	1.5	10.4
Burundi	190	23.3	8.0	18.2	3.0	8.4	22.8	1.8	14.6
Cameroon	149	24.8	7.2	17.3	4.1	22.8	21.5	2.2	0.0
Cape Verde	36	25.9	3.7	12.2	4.4	4.3	13.3	3.5	32.6
Central African Republic	193	27.2	12.4	14.7	6.5	18.5	18.7	2.0	0.0
Chad	200	24.0	4.1	18.1	7.0	22.3	22.8	1.8	0.1
Comoros	70	37.3	3.7	13.6	5.9	19.4	16.3	3.4	0.5
Congo	108	30.9	9.3	11.2	6.6	25.7	13.6	2.6	0.0
Côte d'Ivoire	194	34.9	5.6	14.8	2.5	20.5	19.6	2.2	0.0
Democratic Republic of the Congo	205	25.7	3.7	18.1	4.7	16.9	23.1	1.6	6.3
Equatorial Guinea	204	27.5	7.4	13.6	7.4	24.0	17.3	2.5	0.3
Eritrea	82	27.4	6.2	15.6	2.5	13.6	18.6	3.0	13.0
Ethiopia	166	30.2	3.8	17.3	4.2	6.1	22.3	1.7	14.3
Gabon	91	35.1	10.1	8.8	4.4	28.3	10.7	2.5	0.0
Gambia	122	36.6	1.3	12.2	2.5	29.4	15.5	2.6	0.0
Ghana	112	28.5	5.7	12.2	2.9	33.0	14.6	3.0	0.0
Guinea	155	28.8	2.3	16.5	5.5	24.5	20.9	1.4	0.0
Guinea-Bissau	203	24.1	2.6	18.6	3.4	21.0	23.4	1.4	5.5
Kenya	120	24.2	14.6	16.5	3.2	13.6	19.9	2.7	5.3

continues

281

DATA TABLE 13 *continued*

WHO Regions/Country	Probability of Dying per 1,000 Live Births Under 5 Years (Under 5 Mortality)	Neonatal Causes	HIV/AIDS	Diarrheal Diseases	Measles	Malaria	Pneumonia	Injuries	Other
	2004	2000	2000	2000	2000	2000	2000	2000	2000
Lesotho	82	32.8	56.2	3.9	0.1	0.0	4.7	2.2	0.0
Liberia	235	29.1	3.6	17.3	6.0	18.9	23.0	1.7	0.3
Madagascar	123	25.6	1.3	16.9	5.0	20.1	20.7	2.4	8.0
Malawi	175	21.7	14.0	18.1	0.3	14.1	22.6	1.7	7.6
Mali	219	25.9	1.6	18.3	6.1	16.9	23.9	1.4	5.9
Mauritania	125	39.4	0.3	16.2	1.7	12.2	22.3	1.9	5.9
Mauritius	15	66.0	0.0	1.2	0.0	0.0	3.9	5.2	23.6
Mozambique	152	29.0	12.9	16.5	0.3	18.9	21.2	1.0	0.1
Namibia	63	38.5	53.0	2.5	0.1	0.0	3.0	3.0	0.0
Niger	259	16.7	0.6	19.8	7.3	14.3	25.1	1.4	14.8
Nigeria	197	26.1	5.0	15.7	6.3	24.1	20.1	1.9	0.8
Rwanda	203	21.7	5.0	18.5	1.6	4.6	23.2	1.8	23.7
Sao Tome and Principe	118	32.1	3.7	16.0	4.8	0.6	21.2	3.5	18.1
Senegal	137	22.8	1.0	17.1	8.1	27.6	20.7	2.6	0.2
Seychelles	14	27.2	0.0	0.0	0.0	0.0	10.1	12.3	50.3
Sierra Leone	283	21.9	1.3	19.7	5.3	12.4	25.5	1.2	12.7
South Africa	67	35.1	57.1	0.8	0.0	0.0	0.9	5.0	1.1
Swaziland	156	26.8	47.0	9.6	0.2	0.2	11.8	3.8	0.5
Togo	140	29.0	5.8	13.8	6.6	25.3	17.1	2.5	0.0
Uganda	138	23.6	7.7	17.2	3.0	23.1	21.1	2.2	2.1
United Republic of Tanzania	126	26.9	9.3	16.8	1.3	22.7	21.1	2.0	0.0
Zambia	182	22.9	16.1	17.5	1.2	19.4	21.8	1.0	0.1
Zimbabwe	129	28.1	40.6	12.1	2.9	0.2	14.7	1.2	0.3

Region of the Americas	25	43.7	1.4	10.1	0.1	0.4	11.6	4.9	27.9
Antigua and Barbuda	12	25.3	1.0	2.4	0.0	0.0	1.5	2.4	67.4
Argentina	18	56.5	0.2	1.3	0.0	0.0	3.4	7.7	30.8
Bahamas	13	43.5	5.3	0.8	0.0	0.0	5.3	13.0	32.1
Barbados	12	63.8	1.7	0.0	0.0	0.0	0.0	1.7	32.8
Belize	39	49.0	1.0	3.5	0.0	0.0	6.9	9.8	29.9
Bolivia	69	37.9	0.1	14.3	0.1	0.7	17.1	5.1	24.7
Brazil	34	38.0	0.3	12.0	0.0	0.5	13.2	3.2	32.8
Canada	6	58.5	0.0	0.2	0.0	0.0	1.1	7.2	32.9
Chile	9	52.8	0.1	0.5	0.0	0.0	6.2	9.1	31.2
Colombia	21	62.1	1.4	10.3	0.0	0.2	10.4	4.6	11.0
Costa Rica	13	58.7	0.2	3.0	0.0	0.0	4.0	3.9	30.1
Cuba	7	49.9	0.0	1.3	0.0	0.0	4.1	7.9	36.9
Dominica	14	99.9	0.0	0.0	0.0	0.0	0.0	0.0	0.1
Dominican Republic	32	47.2	3.9	11.7	0.1	0.6	13.0	2.9	20.6
Ecuador	26	49.8	1.1	11.0	0.1	0.5	12.0	4.6	20.9
El Salvador	28	39.9	1.7	12.4	0.0	0.5	13.4	3.7	28.4
Grenada	21	43.8	2.6	1.6	0.0	0.0	9.5	5.2	37.3
Guatemala	45	37.3	2.7	13.1	0.1	0.4	15.0	1.5	29.8
Guyana	64	33.7	7.7	21.4	0.0	0.7	5.2	6.2	25.2
Haiti	117	26.4	8.3	16.5	0.5	0.7	20.2	0.4	27.0
Honduras	41	43.1	6.3	12.2	0.0	0.4	13.8	4.2	20.1
Jamaica	20	52.1	6.1	9.6	0.0	0.0	9.3	2.4	20.6
Mexico	28	52.5	0.1	5.1	0.0	0.0	8.5	7.0	26.8
Nicaragua	38	42.4	0.5	12.2	0.0	0.4	13.7	3.0	27.7
Panama	24	42.4	2.4	10.7	0.0	0.2	10.8	3.8	29.6
Paraguay	24	53.5	0.2	10.7	0.1	0.3	11.9	3.8	19.6
Peru	29	38.5	0.9	12.2	0.0	0.4	13.6	9.5	24.9
Saint Kitts and Nevis	21	2.8	0.0	14.4	0.0	0.0	0.0	7.9	74.9
Saint Lucia	14	30.9	1.3	1.3	0.0	0.0	1.3	4.7	60.4

continues

283

WHO Regions/Country	Probability of Dying per 1,000 Live Births Under 5 Years (Under 5 Mortality)	Neonatal Causes	HIV/AIDS	Diarrheal Diseases	Measles	Malaria	Pneumonia	Injuries	Other
	2004	2000	2000	2000	2000	2000	2000	2000	2000
Saint Vincent and the Grenadines	22	49.6	2.9	0.5	0.0	0.0	10.5	4.0	32.4
Suriname	39	40.5	2.5	13.1	0.3	2.4	11.5	5.8	23.9
Trinidad and Tobago	20	46.3	4.7	1.3	0.0	0.0	2.0	3.1	42.5
United States of America	8	56.9	0.1	0.1	0.0	0.0	1.3	10.3	31.3
Uruguay	14	48.1	0.2	2.3	0.0	0.0	5.4	7.0	36.9
Venezuela (Bolivarian Republic of)	19	52.6	0.2	9.9	0.0	0.0	5.9	6.5	24.8
Eastern Mediterranean Region	**94**	**43.4**	**0.4**	**14.6**	**3.0**	**2.9**	**19.0**	**3.2**	**13.5**
Afghanistan	257	26.0	0.3	18.9	5.9	1.0	24.8	1.1	22.1
Bahrain	11	46.0	0.2	0.7	0.0	0.0	1.4	10.2	41.5
Djibouti	126	27.0	2.7	16.6	4.4	0.8	20.4	1.8	26.2
Egypt	36	44.3	0.0	12.8	0.1	0.4	14.6	2.1	25.7
Iran (Islamic Republic of)	38	62.9	0.1	5.5	0.0	0.2	6.4	12.8	12.1
Iraq	125	50.8	0.3	13.2	0.5	0.7	17.6	5.7	11.2
Jordan	27	55.4	0.1	10.7	0.0	0.3	11.7	2.3	19.5
Kuwait	12	35.5	0.0	0.7	0.0	0.0	4.4	7.9	51.5
Lebanon	31	64.9	0.0	1.0	0.0	0.0	1.1	11.0	22.0
Libyan Arab Jamahiriya	20	55.6	0.1	8.4	0.1	0.0	8.5	2.6	24.8
Morocco	43	44.7	0.3	12.2	0.2	0.4	14.0	4.0	24.1
Oman	13	42.3	0.3	8.1	0.0	0.1	7.2	4.1	37.9
Pakistan	101	55.7	0.0	14.0	2.4	0.7	19.3	2.1	5.7
Qatar	12	29.6	0.1	8.4	0.0	0.0	7.7	5.2	48.9
Saudi Arabia	27	40.2	0.1	6.2	0.0	0.2	6.6	14.5	32.2
Somalia	225	23.3	0.8	18.7	6.8	4.5	23.9	2.6	19.5
Sudan	91	31.4	2.9	12.9	5.4	21.2	15.5	4.6	6.2

Syrian Arab Republic	16	42.7	0.0	9.6	0.0	0.2	9.9	3.4	34.1
Tunisia	25	52.7	0.0	7.0	0.0	0.2	7.6	9.7	22.8
United Arab Emirates	8	55.7	0.1	6.3	0.0	0.0	4.7	15.0	18.2
Yemen	111	33.3	0.3	16.1	2.2	7.5	19.8	3.7	17.1
European Region	**22**	**44.3**	**0.2**	**10.2**	**0.1**	**0.5**	**13.1**	**6.2**	**25.4**
Albania	19	52.8	0.0	10.5	0.1	0.4	10.6	4.4	21.2
Andorra	7	...	...	...	...	...	...	...	22.7
Armenia	32	48.4	0.2	10.5	0.1	0.5	11.8	5.8	34.9
Austria	5	56.0	0.0	0.0	0.0	0.0	0.7	8.4	19.7
Azerbaijan	90	44.1	0.0	15.3	0.1	1.0	18.4	1.3	30.8
Belarus	10	37.5	3.2	1.5	0.0	0.0	9.0	18.1	38.7
Belgium	5	50.1	0.5	0.3	0.0	0.0	0.8	9.7	40.5
Bosnia and Herzegovina	15	52.7	0.0	0.6	0.0	0.0	2.5	3.7	29.1
Bulgaria	15	47.3	0.0	2.3	0.0	0.0	16.1	5.2	24.6
Croatia	7	65.3	0.0	0.3	0.0	0.0	1.3	8.5	28.2
Cyprus	5	61.5	0.1	3.2	0.0	0.0	1.7	5.4	34.7
Czech Republic	5	48.9	0.0	0.2	0.0	0.0	3.6	12.5	19.4
Denmark	5	73.8	0.0	0.3	0.0	0.0	0.9	5.5	24.3
Estonia	8	54.3	0.0	1.4	0.0	0.0	2.1	17.9	36.0
Finland	4	55.1	0.0	0.8	0.0	0.0	1.2	6.9	37.5
France	5	52.6	0.0	0.9	0.0	0.0	0.6	8.3	22.3
Georgia	45	52.1	0.0	11.5	0.1	0.3	12.5	1.2	41.8
Germany	5	50.7	0.1	0.2	0.0	0.0	0.7	6.6	28.6
Greece	5	63.0	0.0	0.0	0.0	0.0	2.6	5.8	33.6
Hungary	8	56.9	0.0	0.1	0.0	0.0	3.9	5.6	34.1
Iceland	3	61.0	0.0	0.0	0.0	0.0	0.0	4.9	34.2
Ireland	6	61.1	0.0	0.0	0.5	0.0	1.3	2.9	40.3
Israel	6	52.8	0.0	0.6	0.0	0.0	0.4	5.9	32.8
Italy	5	62.0	0.2	0.0	0.0	0.0	1.0	4.0	17.9
Kazakhstan	73	43.1	0.0	14.5	0.1	0.8	16.9	6.8	

continues

DATA TABLE 13 *continued*

WHO Regions/Country	Probability of Dying per 1,000 Live Births Under 5 Years (Under 5 Mortality) 2004	Neonatal Causes 2000	HIV/AIDS 2000	Diarrheal Diseases 2000	Measles 2000	Malaria 2000	Pneumonia 2000	Injuries 2000	Other 2000
Kyrgyzstan	68	43.8	0.0	14.1	0.1	0.9	16.7	6.6	17.9
Latvia	11	53.2	0.0	0.0	0.0	0.0	1.2	11.3	34.3
Lithuania	10	41.4	0.0	0.3	0.0	0.0	5.3	17.4	35.6
Luxembourg	6	54.0	0.0	0.0	0.0	0.0	1.1	14.9	29.9
Malta	6	66.7	0.0	0.0	0.0	0.0	0.0	6.0	27.4
Monaco	4	...	...	...	...	...	...	...	...
Netherlands	5	63.1	0.0	0.0	0.0	0.0	1.1	5.2	30.6
Norway	4	54.0	0.0	0.3	0.0	0.0	1.4	6.2	38.1
Poland	8	59.1	0.0	0.1	0.0	0.0	2.7	5.6	32.5
Portugal	5	47.9	0.1	0.1	0.0	0.1	1.8	9.0	41.0
Republic of Moldova	28	46.1	0.0	2.0	0.0	0.0	15.5	13.3	23.1
Romania	20	41.4	0.1	2.5	0.0	0.0	27.1	8.6	20.3
Russian Federation	16	40.8	0.4	2.5	0.0	0.0	6.3	12.0	38.0
San Marino	4	...	...	...	...	...	...	...	...
Serbia and Montenegro	15	57.1	0.1	6.0	0.1	0.0	9.1	2.9	24.7
Slovakia	8	52.7	0.0	1.4	0.0	0.0	9.4	6.0	30.5
Slovenia	4	64.4	0.0	0.0	0.0	0.0	0.0	5.9	29.7
Spain	5	52.4	0.0	0.1	0.0	0.0	1.3	6.5	39.6
Sweden	4	59.4	0.0	0.0	0.0	0.0	0.8	3.4	36.3
Switzerland	5	62.1	0.0	0.2	0.0	0.0	0.7	7.5	29.5
Tajikistan	118	29.7	0.0	16.4	0.2	0.8	19.9	2.6	30.4
The former Yugoslav Republic of Macedonia	14	63.1	0.0	5.0	0.0	0.0	4.3	2.5	25.1
Turkey	32	49.1	0.0	12.2	0.3	0.5	14.0	4.0	19.8
Turkmenistan	103	37.8	0.0	15.6	0.1	0.9	18.8	4.8	22.0

Ukraine	18	42.3	4.9	1.2	0.0	0.0	6.3	14.5	30.7
United Kingdom	6	59.1	0.0	0.9	0.0	0.0	2.2	4.4	33.4
Uzbekistan	69	38.1	0.0	14.8	0.1	0.8	16.8	7.0	22.4
South-East Asia Region	**77**	**44.4**	**0.6**	**20.1**	**3.5**	**1.1**	**18.1**	**2.3**	**9.9**
Bangladesh	77	45.4	0.0	20.0	2.0	0.7	17.6	2.7	11.4
Bhutan	80	38.9	0.7	20.9	1.2	0.8	18.8	2.4	16.3
Democratic People's Republic of Korea	55	41.8	0.7	18.9	0.8	0.7	15.2	3.0	18.9
India	85	45.2	0.7	20.3	3.7	0.9	18.5	2.2	8.5
Indonesia	38	37.6	0.0	18.3	4.7	0.5	14.4	2.8	21.8
Maldives	46	45.1	0.7	20.3	0.1	0.6	17.5	2.5	13.1
Myanmar	105	39.1	0.9	21.1	2.4	9.0	19.3	2.0	6.2
Nepal	76	43.5	0.2	20.5	2.7	0.8	18.5	2.3	11.5
Sri Lanka	14	59.5	0.0	13.5	1.7	0.4	8.5	5.4	10.9
Thailand	21	44.9	6.2	16.2	0.1	0.3	11.5	4.8	16.0
Timor-Leste	80	32.3	0.7	21.9	3.5	0.4	19.6	1.9	19.7
Western Pacific Region	**31**	**47.0**	**0.3**	**12.0**	**0.8**	**0.4**	**13.8**	**7.3**	**18.4**
Australia	5	55.6	0.0	0.1	0.0	0.0	1.2	10.6	32.5
Brunei Darussalam	9	63.7	0.0	1.1	0.0	0.0	0.7	9.2	25.4
Cambodia	141	29.8	2.0	16.6	2.3	0.9	20.6	1.7	26.1
China	31	49.2	0.1	11.8	0.4	0.4	13.4	8.4	16.3
Cook Islands	21	96.1	0.0	0.7	0.5	0.0	1.1	0.2	1.4
Fiji	20	41.2	0.2	10.6	0.0	0.0	9.2	2.9	36.0
Japan	4	40.0	0.0	0.4	0.2	0.0	3.9	11.6	43.9
Kiribati	65	22.1	0.0	21.9	2.6	0.7	11.5	1.3	39.9
Lao People's Democratic Republic	83	34.5	0.0	15.6	5.9	0.7	19.1	2.3	21.9
Malaysia	12	61.8	1.4	5.4	0.9	0.1	4.0	7.7	18.7
Marshall Islands	59	37.1	0.3	14.1	0.5	0.0	13.5	3.1	31.4
Micronesia (Federated States of)	23	49.2	0.3	8.0	1.5	0.0	11.3	2.7	26.9

continues

DATA TABLE 13 *continued*

WHO Regions/Country	Probability of Dying per 1,000 Live Births Under 5 Years (Under 5 Mortality)	Neonatal Causes	HIV/AIDS	Diarrheal Diseases	Measles	Malaria	Pneumonia	Injuries	Other
	2004	2000	2000	2000	2000	2000	2000	2000	2000
Mongolia	52	34.1	0.3	14.5	0.3	1.0	17.1	4.4	28.3
Nauru	30	7.0	0.0	37.8	5.5	0.0	30.3	19.4	0.1
New Zealand	6	48.3	0.0	0.2	0.0	0.0	2.7	11.4	37.4
Niue	36	...	...	...	...	...	...	...	...
Palau	27	47.0	0.3	9.7	0.7	0.0	12.4	2.5	27.4
Papua New Guinea	93	35.4	0.3	15.3	2.1	0.8	18.5	2.3	25.4
Philippines	34	36.9	0.0	12.0	1.2	0.4	13.4	2.7	33.5
Republic of Korea	6	71.5	0.0	0.4	0.2	0.0	1.8	11.2	15.0
Samoa	30	49.2	0.3	9.7	0.1	0.1	10.2	2.9	27.4
Singapore	3	40.0	0.0	0.4	0.0	0.0	9.0	7.1	43.5
Solomon Islands	56	49.5	0.3	8.8	0.5	0.1	9.5	2.5	28.7
Tonga	25	57.2	0.0	10.0	1.8	1.3	7.3	2.0	20.4
Tuvalu	51	40.0	0.3	13.2	1.2	0.0	13.5	3.0	28.8
Vanuatu	40	42.3	0.3	11.5	0.3	0.6	13.0	2.7	29.4
Viet Nam	23	56.4	1.0	10.4	3.4	0.4	11.5	4.9	11.9

. . . Data not available or not applicable.

Rates are age-standardized to the WHO world standard population.

Sum of individual proportions may not add up to 100% due to rounding.

International River Basins of Africa, Asia, Europe, North America, and South America

Description

This set of tables lists all international river basins in Africa (Table 14A), Asia (Table 14B), Europe (Table 14C), North America (Table 14D), and South America (Table 14E) from an updated assessment prepared by Dr. Aaron Wolf and his colleagues at Oregon State University published in 2002 (see sources, below). A "river basin" is defined as the area that contributes hydrologically to a first-order stream that ends in the ocean or a terminal (closed) lake or inland sea. "River basin" is thus synonymous with "watershed" or "catchment." A basin is considered "international" if any perennial tributary crosses the political boundaries of two or more nations or states. The total area of each basin is provided in square kilometers, along with the countries that share each basin and the area (and percentage of area) of each country within the basin. All together there are around 260 such international river basins, comprising just under half the land area of the earth, excluding Antarctica. See the original on-line source for up-to-date footnotes, as noted in the Tables.

Limitations

Changing political boundaries constantly alter the number and extent of international river basins – each table includes some relevant footnotes that expound on ongoing political discussions and disputes over borders and watershed boundaries. In the first comprehensive assessment done, in 1978, 214 international river basins were identified. The increase from that number to the current estimate of over 260 reflects both better mapping of world's hydrologic boundaries and major changes in political borders in the former Soviet Union, Eastern European countries, and elsewhere. Political borders will continue to shift, and the groupings and numbers in this table will continue to change.

Major rivers that join above an outlet are grouped together here. For example, the Tigris and Euphrates rivers are distinct and separate rivers for most of their extent, but they merge upstream of the Arabian Gulf and hence are grouped in this table. Similarly, the Ganges, Brahmaputra, and Meghna rivers – each a significant river in its own right – join before they reach the ocean and are thus grouped here.

Some border areas are disputed. Where such disputes are well known, they are explicitly identified in this table. For example, portions of the Indus and the Ganges-Brahmaputra-Meghna basins are under Chinese control but claimed by India, while other portions are under Indian control but claimed by China.

Groundwater basins also often extend across international boundaries, but these are not included here. Indeed, no assessment of international groundwater basins has been published.

SOURCES

International River Basin Registry, Oregon State University. 2002 update. http://www.trans-boundarywaters.orst.edu/database/interriverbasinreg.html, with permission.

Wolf, A.T., Natharius, J.A., Danielson, J.J., Ward, B.S., Pender, J.K. 1999. International river basins of the world. *International Journal of Water Resources Development* 15:4 (December).

DATA TABLE 14A   Africa

Basin Name	Area of Basin (km²) (1)	Countries Sharing the Basin	Area within Country (km²)	Percentage of Watershed Within Country
Akpa (2)	4,900	Cameroon	3,000	61.65
		Nigeria	1,900	38.17
Atui (3)	32,600	Mauritania	20,500	62.91
		Western Sahara	11,200	34.24
Awash	154,900	Ethiopia	143,700	92.74
		Djibouti	11,000	7.09
		Somalia	300	0.16
Baraka	66,200	Eritrea	41,500	62.57
		Sudan	24,800	37.43
Benito/Ntem	45,100	Cameroon	18,900	41.87
		Equatorial Guinea	15,400	34.11
		Gabon	10,800	23.86
Bia	11,100	Ghana	6,400	57.58
		Ivory Coast	4,500	40.28
Buzi	27,700	Mozambique	24,500	88.35
		Zimbabwe	3,200	11.65
Cavally	30,600	Ivory Coast	16,600	54.12
		Liberia	12,700	41.66
		Guinea	1,300	4.22
Cestos	15,000	Liberia	12,800	84.99
		Ivory Coast	2,200	14.91
		Guinea	20	0.11
Chiloango	11,600	Congo, Democratic Republic of (Kinshasa)	7,500	64.6
		Angola	3,800	32.71
		Congo, Republic of the (Brazzaville)	300	2.69
Congo/Zaire (4, 5)	3,691,000	Congo, Democratic Republic of (Kinshasa)	2,302,800	62.39
		Central African Republic	400,800	10.86
		Angola	290,600	7.87
		Congo, Republic of the (Brazzaville)	248,100	6.72
		Zambia	176,000	4.77
		Tanzania, United Republic of	166,300	4.51
		Cameroon	85,200	2.31
		Burundi	14,400	0.39
		Rwanda	4,500	0.12
		Sudan	1,400	0.04
		Gabon	500	0.01
		Malawi	100	0
		Uganda	70	0
Corubal	24,000	Guinea	17,500	72.71
		Guinea-Bissau	6,500	27.02
Cross	52,800	Nigeria	40,300	76.34
		Cameroon	12,500	23.66

continues

DATA TABLE 14A   *continued*

Basin Name	Area of Basin (km²) (1)	Countries Sharing the Basin	Area within Country (km²)	Percentage of Watershed Within Country
Cuvelai/Etosha	167,400	Namibia	114,100	68.15
		Angola	53,300	31.85
Daoura	34,500	Morocco	18,200	52.72
		Algeria	16,300	47.28
Dra	96,400	Morocco	75,800	78.65
		Algeria	20,600	21.33
Gambia	69,900	Senegal	50,700	72.48
		Guinea	13,200	18.92
		Gambia	5,900	8.51
Gash	40,000	Eritrea	21,400	53.39
		Sudan	9,600	24.09
		Ethiopia	9,000	22.52
Geba	12,800	Guinea-Bissau	8,700	67.69
		Senegal	4,100	31.88
		Guinea	50	0.42
Great Scarcies	12,100	Guinea	9,000	74.96
		Sierra Leone	3,000	25.04
Guir	78,900	Algeria	61,200	77.53
		Morocco	17,700	22.47
Incomati (6)	46,700	South Africa	29,200	62.47
		Mozambique	14,600	31.2
		Swaziland	3,000	6.33
Juba-Shibeli	803,500	Ethiopia	367,400	45.72
		Somalia	220,900	27.49
		Kenya	215,300	26.79
Komoe	78,100	Ivory Coast	58,300	74.67
		Burkina Faso	16,900	21.66
		Ghana	2,200	2.85
		Mali	600	0.82
Kunene	110,000	Angola	95,300	86.68
		Namibia	14,700	13.32
		Niger	674,200	28.23
		Central African Republic	218,600	9.15
		Nigeria	180,200	7.54
		Algeria	90,000	3.77
		Sudan	82,800	3.47
		Cameroon	46,800	1.96
		Chad, claimed by Libya	12,300	0.51
		Libya	4,600	0.19
Lake Natron	55,400	Tanzania, United Republic of	37,100	67
		Kenya	18,300	33
Lake Turkana (8)	206,900	Ethiopia	113,200	54.69
		Kenya	89,700	43.36
		Uganda	2,500	1.21
		Sudan	1,500	0.7
		Sudan, administered by Kenya	70	0.03

Basin Name	Area of Basin (km²) (1)	Countries Sharing the Basin	Area within Country (km²)	Percentage of Watershed Within Country
Limpopo	414,800	South Africa	183,500	44.25
		Mozambique	87,200	21.02
		Botswana	81,500	19.65
		Zimbabwe	62,600	15.08
Little Scarcies	18,900	Sierra Leone	13,000	69.12
		Guinea	5,800	30.88
Loffa	11,400	Liberia	10,100	88.56
		Guinea	1,300	11.38
Lotagipi Swamp (8)	38,700	Kenya	20,300	52.33
		Sudan	9,900	25.54
		Sudan, administered by Kenya	3,300	8.52
		Ethiopia	3,200	8.32
		Uganda	2,100	5.29
Mana-Morro	6,800	Liberia	5,700	82.84
		Sierra Leone	1,200	17.16
Maputo (6)	30,700	South Africa	18,500	60.31
		Swaziland	10,600	34.71
		Mozambique	1,500	4.98
Mbe	7,000	Gabon	6,500	92.97
		Equatorial Guinea	500	7.02
Medjerda	23,100	Tunisia	15,600	67.53
		Algeria	7,600	32.9
Moa	22,500	Sierra Leone	10,800	47.79
		Guinea	8,800	39.2
		Liberia	2,900	13.01
Mono	23,400	Togo	22,300	95.19
		Benin	1,100	4.81
Niger	2,113,200	Nigeria	561,900	26.59
		Mali	540,700	25.58
		Niger	497,900	23.56
		Algeria	161,300	7.63
		Guinea	95,900	4.54
		Cameroon	88,100	4.17
		Burkina Faso	82,900	3.93
		Benin	45,300	2.14
		Ivory Coast	22,900	1.08
		Chad	16,400	0.78
		Sierra Leone	50	0
Nile (9)	3,031,700	Sudan	1,927,300	63.57
		Ethiopia	356,000	11.74
		Egypt	272,600	8.99
		Uganda	238,500	7.87
		Tanzania, United Republic of	120,200	3.96
		Kenya	50,900	1.68
		Congo, Democratic Republic of (Kinshasa)	21,400	0.71

continues

DATA TABLE 14A *continued*

Basin Name	Area of Basin (km²) (1)	Countries Sharing the Basin	Area within Country (km²)	Percentage of Watershed Within Country
Nile (*continued*)		Rwanda	20,700	0.68
		Burundi	12,900	0.43
		Egypt, administered by Sudan	4,400	0.15
		Eritrea	3,500	0.12
		Sudan, administered by Egypt	2,000	0.07
		Central African Republic	1,200	0.04
Nyanga	12,300	Gabon	11,500	93.56
		Congo, Republic of the (Brazzaville)	800	6.44
Ogooue	223,000	Gabon	189,500	84.98
		Congo, Republic of the (Brazzaville)	26,300	11.79
		Cameroon	5,200	2.34
		Equatorial Guinea	2,000	0.89
Okavango	706,900	Botswana	358,000	50.65
		Namibia	176,200	24.93
		Angola	150,100	21.23
		Zimbabwe	22,600	3.19
Orange (6, 10, 11)	945,500	South Africa	563,900	59.65
		Namibia	240,200	25.4
		Botswana	121,400	12.85
		Lesotho	19,900	2.1
Oued Bon Naima	500	Morocco	300	65.08
		Algeria	200	34.92
Oueme	59,500	Benin	49,400	82.98
		Nigeria	9,700	16.29
		Togo	400	0.73
Ruvuma (12)	151,700	Mozambique	99,000	65.27
		Tanzania, United Republic of	52,200	34.43
		Malawi	400	0.3
Sabi	115,700	Zimbabwe	85,400	73.85
		Mozambique	30,300	26.15
Sassandra	68,200	Ivory Coast	59,800	87.64
		Guinea	8,400	12.36
Senegal	436,000	Mauritania	219,100	50.25
		Mali	150,800	34.59
		Senegal	35,200	8.08
		Guinea	30,800	7.07
St. John (Africa)	15,600	Liberia	12,900	83.04
		Guinea	2,600	16.96
St. Paul	21,200	Liberia	11,800	55.75
		Guinea	9,400	44.25
Tafna	9,500	Algeria	7,000	74.39
		Morocco	2,400	25.6

Basin Name	Area of Basin (km^2) (1)	Countries Sharing the Basin	Area within Country (km^2)	Percentage of Watershed Within Country
Tano	15,600	Ghana	13,700	87.96
		Ivory Coast	1,700	11.21
Umba	8,200	Tanzania, United Republic of	6,800	83.58
		Kenya	1,300	16.41
Umbeluzi (6)	10,900	Mozambique	7,200	65.87
		Swaziland	3,500	32.44
		South Africa	30	0.27
Utamboni	7,700	Gabon	4,500	58.65
		Equatorial Guinea	3,100	40.4
Volta	412,800	Burkina Faso	173,500	42.04
		Ghana	166,000	40.21
		Togo	25,800	6.26
		Mali	18,800	4.56
		Benin	15,000	3.63
		Ivory Coast	13,500	3.27
Zambezi (13, 14)	1,385,300	Zambia	576,900	41.64
		Angola	254,600	18.38
		Zimbabwe	215,500	15.55
		Mozambique	163,500	11.81
		Malawi	110,400	7.97
		Tanzania, United Republic of	27,200	1.97
		Botswana	18,900	1.37
		Namibia	17,200	1.24
		Congo, Democratic Republic of (Kinshasa)	1,100	0.08

1) The numbers referring to basin areas have been rounded to significant digits and, as a result, the numbers for area within each basin do not necessarily add up to the total area for that basin. Also, the percentages were calculated based on raw data, and therefore do not reflect the rounding of the areas.

2) The dispute between Nigeria and Cameroon, over land and maritime boundaries in the vicinity of the oil rich Bakasi Peninsula, was referred to the International Court of Justice, which gave a ruling in 1998. Nigeria has filed an appeal on the ruling and the dispute has yet to be resolved. The Bakasi Peninsula, in the southwest province of Cameroon, is divided by the Akpa Yafi river and lies to the west of Cameroon's Rio del Ray. (CIA World Factbook, 1998; Columbia Gazetteer, 1998).

3) Morocco claims and administers Western Sahara, but the region's sovereignty is unresolved and the UN is attempting to hold a referendum on the issue. A UN-administered cease-fire remains in effect since September 1991. (Encyclopedia of International Boundaries, 1995; CIA World Factbook, 1998).

4) It has been informally reported that the indefinite segment of the Democratic Republic of the Congo (Kinshasa)-Zambia boundary has been settled. Therefore, the Democratic Republic of the Congo (Kinshasa)-Tanzania-Zambia tripoint in Lake Tanganyika also may no longer be indefinite. (CIA World Factbook, 1998).

5) A long segment of the boundary between the Democratic Republic of the Congo (Kinshasa) and the Republic of the Congo (Brazzaville) along the Congo River remains indefinite, as no division of the river or its islands has been made. (CIA World Factbook, 1998).

6) Swaziland has asked South Africa to open negotiations on reincorporating some nearby South African territories that are populated by ethnic Swazis or that were long ago part of the Swazi Kingdom. (CIA World Factbook, 1998).

7) Lake Chad varies in extent between rainy and dry seasons - from 50,000 to 20,000 km^2. Demarcation of international boundaries in the vicinity of Lake Chad is complete and awaits ratification by Cameroon, Chad, Niger, and Nigeria. Determining the boundaries of sectors involving rivers draining into Lake Chad is complicated by flooding and the uncovering or covering of islands. The lack of demarcated boundaries has led to border incidents in the past. (Encyclopedia of International Boundaries, 1995; The CIA World Factbook, 1998).

8) The administrative boundary between Kenya and Sudan does not coincide with the international boundary. (CIA World Factbook, 1998).

continues

9) Egypt's administrative boundary with Sudan does not coincide with the international boundary and creates the "Hala'ib Triangle," a barren area of 20,580 km² north of the 22nd parallel. (CIA World Factbook, 1998).

10) Although topographically Botswana is riparian to the Orange River basin, it is unknown whether Botswana territory contributes water to the Orange River. Botswana's political status as riparian to the Orange River basin remains to be clarified among the basin states. (Conley and van Niekerk, 1998).

11) Namibia and South Africa are undergoing negotiations to confirm the exact positions of their boundary along the Orange River. (Conley and van Niekerk, 1998).'

12) Malawi is in dispute with Tanzania over the boundary in Lake Nyasa (Lake Malawi). (CIA World Factbook, 1998).

13) The quadripoint between Botswana, Namibia, Zambia and Zimbabwe is in disagreement. (CIA World Factbook, 1998).

14) The dispute between Botswana and Namibia over the uninhabited Kasikili (Sidudu) Island in the Linyanti (Chobe) River is presently before the International Court of Justice. Botswana and Namibia are also contesting at least one other island in Linyanti River. (CIA World Factbook, 1998).

Data Table 14B   Asia

Basin Name	Area of Basin (km²) (1)	Countries Sharing the Basin	Area within Country (km²)	Percentage of Watershed Within Country
Amur (15)	2,085,900	Russia	1,006,100	48.23
		China	889,100	42.62
		Mongolia	190,600	9.14
		Korea, Democratic People's Republic of (North)	100	0.01
An Nahr Al Kabir	1,300	Syria	900	67.6
		Lebanon	400	31.7
Aral Sea (16, 17)	1,231,400	Kazakhstan	424,400	34.46
		Uzbekistan	382,600	31.07
		Tajikistan	135,700	11.02
		Kyrgyzstan	111,700	9.07
		Afghanistan	104,900	8.52
		Turkmenistan	70,000	5.68
		China	1,900	0.15
		Pakistan	200	0.01
Asi/Orontes	37,900	Turkey	18,900	49.94
		Syria	16,800	44.32
		Lebanon	2,200	5.74
Astara Chay (18)	600	Iran	500	81.64
		Azerbaijan	100	18.36
Atrak (17)	34,200	Iran	23,600	68.86
		Turkmenistan	10,700	31.14
BahuKalat/ Rudkhanehye	18,000	Iran	18,000	99.83
		Pakistan	30	0.17
Bangau (19)	60	Brunei	30	46.03
		Malaysia	30	49.21
Bei Jiang/Hsi (20)	417,800	China	407,900	97.63
		Vietnam	9,800	2.35
Beilun (20)	900	China	800	84.92
		Vietnam	100	15.08
Ca/Song Koi	31,000	Vietnam	20,100	64.91
		Laos, People's Democratic Republic of	10,900	35.09

Basin Name	Area of Basin (km^2) (1)	Countries Sharing the Basin	Area within Country (km^2)	Percentage of Watershed within Country
Coruh (18)	22,100	Turkey	20,000	90.85
		Georgia	2,000	9.01
Dasht	33,400	Pakistan	26,200	78.42
		Iran	7,200	21.58
Fenney	2,800	India	1,800	65.83
		Bangladesh	1,000	34.17
Fly	64,600	Papua New Guinea	60,400	93.4
		Indonesia	4,300	6.6
Ganges-Brahmaputra-Meghna (21, 22)	1,634,900	India	948,400	58.01
		China	321,300	19.65
		Nepal	147,400	9.01
		Bangladesh	107,100	6.55
		India, claimed by China	67,100	4.11
		Bhutan	39,900	2.44
		India control, claimed by China	1,200	0.07
		Myanmar (Burma)	80	0
Golok	1,800	Thailand	1,000	56.62
		Malaysia	800	43.38
Han (23, 24)	35,300	Korea, Republic of (South)	25,100	71.22
		Korea, Democratic People's Republic of (North)	10,100	28.67
Har Us Nur	185,300	Mongolia	179,300	96.81
		Russia	5,600	3.04
		China	300	0.15
Hari/Harirud	92,600	Afghanistan	41,000	44.31
		Iran	35,400	38.27
		Turkmenistan	16,100	17.41
Helmand	353,500	Afghanistan	288,200	81.53
		Iran	54,900	15.52
		Pakistan	10,400	2.95
Ili/Kunes He	161,200	Kazakhstan	97,100	60.24
		China	55,300	34.32
		Kyrgyzstan	8,800	5.44
Indus (25, 26)	1,138,800	Pakistan	597,700	52.48
		India	381,600	33.51
		China	76,200	6.69
		Afghanistan	72,100	6.33
		Chinese control, claimed by India	9,600	0.84
		Indian control, claimed by China	1,600	0.14
		Nepal	10	0

continues

DATA TABLE 14B   *continued*

Basin Name	Area of Basin (km²) (1)	Countries Sharing the Basin	Area within Country (km²)	Percentage of Watershed within Country
Irrawaddy	404,200	Myanmar (Burma)	368,600	91.2
		China	18,500	4.58
		India	14,100	3.49
		India, claimed by China	1,200	0.3
Jenisej/Yenisey	2,557,800	Russia	2,229,800	87.17
		Mongolia	327,900	12.82
Jordan (27, 28, 29)	42,800	Jordan	20,600	48.13
		Israel	9,100	21.26
		Syria	4,900	11.45
		West Bank	3,200	7.48
		Egypt	2,700	6.31
		Golan Heights	1,500	3.5
		Lebanon	600	1.33
Kaladan	30,500	Myanmar (Burma)	22,900	74.91
		India	7,300	23.84
Karnaphuli	12,500	Bangladesh	7,400	58.78
		India	5,100	41.14
		Myanmar (Burma)	10	0.09
Kowl E Namaksar	36,500	Iran	25,900	71.13
		Afghanistan	10,500	28.87
Kura-Araks (18)	193,200	Azerbaijan	56,600	29.28
		Iran	39,700	20.55
		Armenia	34,800	18.03
		Georgia	34,300	17.77
		Turkey	27,700	14.32
		Russia	60	0.03
Lake Ubsa-Nur	62,800	Mongolia	47,600	75.78
		Russia	15,200	24.22
Ma	30,300	Vietnam	17,100	56.48
		Laos, People's Democratic Republic of	13,200	43.52
Mekong (30, 31)	787,800	Laos, People's Democratic Republic of	198,000	25.14
		Thailand	193,900	24.62
		China	171,700	21.79
		Cambodia (Kampuchea)	158,400	20.1
		Vietnam	38,200	4.84
		Myanmar (Burma)	27,600	3.51
Murgab	60,900	Afghanistan	36,400	59.79
		Turkmenistan	24,500	40.21
Nahr El Kebir	1,500	Syria	1,300	85.61
		Turkey	200	13.87

Basin Name	Area of Basin (km²) (1)	Countries Sharing the Basin	Area within Country (km²)	Percentage of Watershed within Country
Ob (18)	2,950,800	Russia	2,192,700	74.31
		Kazakhstan	743,800	25.21
		China	13,900	0.47
		Mongolia	200	0.01
Oral/Ural (18)	311,000	Kazakhstan	175,500	56.43
		Russia	135,500	43.57
Pakchan	3,900	Myanmar (Burma)	1,900	49.11
		Thailand	1,800	47.24
Pandaruan (19)	400	Brunei	200	60.65
		Malaysia	100	39.08
Pu Lun T'o	89,000	China	77,800	87.39
		Mongolia	11,100	12.48
		Russia	80	0.09
		Kazakhstan	30	0.04
Red/Song Hong (20)	157,100	China	84,500	53.75
		Vietnam	71,500	45.5
		Laos, People's Democratic Republic of	1,200	0.74
Saigon	25,100	Vietnam	24,800	98.67
		Cambodia (Kampuchea)	200	0.99
Salween	244,000	China	127,900	52.4
		Myanmar (Burma)	107,000	43.85
		Thailand	9,100	3.73
Samur (18)	6,800	Russia	6,300	93.75
		Azerbaijan	400	6.22
Sembakung (19)	15,300	Indonesia	8,100	52.86
		Malaysia	7,200	47.14
Sepik	73,400	Papua New Guinea	71,000	96.81
		Indonesia	2,300	3.19
Song Vam Co Dong	15,300	Vietnam	7,800	50.68
		Cambodia (Kampuchea)	7,500	49.23
Sujfun	18,300	China	11,800	64.46
		Russia	6,500	35.54
Sulak (18)	15,100	Russia	13,900	92.38
		Georgia	1,100	7.24
		Azerbaijan	60	0.38
Tami	89,900	Indonesia	87,700	97.55
		Papua New Guinea	2,200	2.45
Tarim (16, 17, 25, 26)	1,051,600	China	1,000,300	95.12
		Chinese control, claimed by India	21,500	2.04
		Kyrgyzstan	21,100	2
		Tajikistan	6,600	0.63
		Pakistan	2,000	0.19
		Afghanistan	60	0.01

continues

DATA TABLE 14B *continued*

Basin Name	Area of Basin (km²) (1)	Countries Sharing the Basin	Area within Country (km²)	Percentage of Watershed within Country
Terek (18)	38,700	Russia	37,000	95.39
		Georgia	1,800	4.61
Tigris-Euphrates/ Shatt al Arab (32)	789,000	Iraq	319,400	40.48
		Turkey	195,700	24.8
		Iran	155,400	19.7
		Syria	116,300	14.73
		Jordan	2,000	0.25
		Saudi Arabia	80	0.01
Tjeroaka-Wanggoe	6,600	Indonesia	4,000	61.57
		Papua New Guinea	2,500	38.43
Tumen	29,100	China	20,300	69.75
		Korea, Democratic People's Republic of (North)	8,300	28.59
		Russia	500	1.66
Wadi Al Izziyah	600	Lebanon	400	68.23
		Israel	200	31.6
Yalu	50,900	China	26,800	52.65
		Korea, Democratic People's Republic of (North)	23,800	46.82

1) The numbers referring to basin areas have been rounded to significant digits and, as a result, the numbers for area within each basin do not necessarily add up to the total area for that basin. Also, the percentages were calculated based on raw data, and therefore do not reflect the rounding of the areas.

15) Two disputed sections of the boundary between China and Russia remain to be settled. China holds that the main channel of the Amur River is followed northeast to a point opposite the city of Khabarovsk. Russia claims that the line follows the Kazakevicheva channel southeastward to the Ussuri River. The two countries dispute control of islands in the Amur and Ussuri Rivers, despite a 1987 agreement that established the line as running through the median lines of the main navigable and unnavigable channels. The five disputed islands in the Amur ••• Popov, Savelyev, Evrasikha, Nizhne-Petrovskiy and Lugovskoy ••• amount to 3,000 km² of territory. Also in dispute are the Tarbarov and Bolshoy Ussuriyskiy islands, located in a 30 km section of the boundary at the confluence of the Amur and Ussuri rivers, and the Bolshoy Island, located in the upper reaches of the Argun river. (Encyclopedia of International Boundaries, 1995; CIA World Factbook, 1998; IBRU, 1999).

16) Most of the boundary shared between China and Tajikistan is in dispute, including in the Pamir mountain region. (CIA World Factbook, 1998; IBRU, 1999).

17) Kyrgyzstan and Tajikistan have a territorial dispute regarding their boundary in the Isfara Valley area. (CIA World Factbook, 1998).

18) The boundaries of the Caspian Sea remain to be determined among Azerbaijan, Iran, Kazakhstan, Russia, and Turkmenistan. (CIA World Factbook, 1998).

19) Brunei may wish to purchase the Malaysian salient that divides the country. (CIA World Factbook, 1998).

20) Sections of the land boundary between China and Vietnam are indefinite. (CIA World Factbook, 1998).

21) India and China dispute approximately 83,000 km², including three of the four political divisions of the Northeast Frontier Agency ••• the Sumdurong Cho sector. This region falls in the Ganges-Brahmaputra basin. (Conflict and Border Disputes, 1993; Columbia Gazetteer, 1998; IBRU 1999).

22) Portions of the boundary between Bangladesh and India are indefinite. Much of the boundary between the two countries is based on administrative units that do not shift with the rivers as they change course or level over time. Alluvial or "char" land that is exposed as a river shifts often leads to dispute, as the land is highly valued for agriculture. (CIA World Factbook, 1998; IBRU, 1999).

23) A 33-km section of the boundary between China and North Korea in the Paektu-san (mountain) area is indefinite. North Korea claims territorial rights to two thirds of Chonji, the crater lake on Mount Paektu. (CIA World Factbook, 1998; IBRU, 1999).

24) The Demarcation Line between North Korea and South Korea is in dispute. (CIA World Factbook, 1998).

25) Disputed boundaries between China and India include approximately 25,900 km² in the regions of Sang, Demchok, and Aksai, China. (Encyclopedia of International Boundaries, 1995; Columbia Gazetteer, 1998).

26) India and Pakistan dispute the status of the Jammu and Kashmir region, an area of approximately 220,000 km². (Encyclopedia of International Boundaries, 1995; CIA World Factbook, 1998).

27) The West Bank and Gaza Strip are Israeli-occupied with the exception of territories under control of the Palestinian Authority, as delineated in the 1995 •••Israeli-Palestinian Interim Agreement on the West Bank and the Gaza Strip,••• commonly referred to as •••Oslo II•••, and in the 1998 agreement signed at Wye. Permanent status is to be determined during further negotiation. (CIA World Factbook, 1998).

28) Israel and Syria dispute the Golan Heights, which is currently administered by Israel. (CIA World Factbook, 1998).

29) Topographically, Egypt is riparian to the Jordan River basin, however Egyptian territory does not contribute water to the basin, except for the possibility of intermittent, seasonal wadis.

30) Parts of the boundary between Cambodia and Thailand are indefinite, including overlapping claims in the Gulf of Thailand, an area potentially containing oil and gas deposits, and an island located near the boundary between Cambodian Koh Kong and the Thai province of Trat. (CIA World Factbook, 1998; IBRU, 1999).

31) Parts of the boundary between People's Democratic Republic of Laos and Thailand are indefinite. The two countries have an agreement to demarcate their boundary, but demarcation was suspended in February, 1998. (CIA World Factbook, 1998; IBRU, 1999).

32) Iran and Iraq restored diplomatic relations in 1990, but work continues on developing written agreements to settle outstanding disputes from their eight-year war, including boundary demarcation, prisoners-of-war, and freedom of navigation and sovereignty over the Shatt al Arab waterway. (CIA World Factbook, 1998).

DATA TABLE 14C   Europe

Basin Name	Area of Basin (km²) (1)	Countries Sharing the Basin	Area within Country (km²)	Percentage of Watershed within Country
Bann	5,600	United Kingdom	5,400	97.14
		Ireland	200	2.86
Barta	1,800	Latvia	1,100	60.87
		Lithuania	700	37.71
Bidasoa	500	Spain	500	89.33
		France	60	10.67
Castletown	400	United Kingdom	300	76.12
		Ireland	90	23.88
Danube (33, 34, 35, 36, 37)	790,100	Romania	228,500	28.93
		Hungary	92,800	11.74
		Austria	81,600	10.32
		Yugoslavia (Serbia and Montenegro)	81,500	10.31
		Germany	59,000	7.47
		Slovakia	45,600	5.77
		Bulgaria	40,900	5.17
		Bosnia and Herzegovina	38,200	4.83
		Croatia	35,900	4.54
		Ukraine	29,600	3.75
		Czech Republic	20,500	2.59
		Slovenia	17,200	2.18
		Moldova	13,900	1.76
		Switzerland	2,500	0.32
		Italy	1,200	0.15
		Poland	700	0.09
		Albania	200	0.03
Daugava (38, 39)	58,700	Byelarus	28,300	48.14
		Latvia	20,200	34.38
		Russia	9,500	16.11
		Lithuania	800	1.38

continues

DATA TABLE 14C *continued*

Basin Name	Area of Basin (km²) (1)	Countries Sharing the Basin	Area within Country (km²)	Percentage of Watershed within Country
Dnieper	516,300	Ukraine	299,300	57.97
		Byelarus	124,900	24.19
		Russia	92,100	17.83
Dniester (37)	62,000	Ukraine	46,800	75.44
		Moldova	15,200	24.52
		Poland	30	0.05
Don	425,600	Russia	371,200	87.23
		Ukraine	54,300	12.76
Douro/Duero	98,900	Spain	80,700	81.63
		Portugal	18,200	18.37
Drin (36)	17,900	Albania	8,100	45.39
		Yugoslavia (Serbia and Montenegro)	7,400	41.4
		Macedonia	2,200	12.18
Ebro	85,800	Spain	85,200	99.36
		Andorra	400	0.48
		France	100	0.16
Elancik	900	Russia	700	71.32
		Ukraine	300	28.68
Elbe	132,200	Germany	83,100	62.86
		Czech Republic	47,600	36.02
		Austria	700	0.54
		Poland	700	0.56
Erne	4,800	Ireland	2,800	59.28
		United Kingdom	1,900	40.72
Fane	200	Ireland	200	96.46
		United Kingdom	10	3.54
Flurry	60	United Kingdom	50	73.77
		Ireland	20	26.23
Foyle	2,900	United Kingdom	2,000	67.3
		Ireland	1,000	32.7
Garonne	55,800	France	55,100	98.83
		Spain	600	1.07
		Andorra	40	0.08
Gauja	11,600	Latvia	10,400	90.42
		Estonia	1,100	9.58
Glama	43,000	Norway	42,600	99
		Sweden	400	0.99
Guadiana	67,900	Spain	54,900	80.82
		Portugal	13,000	19.18
Isonzo	3,000	Slovenia	1,800	59.48
		Italy	1,200	40.09
Jacobs	400	Norway	300	68.1
		Russia	100	31.9
Kemi	55,700	Finland	52,700	94.52
		Russia	3,000	5.41
		Norway	10	0.01

Basin Name	Area of Basin (km²) (1)	Countries Sharing the Basin	Area within Country (km²)	Percentage of Watershed within Country
Klaralven	51,000	Sweden	43,100	84.54
		Norway	7,900	15.46
Kogilnik (37)	6,100	Moldova	3,600	57.82
		Ukraine	2,600	42.18
Krka	1,300	Croatia	1,100	89.55
		Bosnia and Herzegovina	100	8.93
		Yugoslavia (Serbia and Montenegro)	10	0.4
Lake Prespa	9,000	Albania	8,000	88.17
		Macedonia	800	8.5
		Greece	300	3.32
Lava/Pregel	8,600	Russia	6,300	74
		Poland	2,000	23.84
Lielupe	14,400	Latvia	9,600	66.76
		Lithuania	4,800	33.22
Lima	2,300	Spain	1,200	50.88
		Portugal	1,100	49.04
Maritsa	49,600	Bulgaria	33,000	66.49
		Turkey	12,800	25.69
		Greece	3,700	7.55
Mino	15,100	Spain	14,500	96.18
		Portugal	600	3.7
Mius	2,800	Russia	1,900	69.82
		Ukraine	800	30.07
Naatamo	1,000	Norway	600	57.73
		Finland	400	41.97
Narva (40, 41)	53,000	Russia	28,200	53.2
		Estonia	18,100	34.09
		Latvia	5,900	11.13
		Byelarus	800	1.57
Neman (38, 39)	90,300	Byelarus	41,700	46.13
		Lithuania	39,700	43.97
		Russia	4,800	5.3
		Poland	3,800	4.21
		Latvia	300	0.36
Neretva	5,500	Bosnia and Herzegovina	5,300	95.98
		Croatia	200	3.47
Nestos	10,200	Bulgaria	5,500	53.63
		Greece	4,700	46.36
Oder/Odra	122,400	Poland	103,100	84.2
		Czech Republic	10,300	8.38
		Germany	7,800	6.33
		Slovakia	1,300	1.09
Olanga	18,800	Russia	16,800	89.37
		Finland	2,000	10.62

continues

DATA TABLE 14C   *continued*

Basin Name	Area of Basin (km^2) (1)	Countries Sharing the Basin	Area within Country (km^2)	Percentage of Watershed within Country
Oulu	28,700	Finland	26,700	93.2
		Russia	1,900	6.78
Parnu	5,800	Estonia	5,800	99.85
		Latvia	10	0.15
Pasvik	16,000	Finland	12,400	77.46
		Russia	2,600	16.15
		Norway	1,000	6.39
Po	87,100	Italy	82,200	94.44
		Switzerland	4,300	4.92
		France	500	0.54
		Austria	90	0.1
Prohladnaja	600	Russia	500	76.9
		Poland	100	23.1
Rezvaya	700	Turkey	500	74.66
		Bulgaria	200	25.34
Rhine (42)	172,900	Germany	97,700	56.49
		Switzerland	24,300	14.05
		France	23,100	13.34
		Belgium	13,900	8.03
		Netherlands	9,900	5.75
		Luxembourg	2,500	1.46
		Austria	1,300	0.76
		Liechtenstein	200	0.09
		Italy	70	0.04
Rhone	100,200	France	90,100	89.88
		Switzerland	10,100	10.05
		Italy	50	0.05
Roia	600	France	400	67.39
		Italy	200	30.45
Salaca	2,100	Latvia	1,600	78.52
		Estonia	100	5.7
Sarata (37)	1,800	Ukraine	1,100	63.9
		Moldova	600	36.05
Schelde	17,100	France	8,600	50.03
		Belgium	8,400	49.28
		Netherlands	80	0.47
Seine	85,700	France	83,800	97.78
		Belgium	1,800	2.14
		Luxembourg	70	0.08
Struma (36)	15,000	Bulgaria	8,600	57.66
		Greece	3,900	25.88
		Macedonia	1,800	12.22
		Yugoslavia (Serbia and Montenegro)	600	4.19
Tagus/Tejo	77,900	Spain	51,400	66.06
		Portugal	26,100	33.5
Tana	15,600	Norway	9,300	59.71
		Finland	6,300	40.23

Basin Name	Area of Basin (km^2) (1)	Countries Sharing the Basin	Area within Country (km^2)	Percentage of Watershed within Country
Torne/Tornealven	37,300	Sweden	25,400	67.98
		Finland	10,400	28
		Norway	1,500	4.03
Tuloma	25,800	Russia	23,700	91.85
		Finland	2,000	7.93
Vardar (36)	32,400	Macedonia	20,300	62.83
		Yugoslavia (Serbia and Montenegro)	8,200	25.22
		Greece	3,900	11.94
Velaka	700	Bulgaria	700	95.25
		Turkey	30	3.74
Venta	9,500	Latvia	6,200	65.15
		Lithuania	3,300	34.72
Vijose	7,200	Albania	4,600	64.83
		Greece	2,500	34.66
Vistula/Wista	194,000	Poland	169,700	87.45
		Ukraine	12,700	6.55
		Byelarus	9,800	5.03
		Slovakia	1,900	0.96
		Czech Republic	20	0.01
Volga (18)	1,554,900	Russia	1,551,300	99.77
		Kazakhstan	2,200	0.14
		Byelarus	1,300	0.08
Vuoksa	62,700	Finland	54,300	86.48
		Russia	8,500	13.52
Wiedau	1,100	Denmark	1,000	86.23
		Germany	200	13.32
Yser	900	France	500	53.63
		Belgium	400	46.37

1) The numbers referring to basin areas have been rounded to significant digits and, as a result, the numbers for area within each basin do not necessarily add up to the total area for that basin. Also, the percentages were calculated based on raw data, and therefore do not reflect the rounding of the areas.

33) Disputes are ongoing between Bosnia-Herzegovina and Serbia, over Serbian populated areas. According to the Serbian Republic of Bosnia-Herzegovina (SRBH), the external boundaries are marked by the Una river in the west, the Sava river in the north, the state boundary with the Federal Republic of Yugoslavia in the east, and Croatia and the Serbian Republic Krajina in the south. (CIA World Factbook, 1998; IBRU, 1999).

34) Eastern Slavonia, which was held by Serbs during the ethnic conflict in the former Yugoslavia, was returned to Croatian control by the UN Transitional Administration for Eastern Slavonia on January 15, 1998. (CIA World Factbook, 1998).

35) Under an International Court of Justice (ICJ) ruling, Hungary and Slovakia were to agree on the future of the Gabcikovo Dam complex by March 1998. The dispute, however, has yet to be resolved. Completion of the dam system would alter the boundaries between Hungary and Slovakia established under the 1920 Treaty of Trianon. (CIA World Factbook, 1998; IBRU, 1999).

36) The boundary commission formed by Serbia and Montenegro, and the Former Yugoslav Republic of Macedonia in April 1996 to resolve differences in delineation of their mutual boundary has made no progress so far. (CIA World Factbook, 1998).

37) Romania considers certain territories of Moldova and Ukraine-including Bessarabia (45,600 km^2) and Northern Bukovina-as historically part of Romania. This territory was incorporated into the former Soviet Union following the Molotov-Ribbentrop Pact in 1940. (CIA World Factbook, 1998; Columbia Gazetteer, 1998).

38) Border problems between Byelarus and Lithuania in part lie in the fact that the new boundary is different from the old Soviet administrative division between the two republics. Areas of dispute include the land around the Adutiskis railway station and the Druskininkai resort claimed by Byelarus. Demarcation of the boundary between Byelarus and Lithuania is underway. (CIA World Factbook, 1998; IBRU 1999).

39) The 1997 boundary agreement Lithuania and Russia remains to be ratified. (CIA World Factbook, 1998).

continues

40) The December 1996 technical boundary agreement reached between Estonian and Russian negotiators remains to be ratified. Estonia claimed over 2,000 km² of territory in the Narva and Pechory regions of Russia-based on the boundary established under the 1920 Peace Treaty of Tartu. (CIA World Factbook, 1998).

41) Latvia claimed the Abrene/Pytalovo section of the border ceded by the Latvian Soviet Socialist Republic to Russia in 1944, based on the 1920 Treaty of Riga. A draft treaty delimiting the boundary between Latvia and Russia has not been signed. The Abrene/Pytalovo region is crossed by the Utroya River, a tributary of the Vclikaya river. (CIA World Factbook, 1998; Columbia Gazetteer, 1998).

42) While the Meuse basin is topographically part of the Rhine basin, European nations treat it as a politically separate basin. (Huisman, de Jong, and Wieriks, 1998).

DATA TABLE 14D   North America

Basin Name	Area of Basin (km²) (1)	Countries Sharing the Basin	Area within Country (km²)	Percentage of Watershed within Country
Alsek	28,400	Canada	26,500	93.5
		United States of America	1,800	6.5
Artibonite	8,800	Haiti	6,600	74.37
		Dominican Republic	2,300	25.55
Belize (43)	11,500	Belize	7,000	60.86
		Guatemala	4,500	39.14
Candelaria	12,800	Mexico	11,300	88.24
		Guatemala	1,500	11.74
Changuinola	3,200	Panama	2,900	91.29
		Costa Rica	300	8.33
Chilkat	3,800	United States of America	2,100	56.59
		Canada	1,600	43.35
Chiriqui	1,700	Panama	1,500	86.17
		Costa Rica	200	13.83
Choluteca	7,400	Honduras	7,200	97.68
		Nicaragua	200	2.32
Coatan Achute	2,000	Mexico	1,700	86.27
		Guatemala	300	13.73
Coco/Segovia	25,400	Nicaragua	17,900	70.52
		Honduras	7,500	29.48
Colorado	655,000	United States of America	644,600	98.41
		Mexico	10,400	1.59
Columbia	668,400	United States of America	566,500	84.75
		Canada	101,900	15.24
Firth	6,000	Canada	3,800	63.6
		United States of America	2,200	36.4
Fraser	239,700	Canada	239,100	99.74
		United States of America	600	0.26
Goascoran	2,800	Honduras	1,500	53.36
		El Salvador	1,300	46.64
Grijalva (43)	126,800	Mexico	78,900	62.25
		Guatemala	47,800	37.72
		Belize	20	0.02

Basin Name	Area of Basin (km²) (1)	Countries Sharing the Basin	Area within Country (km²)	Percentage of Watershed within Country
Hondo (43)	14,600	Mexico	8,900	61.14
		Guatemala	4,200	28.5
		Belize	1,500	10.36
Lempa	18,000	El Salvador	9,500	52.45
		Honduras	5,800	32.01
		Guatemala	2,800	15.54
Massacre	800	Haiti	500	62.03
		Dominican Republic	300	35.96
Mississippi	3,226,300	United States of America	3,176,500	98.46
		Canada	49,800	1.54
Motaqua	16,100	Guatemala	14,600	90.85
		Honduras	1,500	9.11
Negro	5,800	Nicaragua	4,800	83.87
		Honduras	900	15.68
Nelson-Saskatchewan	1,109,400	Canada	952,000	85.81
		United States of America	157,400	14.19
Paz	2,200	Guatemala	1,400	64.47
		El Salvador	800	35.53
Pedernales	400	Haiti	200	67.32
		Dominican Republic	100	32.68
Rio Grande (North America)	656,100	United States of America	341,800	52.1
		Mexico	314,300	47.9
San Juan	42,200	Nicaragua	30,400	72.02
		Costa Rica	11,800	27.93
Sarstun (43)	2,100	Guatemala	1,800	87.63
		Belize	300	12.37
Sixaola	2,900	Costa Rica	2,500	88.65
		Panama	300	9.99
Skagit	8,000	United States of America	7,100	88.46
		Canada	900	11.54
St. Croix	4,600	United States of America	3,300	70.86
		Canada	1,400	29.14
St. John (North America)	47,700	Canada	30,300	63.5
		United States of America	17,300	36.22
St. Lawrence	1,055,200	Canada	559,000	52.98
		United States of America	496,100	47.02
Stikine	50,900	Canada	50,000	98.32
		United States of America	900	1.68
Suchiate	1,600	Guatemala	1,100	68.79
		Mexico	500	31.21

continues

DATA TABLE 14D   *continued*

Basin Name	Area of Basin (km²) (1)	Countries Sharing the Basin	Area within Country (km²)	Percentage of Watershed within Country
Taku	18,100	Canada	16,300	90.09
		United States of America	1,700	9.13
Tijuana	4,400	Mexico	3,100	70.57
		United States of America	1,300	29.43
Whiting	2,600	Canada	2,000	80.06
		United States of America	500	19.94
Yaqui	74,700	Mexico	70,100	93.87
		United States of America	4,600	6.13
Yukon	829,700	United States of America	496,400	59.83
		Canada	333,300	40.17

1) The numbers referring to basin areas have been rounded to significant digits and, as a result, the numbers for area within each basin do not necessarily add up to the total area for that basin. Also, the percentages were calculated based on raw data, and therefore do not reflect the rounding of the areas.

43) The boundary between Belize and Guatemala is in dispute. Talks to resolve the dispute are ongoing. Changes in the boundary between Guatemala and Belize could impact the Hondo, Belize, Grijalva, and/or Sarstun basins. (Until 1991, Guatemala claimed all of Belize). (CIA World Factbook, 1998; Columbia Gazetteer, 1998; IBRU, 1999).

DATA TABLE 14E   South America

Basin Name	Area of Basin (km²) (1)	Countries Sharing the Basin	Area within Country (km²)	Percentage of Watershed within Country
Amacuro	5,600	Venezuela	4,900	86.89
		Guyana	700	13.11
Amazon (44)	5,883,400	Brazil	3,670,300	62.38
		Peru	956,500	16.26
		Bolivia	706,700	12.01
		Colombia	367,800	6.25
		Ecuador	123,800	2.1
		Venezuela	40,300	0.68
		Guyana	14,500	0.25
		Suriname	1,400	0.02
		French Guiana	30	0
Aviles	300	Argentina	200	88.72
		Chile	30	11.28
Aysen	13,600	Chile	13,100	96.07
		Argentina	500	3.93
Baker	30,800	Chile	21,000	68.15
		Argentina	9,800	31.83
Barima	2,100	Guyana	1,100	51.05
		Venezuela	1,000	47.84
Cancoso/Lauca	23,500	Bolivia	20,200	85.72
		Chile	3,400	14.28

Basin Name	Area of Basin (km^2) (1)	Countries Sharing the Basin	Area within Country (km^2)	Percentage of Watershed within Country
Carmen Silva/ Chico	1,700	Argentina	1,000	59.7
		Chile	700	40.3
Catatumbo	31,000	Colombia	19,600	63.15
		Venezuela	11,400	36.75
Chira (44)	15,700	Peru	9,800	62.23
		Ecuador	5,800	37.23
Chuy	200	Brazil	100	64.57
		Uruguay	60	32.57
Comau	900	Chile	900	91.36
		Argentina	80	8.64
Corantijn/ Courantyne (45)	41,800	Guyana	21,700	52.06
		Suriname	19,900	47.75
		Brazil	80	0.19
Cullen	600	Chile	500	83
		Argentina	100	17
Essequibo (46)	239,500	Guyana	162,100	67.67
		Venezuela	52,400	21.87
		Suriname	24,300	10.13
		Brazil	200	0.07
Gallegos-Chico	11,600	Argentina	7,000	60.15
		Chile	4,600	39.85
Jurado	700	Colombia	500	82.11
		Panama	100	17.89
La Plata (47, 48)	2,954,500	Brazil	1,379,300	46.69
		Argentina	817,900	27.68
		Paraguay	400,100	13.54
		Bolivia	245,100	8.3
		Uruguay	111,600	3.78
Lagoon Mirim	55,000	Uruguay	31,200	56.69
		Brazil	23,800	43.24
Lake Fagnano (49)	3,200	Argentina	2,700	85.17
		Chile	500	14.83
Lake Titicaca- Poopo System	111,800	Bolivia	63,000	56.32
		Peru	48,000	42.94
		Chile	800	0.74
Maroni (50)	65,000	Suriname	37,500	57.64
		French Guiana	27,200	41.9
		Brazil	200	0.27
Mataje	700	Ecuador	500	73.98
		Colombia	200	26.02
Mira	12,100	Colombia	6,200	50.87
		Ecuador	5,800	47.97
Oiapoque/ Oyupock	23,300	French Guiana	13,700	58.92
		Brazil	9,500	41

continues

Data Table 14E   *continued*

Basin Name	Area of Basin (km²) (1)	Countries Sharing the Basin	Area within Country (km²)	Percentage of Watershed within Country
Orinoco	927,400	Venezuela	604,500	65.18
		Colombia	321,700	34.68
		Brazil	800	0.08
Palena	13,300	Chile	7,300	54.87
		Argentina	6,000	45.13
Pascua	13,700	Chile	7,300	53.51
		Argentina	6,400	46.46
Patia	21,300	Colombia	20,800	97.61
		Ecuador	500	2.38
Puelo	8,400	Argentina	5,500	66.03
		Chile	2,900	33.97
Rio Grande (South America)	8,000	Argentina	4,000	49.74
		Chile	4,000	50.26
San Martin	700	Chile	600	87.44
		Argentina	80	12.56
Seno Union/ Serrano	6,500	Chile	5,700	87.93
		Argentina	700	10.34
Tumbes-Poyango (44)	5,000	Ecuador	3,600	71.62
		Peru	1,400	28.38
Valdivia	15,000	Chile	14,700	98.39
		Argentina	100	0.69
Yelcho	11,100	Argentina	6,900	62.14
		Chile	4,200	37.86
Zapaleri (51)	2,600	Chile	1,600	59.6
		Argentina	500	19.65
		Bolivia	500	20.75
Zarumilla (44)	4,300	Ecuador	3,400	78.71
		Peru	900	20.51

1) The numbers referring to basin areas have been rounded to significant digits and, as a result, the numbers for area within each basin do not necessarily add up to the total area for that basin. Also, the percentages were calculated based on raw data, and therefore do not reflect the rounding of the areas.

44) Three sections of the boundary between Ecuador and Peru have been in dispute. The areas cover over 324,000 km² and include portions of the Amazon and Maranon rivers. The districts of Tumbes, Jaen, and Maynas are claimed by Ecuador and administered by Peru. In December 1998, Peru and Ecuador signed a joint agreement on the implementation of a permanent development policy for the border region. A joint commission was created to determine their common land boundary. (Encyclopedia of International Boundaries, 1995; CIA World Factbook, 1998; Columbia Gazetteer, 1998; BBC Summary of World Broadcasts, 12/3/98; Xinhua News Agency, 12/11/1998).

45) The boundary upstream from the confluence of the Courantyne/Koetari (Kutari) River with the New (Upper Courantyne) River remains unsettled. Guyana administers the triangle formed by the two rivers, while Brazil and Suriname continue to claim the area. Suriname also claims the west bank of the Courantyne River below the New River as the boundary, but de facto the boundary continues to follow the thalweg. (Encyclopedia of International Boundaries, 1995; CIA World Factbook, 1998).

46) Talks are ongoing between Guyana and Venezuela regarding their boundary dispute. Venezuela claims all of the area west of the Essequibo River. (CIA World Factbook, 1998; IBRU, 1999).

47) A short section of the boundary between Brazil and Paraguay, just west of Salto das Sete Quedas (Guaira Falls) on the Rio Parana, has yet to be precisely delimited. (CIA World Factbook, 1998).

48) Two short sections of the boundary between Brazil and Uruguay are in dispute - the Arroio Invernada (Arroyo de la Invernada) area of the Rio Quarai (Rio Cuareim) and the islands at the confluence of the Rio Quarai and the Uruguay River. (CIA World Factbook, 1998).

49) A short section of the southeastern boundary of Chile with Argentina, in the area of the Beagle Channel, remains unclear. The 1991 Aylwin-Menem Treaty delineates the boundary between Argentina and Chile in the continental glaciers area. As of March 1999, the treaty has not been ratified by the Congresses of either country. (CIA World Factbook, 1998; IBRU, 1999).

50) Suriname and French Guiana are in dispute over which of the upper tributaries of the Maroni River was originally intended to carry the boundaries down to the Brazilian boundary. The disputed area is administered by France as a region of the overseas department of French Guiana and claimed by Suriname. The area lies between the Riviere Litani and the Riviere Marouini, both headwaters of the Lawa. (Encyclopedia of International Boundaries, 1995; CIA World Factbook, 1998).

51) Bolivia has desired a sovereign corridor to the South Pacific Ocean, since the Atacama desert area was lost to Chile in 1884. The creation of such a corridor could impact territory in the Zapaleri basin or create a new international basin. (CIA World Factbook, 1998; IBRU, 1999).

OECD Water Tariffs

Description

The cost to provide water and wastewater services to consumers, and the prices charged those consumers, vary enormously from region to region, city to city, and country to country. Very little comprehensive data have traditionally been available on water prices and costs, but this is beginning to change. The Global Water Intelligence and the Organization for Economic Cooperation and Development (OECD) has begun to collect and publish information on water tariffs in the water and wastewater area by city. This table lists water and wastewater tariffs, as of approximately mid-2007, for approximately 170 different cities, in 29 different OECD countries. The data are reported as US$ per cubic meters, for the first 15 cubic meters of water used, or wastewater generated, per month. This table also shows the fraction of income (reported as GDP per person) typically spent on the combination of water and wastewater services. The data are sorted alphabetically by country.

The average price in the OECD countries for water was around $1.40 per cubic meter. Wastewater prices were around $0.91 per cubic meter. And total water and wastewater costs had risen just under 4 percent from the previous year. These data come from surveys of water agencies in each city and substantially more detailed data on the tariff structures, differential rates and subsidies, and block rates are provided in the full tables from OECD.

Limitations

Water rates and rate structures are highly variable among service providers. Rate designs range from fixed monthly rates independent of the volume used to highly sophisticated time-of-use rates for different customer categories. The data in this table are thus gross simplifications of complex water rates, and assume a fixed amount of water used per month – 15 cubic meters. A total monthly bill is then computed based on the specifics of the rate structure and a total monthly cost is presented for water and wastewater. As a result of this approach, a single value can be shown, but actual rates will vary enormously across end users depending on volume of use, location, and other factors. Greater detail is provided in the original survey data available from the GWI/OECD survey. The reported values are in US dollars per cubic meter, rather than the country currency and thus will fluctuate as exchange rates fluctuate.

SOURCE

GWI/OECD Water Tariff Survey 2007. www.globalwaterintel.com/survey2007.xls.
Full data provided to the Pacific Institute by Chris Gasson of the Global Water Intelligence.

DATA TABLE 15 OECD Water Tariffs

City name	Country	Water $/m³ for 15 m³/month (June 2007)	Wastewater $/m³ for 15 m³/month (June 2007)	Total Water and Wastewater US $/m³ for 15 m³/month	Water/Wastewater bill as % of GDP/head
Brisbane	Australia	$ 1.74	$ 1.84	$ 3.58	4.3
Sydney	Australia	$ 1.46	$ 2.00	$ 3.45	4.2
Melbourne (City West)	Australia	$ 1.29	$ 1.54	$ 2.82	3.4
Adelaide	Australia	$ 1.39	$ 1.39	$ 2.78	3.4
Perth	Australia	$ 1.30	$ 1.31	$ 2.61	3.2
Graz	Austria	$ 2.55	$ 2.07	$ 4.62	2.6
Linz	Austria	$ 2.73	$ 1.15	$ 3.88	2.2
Vienna	Austria	$ 1.85	$ 1.43	$ 3.28	1.8
Gent	Belgium	$ 5.49	$ -	$ 5.49	2.3
Brussels	Belgium	$ 3.76	$ 0.59	$ 4.35	1.8
Antwerp	Belgium	$ 4.18	$ -	$ 4.18	1.7
Calgary	Canada	$ 1.74	$ 1.49	$ 3.23	7.3
Edmonton	Canada	$ 1.65	$ 1.22	$ 2.87	6.5
Ottawa	Canada	$ 1.00	$ 1.45	$ 2.45	5.5
Toronto	Canada	$ 1.50	$ -	$ 1.50	3.4
Vancouver	Canada	$ 0.57	$ 0.91	$ 1.48	3.3
Montreal	Canada	$ 0.42	$ -	$ 0.42	1.0
Brno	Czech Republic	$ 1.19	$ 1.36	$ 2.55	1.7
Prague	Czech Republic	$ 1.30	$ 1.12	$ 2.42	1.6
Ostrava	Czech Republic	$ 1.20	$ 1.02	$ 2.22	1.5
Copenhagen	Denmark	$ 7.71	$ -	$ 7.71	3.5
Aahus	Denmark	$ 7.58	$ -	$ 7.58	3.5
Espoo	Finland	$ 2.00	$ 2.48	$ 4.48	4.3
Helsinki	Finland	$ 1.30	$ 1.69	$ 2.99	2.9
Bordeaux	France	$ 1.57	$ 3.16	$ 4.73	3.3
Toulouse	France	$ 3.80	$ -	$ 3.80	2.6

Marseille	France	$ 2.91	$ 1.21	$ 4.12	2.8
Lyon	France	$ 4.05	$ -	$ 4.05	2.8
Nice	France	$ 2.07	$ 1.91	$ 3.98	2.7
Nantes	France	$ 3.76	$ -	$ 3.76	2.6
Strasbourg	France	$ 3.74	$ -	$ 3.74	2.6
Paris	France	$ 3.72	$ -	$ 3.72	2.6
Frankfurt	Germany	$ 3.10	$ 2.43	$ 5.53	3.0
Stuttgart	Germany	$ 5.32	$ 2.17	$ 7.49	4.1
Berlin	Germany	$ 3.13	$ 3.49	$ 6.62	3.6
Hamburg	Germany	$ 2.28	$ 3.57	$ 5.85	3.2
Köln	Germany	$ 3.10	$ 2.50	$ 5.60	3.0
Düsseldorf	Germany	$ 3.19	$ 1.98	$ 5.16	2.8
Munich	Germany	$ 1.97	$ 2.16	$ 4.12	2.2
Essen	Germany	$ 4.05	$ -	$ 4.05	2.2
Dortmund	Germany	$ 3.49	$ -	$ 3.49	1.9
Bremen	Germany	$ 2.96	$ -	$ 2.96	1.6
Thessaloniki	Greece	$ 0.81	$ 0.53	$ 1.34	1.0
Athens	Greece	$ 0.81	$ 0.53	$ 1.34	1.0
Piraeus	Greece	$ 0.81	$ 0.53	$ 1.34	1.0
Miskolc	Hungary	$ 1.40	$ 1.08	$ 2.49	1.8
Budapest	Hungary	$ 0.98	$ 1.39	$ 2.37	1.7
Debrecen	Hungary	$ 1.00	$ 0.71	$ 1.71	1.3
Reykjavik	Iceland	$ 1.95	$ -	$ 1.95	3.4
Dublin	Ireland	$ -	$ -	$ -	0.0
Cork	Ireland	$ -	$ -	$ -	0.0
Bologna	Italy	$ 1.11	$ 0.52	$ 1.63	1.9
Rome	Italy	$ 0.66	$ 0.61	$ 1.28	1.5
Palermo	Italy	$ 0.75	$ 0.53	$ 1.28	1.5
Naples	Italy	$ 0.64	$ 0.55	$ 1.19	1.4
Venice	Italy	$ 0.61	$ 0.49	$ 1.09	1.3

continues

DATA TABLE 15 *continued*

City name	Country	Water $/m³ for 15 m³/month (June 2007)	Wastewater $/m³ for 15 m³/month (June 2007)	Total Water and Wastewater US $/m³ for 15 m³/month	Water/Wastewater bill as % of GDP/head
Genova	Italy	$ 0.37	$ 0.56	$ 0.92	1.1
Milan	Italy	$ 0.41	$ 0.43	$ 0.84	1.0
Turin	Italy	$ 0.38	$ 0.37	$ 0.75	0.9
Kumamoto	Japan	$ 0.94	$ 1.38	$ 2.31	3.4
Sakai	Japan	$ 1.12	$ 1.13	$ 2.25	3.3
Chiba	Japan	$ 1.00	$ 1.24	$ 2.25	3.3
Fukuoka	Japan	$ 1.06	$ 0.56	$ 1.62	2.4
Okayama	Japan	$ 1.06	$ 1.12	$ 2.17	3.2
Sendai	Japan	$ 1.30	$ 0.69	$ 1.98	2.9
Sapporo	Japan	$ 1.37	$ 0.55	$ 1.92	2.8
Kyoto	Japan	$ 0.99	$ 0.76	$ 1.76	2.6
Yokohama	Japan	$ 0.94	$ 0.71	$ 1.65	2.4
Tokyo	Japan	$ 0.91	$ 0.75	$ 1.65	2.4
Kitakyushu	Japan	$ 0.83	$ 0.79	$ 1.62	2.4
Hiroshima	Japan	$ 0.83	$ 0.74	$ 1.57	2.3
Kawasaki	Japan	$ 0.84	$ 0.78	$ 1.61	2.4
Nagoya	Japan	$ 0.86	$ 0.65	$ 1.51	2.2
Osaka	Japan	$ 0.85	$ 0.50	$ 1.35	2.0
Pusan	Korea	$ 0.71	$ 0.74	$ 1.45	3.0
Ulsan	Korea	$ 0.81	$ 0.21	$ 1.03	2.1
Gwangju	Korea	$ 0.63	$ 0.22	$ 0.85	1.7
Daegu	Korea	$ 0.61	$ 0.22	$ 0.83	1.7
Incheon	Korea	$ 0.69	$ 0.14	$ 0.82	1.7
Daejeon	Korea	$ 0.61	$ 0.16	$ 0.78	1.6
Seoul	Korea	$ 0.58	$ 0.17	$ 0.75	1.5
Luxembourg	Luxembourg	$ 2.91	$ 1.73	$ 4.64	1.5

Leon	Mexico	$ 0.78	$ -	$ 0.78	3.3
Cuernavaca	Mexico	$ 0.64	$ -	$ 0.64	2.7
Morelia	Mexico	$ 0.45	$ -	$ 0.45	1.9
Puebla	Mexico	$ 0.42	$ -	$ 0.42	1.8
Guadalajara	Mexico	$ 0.40	$ -	$ 0.40	1.7
Zapopan	Mexico	$ 0.40	$ -	$ 0.40	1.7
Acapulco	Mexico	$ 0.40	$ -	$ 0.40	1.7
Ciudad Juarez	Mexico	$ 0.39	$ -	$ 0.39	1.7
Chihuahua	Mexico	$ 0.25	$ -	$ 0.25	1.1
Tijuana	Mexico	$ 0.87	$ -	$ 0.87	3.7
Mexico City	Mexico	$ 0.15	$ 0.06	$ 0.10	0.7
Amsterdam	Netherlands	$ 3.15	$ 0.84	$ 2.31	1.5
Rotterdam	Netherlands	$ 3.12	$ 1.14	$ 1.98	1.5
The Hague	Netherlands	$ 2.02	$ -	$ 2.02	1.0
Utrecht	Netherlands	$ 1.67	$ -	$ 1.67	0.8
Aukland	New Zealand	$ 4.13	$ 2.69	$ 1.44	3.8
Christchurch	New Zealand	$ 0.98	$ 0.49	$ 0.49	0.9
Oslo	Norway	$ 3.49	$ -	$ 3.49	2.1
Trondheim	Norway	$ 2.79	$ -	$ 2.79	1.7
Poznan	Poland	$ 2.64	$ 1.52	$ 1.11	1.5
Gdansk	Poland	$ 2.18	$ 1.22	$ 0.96	1.2
Lodz	Poland	$ 2.15	$ 1.15	$ 1.00	1.2
Warsaw	Poland	$ 2.08	$ 1.15	$ 0.93	1.2
Wroclaw	Poland	$ 1.97	$ 0.90	$ 1.07	1.1
Krakow	Poland	$ 1.86	$ 0.90	$ 0.95	1.1
Amadora	Portugal	$ 2.14	$ 0.61	$ 1.53	1.8
Porto	Portugal	$ 2.01	$ 0.57	$ 1.45	1.7
Lisbon	Portugal	$ 1.49	$ 0.52	$ 0.97	1.3
Kosice	Slovakia	$ 2.36	$ 0.89	$ 1.47	2.3
Bratislava	Slovakia	$ 2.02	$ 1.01	$ 1.01	1.9

continues

DATA TABLE 15 *continued*

City name	Country	Water $/m³ for 15 m³/month (June 2007)	Wastewater $/m³ for 15 m³/month (June 2007)	Total Water and Wastewater US $/m³ for 15 m³/month	Water/Wastewater bill as % of GDP/head
Palma-de-Mallorca	Spain	$ 1.55	$ 0.97	$ 2.52	2.8
Barcelona	Spain	$ 1.23	$ 0.73	$ 1.96	2.2
Seville	Spain	$ 0.96	$ 0.84	$ 1.80	2.0
Madrid	Spain	$ 0.97	$ 0.80	$ 1.77	2.0
Bilbao	Spain	$ 0.72	$ 0.67	$ 1.39	1.6
Valencia	Spain	$ 0.45	$ 0.62	$ 1.07	1.2
Malaga	Spain	$ 0.75	$ -	$ 0.75	0.8
Malmo	Sweden	$ 2.80	$ -	$ 2.80	3.6
Gothenburg	Sweden	$ 2.65	$ -	$ 2.65	3.4
Stockholm	Sweden	$ 1.54	$ -	$ 1.54	2.0
Zurich	Switzerland	$ 2.43	$ 2.78	$ 5.21	4.4
Basel	Switzerland	$ 2.30	$ 2.90	$ 5.20	4.4
Geneva	Switzerland	$ 3.09	$ 2.07	$ 5.16	4.4
Istanbul	Turkey	$ 1.40	$ 0.70	$ 2.10	7.3
Izmir	Turkey	$ 1.01	$ 0.77	$ 1.78	6.2
Adana	Turkey	$ 1.05	$ 0.79	$ 1.84	6.4
Bursa	Turkey	$ 1.47	$ 0.22	$ 1.69	5.9
Ankara	Turkey	$ -	$ 0.47	#REF!	
Gaziantep	Turkey	$ 0.94	$ -	$ 0.94	3.3
Diyarbakir	Turkey	$ 0.79	$ 0.37	$ 1.16	4.1
Konya	Turkey	$ 0.88	$ 0.22	$ 1.11	3.9
Antalya	Turkey	$ 0.98	$ 0.27	$ 1.26	4.4
Glasgow	UK	$ 3.31	$ 4.26	$ 7.57	4.3
Cardiff	UK	$ 2.82	$ 3.55	$ 6.37	3.6
Manchester	UK	$ 2.79	$ 2.84	$ 5.62	3.2
Bristol	UK	$ 2.34	$ 3.28	$ 5.62	3.2
Newcastle	UK	$ 2.06	$ 2.39	$ 4.45	2.5
Birmingham	UK	$ 2.58	$ 1.73	$ 4.31	2.4

City	Country				%
London	UK	$ 2.29	$ 1.48	$ 3.77	2.1
Belfast	UK	$ -	$ -	$ -	0.0
Richmond	USA	$ 1.58	$ 2.32	$ 3.90	6.5
San Diego	USA	$ 1.81	$ 1.84	$ 3.65	6.1
Atlanta	USA	$ 3.01	$ -	$ 3.01	5.0
San Francisco	USA	$ 1.11	$ 1.85	$ 2.96	4.9
San Jose	USA	$ 1.06	$ 1.87	$ 2.94	4.9
Philadelphia	USA	$ 1.11	$ 1.73	$ 2.84	4.7
Boston	USA	$ 1.22	$ 1.58	$ 2.80	4.7
Detroit	USA	$ 0.84	$ 1.57	$ 2.41	4.0
Denver	USA	$ 0.71	$ 1.05	$ 1.76	2.9
New York City	USA	$ 0.91	$ 1.44	$ 2.35	3.9
Las Vegas	USA	$ 0.57	$ 1.75	$ 2.31	3.9
Indianapolis	USA	$ 1.19	$ 1.01	$ 2.20	3.7
Fort Worth	USA	$ 0.99	$ 1.19	$ 2.18	3.6
Jacksonville	USA	$ 0.83	$ 1.43	$ 2.26	3.8
Washington DC	USA	$ 0.85	$ 1.12	$ 1.97	3.3
Cleveland	USA	$ 0.64	$ 1.45	$ 2.09	3.5
Louisville	USA	$ 0.92	$ 1.11	$ 2.04	3.4
Columbus	USA	$ 0.84	$ 1.19	$ 2.03	3.4
Los Angeles	USA	$ 0.92	$ 1.08	$ 2.00	3.3
Houston	USA	$ 1.00	$ 0.97	$ 1.97	3.3
Dallas	USA	$ 0.65	$ 1.24	$ 1.89	3.1
Baltimore	USA	$ 0.77	$ 1.03	$ 1.80	3.0
San Antonio	USA	$ 1.06	$ 0.61	$ 1.67	2.8
El Paso	USA	$ 0.80	$ 0.73	$ 1.53	2.6
Phoenix	USA	$ 0.94	$ 0.66	$ 1.60	2.7
Omaha	USA	$ 0.59	$ 0.80	$ 1.39	2.3
Miami-Dade	USA	$ 0.41	$ 0.78	$ 1.19	2.0
Chicago	USA	$ 0.35	$ 0.29	$ 0.64	1.1
Memphis	USA	$ 0.33	$ 0.23	$ 0.57	0.9

Non-OECD Water Tariffs

Description

The cost to provide water and wastewater services to consumers, and the prices charged those consumers, vary enormously from region to region, city to city, and country to country. Very few comprehensive data have traditionally been available on water prices and costs, but this is beginning to change. The Global Water Intelligence and the Organization for Economic Cooperation and Development (OECD) have begun to collect and publish information on water tariffs in the water and wastewater area by city. This table lists water and wastewater tariffs, as of approximately mid-2007, for approximately 85 non-OECD cities and countries. The data are reported as US$ per cubic meters, for the first 15 cubic meters of water used, or wastewater generated, per month. The data are sorted alphabetically by country.

Limitations

Water rates and rate structures are highly variable among service providers. Rate designs range from fixed monthly rates independent of the volume of use to highly sophisticated time-of-use rates for different categories of users. The data in this table are thus gross simplifications of complex water rates, and assume a fixed amount of water used per month – 15 cubic meters. A total monthly bill is then computed based on the specifics of the rate structure and a total monthly cost is presented for water and wastewater. As a result of this approach, a single value can be shown, but actual rates will vary enormously across end users depending on volume of use, location, and other factors. Greater detail is provided in the original survey data available from the GWI/OECD survey. The reported values are in US dollars per cubic meter, rather than the country currency and thus will fluctuate as exchange rates fluctuate.

SOURCE

GWI/OECD Water Tariff Survey 2007. www.globalwaterintel.com/survey2007.xls. Full data provided to the Pacific Institute by Chris Gasson of the Global Water Intelligence.

DATA TABLE 16 Non-OECD Water Tariffs

City name	Country	Water $/m^3 for 15 m^3/month (June 2007)	Wastewater $/m^3 for 15 m^3/month (June 2007)	Total Water and Wastewater $/m^3 for 15 m^3/month
Algiers	Algeria	$ 0.11	$ -	$ 0.11
Buenos Aires	Argentina	$ 0.13	$ 0.13	$ 0.26
Cordoba	Argentina	$ 0.08	$ -	$ 0.08
Yerevan	Armenia	$ 0.43	$ 0.08	$ 0.51
Baku	Azerbaijan	$ 0.15	$ -	$ 0.15
Manama	Bahrain	$ 0.07	$ -	$ 0.07
Minsk	Belarus	$ 0.12	$ 0.06	$ 0.18
Gaberone	Botswana	$ 0.63	$ -	$ 0.63
Sao Paolo	Brazil	$ 0.73	$ 0.73	$ 1.45
Sofia	Bulgaria	$ 0.55	$ 0.20	$ 0.75
Santiago-de-Chile	Chile	$ 0.54	$ 0.49	$ 1.02
Hong Kong	China	$ 0.40	$ 0.12	$ 0.53
Beijing	China	$ 0.37	$ 0.12	$ 0.49
Tianjin	China	$ 0.34	$ 0.11	$ 0.45
Chonqing	China	$ 0.28	$ 0.09	$ 0.37
Hangzhou	China	$ 0.18	$ 0.07	$ 0.24
Guangzhou	China	$ 0.17	$ 0.08	$ 0.26
Shanghai	China	$ 0.14	$ 0.11	$ 0.24
Havana	Cuba	$ 0.01	$ -	$ 0.01
Guayaquil	Ecuador	$ 0.28	$ 0.22	$ 0.50
Cairo	Egypt	$ 0.06	$ -	$ 0.06
Tallinn	Estonia	$ 1.25	$ 1.01	$ 2.26
Addis Ababa	Ethiopia	$ 0.22	$ -	$ 0.22
Tbilisi	Georgia	$ 0.18	$ -	$ 0.18
Tegucigalpa	Honduras	$ 0.05	$ 0.01	$ 0.06
Bangalore	India	$ 0.15	$ 0.02	$ 0.17
Calcutta	India	$ 0.12	$ -	$ 0.12

continues

321

DATA TABLE 16 *continued*

City name	Country	Water $/m^3$ for 15 m^3/month (June 2007)	Wastewater $/m^3$ for 15 m^3/month (June 2007)	Total Water and Wastewater $/m^3$ for 15 m^3/month
New Delhi	India	$ 0.11	$ -	$ 0.11
Mumbai	India	$ 0.09	$ -	$ 0.09
Jakarta	Indonesia	$ 0.75	$ -	$ 0.75
Tehran	Iran	$ -	$ -	$ -
Baghdad	Iraq	$ -	$ -	$ -
Jerusalem	Israel	$ 1.00	$ -	$ 1.00
Tel Aviv	Israel	$ 1.00	$ -	$ 1.00
Astana	Kazakhstan	$ 0.17	$ 0.12	$ 0.28
Almaty	Kazakhstan	$ 0.11	$ 0.07	$ 0.18
Nairobi	Kenya	$ 0.21	$ 0.13	$ 0.35
Riga	Latvia	$ 0.58	$ 0.58	$ 1.16
Vilnius	Lithuania	$ 0.74	$ 0.89	$ 1.63
Antananarivo	Madagascar	$ 0.28	$ -	$ 0.28
Bamako	Mali	$ 0.43	$ -	$ 0.43
Chisinau	Moldova	$ 0.33	$ 0.05	$ 0.38
Ulan Bator	Mongolia	$ 0.02	$ 0.03	$ 0.04
Casablanca	Morocco	$ 0.75	$ 0.12	$ 0.87
Rabat	Morocco	$ 0.28	$ 0.08	$ 0.37
Kathmandu	Nepal	$ 0.40	$ 0.20	$ 0.60
Lagos	Nigeria	$ 0.40	$ -	$ 0.40
Muscat	Oman	$ 1.38	$ -	$ 1.38
Karachi	Pakistan	$ 0.04	$ 0.01	$ 0.05
Ramallah	Palestine	$ 1.00	$ 0.28	$ 1.28
Panama-city	Panama	$ 0.53	$ -	$ 0.53
Lima	Peru	$ 0.22	$ -	$ 0.22
Manila (Maynilad)	Philippines	$ 0.66	$ 0.32	$ 0.99
Manila Water Company	Philippines	$ 0.42	$ 0.04	$ 0.46

Doha	Qatar	$ 1.21	$ -	$ 1.21
Bucharest	Romania	$ 0.99	$ 0.22	$ 1.21
Yekaterinbourg	Russia	$ 0.41	$ 0.15	$ 0.56
Moscow	Russia	$ 0.38	$ 0.31	$ 0.69
Volgograd	Russia	$ 0.32	$ 0.17	$ 0.49
Rostov-on-Don	Russia	$ 0.31	$ 0.20	$ 0.51
St Petersburg	Russia	$ 0.31	$ 0.31	$ 0.61
Saratov	Russia	$ 0.30	$ 0.14	$ 0.44
Krasnoyarsk	Russia	$ 0.28	$ 0.23	$ 0.50
Omsk	Russia	$ 0.26	$ 0.23	$ 0.50
Samara	Russia	$ 0.21	$ 0.09	$ 0.30
Ufa (Russia)	Russia	$ 0.18	$ 0.12	$ 0.30
Novosibirsk	Russia	$ 0.17	$ 0.12	$ 0.29
Jeddah	Saudi Arabia	$ 0.05	$ -	$ 0.05
Riyadh	Saudi Arabia	$ 0.03	$ -	$ 0.03
Belgrade	Serbia	$ 0.49	$ 0.12	$ 0.62
Cape Town	South Africa	$ 0.42	$ 0.78	$ 1.20
Johannesburg	South Africa	$ 0.29	$ 0.23	$ 0.52
Colombo	Sri Lanka	$ 0.01	$ -	$ 0.01
Damascus	Syria	$ 0.07	$ 0.01	$ 0.09
Taipei	Taiwan	$ 0.29	$ 0.11	$ 0.40
Ashgabat	Turkmenistan	$ -	$ -	$ -
Dubai	UAE	$ 0.02	$ -	$ 0.02
Kiev	Ukraine	$ 0.40	$ 0.23	$ 0.63
Kharkiv	Ukraine	$ 0.34	$ 0.06	$ 0.39
Lviv	Ukraine	$ 0.33	$ 0.19	$ 0.51
Odessa	Ukraine	$ 0.22	$ 0.19	$ 0.42
Tashkent	Uzbekistan	$ 0.01	$ 0.01	$ 0.02
Caracas	Venezuela	$ 0.21	$ -	$ 0.21
Ho Chi Min City	Vietnam	$ 0.39	$ -	$ 0.39
Hanoi	Vietnam	$ 0.18	$ -	$ 0.18

Fraction of Arable Land that Is Irrigated, by Country

Description

Irrigated agriculture is critically important for global food production, producing a disproportionate amount of food per unit area. This table shows the fraction of arable land in each country that is actually irrigated, or estimated to be irrigated on the basis of country surveys and land-use reports. The data are grouped by continental area and country and report arable land in thousand hectares, the fraction irrigated, and the area irrigated in thousand hectares. All of the data come from the 2005/2006 Food and Agricultural Organization Statistical Yearbook.

There is a tremendous diversity of experience with irrigation. Many countries report effectively no irrigated area; others, such as Egypt, are almost completely dependent on irrigation.

Limitations

No information is provided in this source on the method of irrigation or on the variations from year to year in actual irrigated area, both of which can fluctuate depending on food markets, water availability, water prices, and other factors. The next table shows crop area "equipped" for irrigation and hence, in theory, capable of being brought into production using irrigated water.

SOURCE

Food and Agriculture Organization of the United Nations (FAO). Statistical Yearbook, 2005/2006. http://www.fao.org/statistics/yearbook/vol_1_2/site_en.asp?page=cp

DATA TABLE 17 Fraction of Arable Land that is Irrigated, by Country

Region and Country	Arable Land (1,000 ha)	Fraction of Arable Land that is Irrigated	Irrigated Land Area (1,000 ha)
AFRICA			
Algeria	7,662	7%	536
Angola	3,000	2%	60
Benin	2,380	0%	-
Botswana	377	0%	-
Burkina Faso	4,040	1%	40
Burundi	960	2%	19
Cameroon	5,960	0%	-
Cape Verde	44	6%	3
Central African Republic	1,930	0%	-
Chad	3,520	1%	35
Comoros	80	0%	
Congo	490	0%	-
Congo, Democratic Republic	6,700	0%	-
Côte d'Ivoire	3,100	1%	31
Egypt	2,801	96%	2,689
Eritrea	560	4%	22
Ethiopia	10,000	3%	300
Gabon	325	1%	3
Gambia	285	1%	3
Ghana	3,950	1%	40
Guinea	975	6%	59
Guinea-Bissau	300	5%	15
Kenya	4,500	2%	90
Lesotho	330	1%	3
Liberia	380	1%	4
Libya	1,815	22%	399
Madagascar	2,900	31%	899
Malawi	2,100	3%	63
Mali	4,634	5%	232
Mauritania	488	10%	49
Mauritius	100	21%	21
Morocco	8,767	15%	1,315
Mozambique	3,900	3%	117
Namibia	816	1%	8
Niger	14,483	1%	145
Nigeria	28,200	1%	282
Rwanda	900	1%	9
Senegal	2,355	5%	118
Sierra Leone	490	5%	25
South Africa	14,753	10%	1,475
Sudan	16,233	11%	1,786
Swaziland	178	26%	46
Tanzania	4,000	4%	160
Togo	2,510	0%	-
Tunisia	2,864	8%	229
Uganda	5,060	0%	-

continues

DATA TABLE 17   *continued*

Region and Country	Arable Land (1,000 ha)	Fraction of Arable Land that is Irrigated	Irrigated Land Area (1,000 ha)
AFRICA *(continued)*			
Zambia	5,260	3%	158
Zimbabwe	3,220	5%	161
ASIA			
Armenia	495	51%	252
Azerbaijan	1,760	73%	1,285
Bangladesh	8,084	56%	4,527
Brunei Darussalam	9	8%	1
Cambodia	3,700	7%	259
China	137,124	37%	50,736
Cyprus	98	29%	28
Georgia	793	44%	349
India	160,555	33%	52,983
Indonesia	20,500	13%	2,665
Iran	14,324	47%	6,732
Israel	338	46%	155
Japan	4,474	54%	2,416
Jordan	242	19%	46
Kazakhstan	21,535	16%	3,446
Korea Republic	1,718	46%	790
Kuwait	10	93%	9
Kyrgyzstan	1,335	77%	1,028
Laos	877	18%	158
Lebanon	190	31%	59
Malaysia	1,820	5%	91
Mongolia	1,174	7%	82
Myanmar	9,909	18%	1,784
Nepal	2,324	48%	1,116
Pakistan	21,302	83%	17,681
Philippines	5,650	15%	848
Saudi Arabia	3,592	43%	1,545
Sri Lanka	895	39%	349
Syrian Arab Republic	4,542	25%	1,136
Tajikistan	930	68%	632
Thailand	15,865	26%	4,125
Turkey	23,826	20%	4,765
Turkmenistan	1,850	94%	1,739
United Arab Emirates	60	n.a.	
Uzbekistan	4,475	89%	3,983
Vietnam	6,200	37%	2,294
Yemen	1,545	33%	510
EUROPE			
Albania	578	51%	295
Austria	1,399	0%	-
Belarus	6,133	2%	123

Region and Country	Arable Land (1,000 ha)	Fraction of Arable Land that is Irrigated	Irrigated Land Area (1,000 ha)
Bosnia and Herzegovina	1,000	0%	-
Bulgaria	3,526	16%	564
Croatia	1,458	1%	15
Czech Republic	3,082	1%	31
Denmark	2,281	20%	456
Estonia	843	0%	-
Finland	2,183	3%	65
France	18,440	13%	2,397
Germany	11,804	4%	472
Greece	2,741	38%	1,042
Hungary	4,602	5%	230
Iceland	7	0%	-
Ireland	1,077	0%	-
Italy	8,479	24%	2,035
Latvia	1,845	1%	18
Lithuania	2,933	0%	-
Macedonia	555	9%	50
Moldova Republic	1,821	14%	255
Malta	8	22%	2
Netherlands	910	60%	546
Norway	883	14%	124
Poland	13,993	1%	140
Portugal	1,800	26%	468
Romania	9,381	31%	2,908
Russian Federation	124,374	4%	4,975
Serbia and Montenegro	3,406	1%	34
Slovakia	1,450	12%	174
Slovenia	173	1%	2
Spain	13,400		-
Sweden	2,706	4%	108
Switzerland	413	6%	25
Ukraine	32,564	7%	2,279
United Kingdom	5,876	3%	176
SOUTH AMERICA			
Argentina	27,800	5%	1,390
Bolivia	2,928	4%	117
Brazil	57,640	4%	2,306
Chile	1,979	83%	1,643
Colombia	2,818	20%	564
Ecuador	1,616	29%	469
Guyana	480	29%	139
Paraguay	2,850	2%	57
Peru	3,700	28%	1,036
Suriname	57	76%	43
Uruguay	1,373	15%	206
Venezuela	2,595	17%	441

continues

Region and Country	Arable Land (1,000 ha)	Fraction of Arable Land that is Irrigated	Irrigated Land Area (1,000 ha)
OCEANIA			
Australia	50,304	5%	2,515
Fiji	200	1%	2
New Zealand	1,500	9%	135
Samoa	59	0%	
NORTH AND CENTRAL AMERICA			
Bahamas	7	9%	1
Barbados	16	29%	5
Belize	64	3%	2
Canada	45,810	2%	916
Costa Rica	225	21%	47
Cuba	3,224	22%	709
Dominican Republic	1,096	17%	186
El Salvador	640	5%	32
Guatemala	1,395	7%	98
Haiti	780	8%	62
Honduras	1,068	6%	64
Jamaica	174	9%	16
Mexico	24,800	23%	5,704
Nicaragua	1,917	3%	58
Panama	540	6%	32
Trinidad and Tobago	75	3%	2
United States of America	176,018	13%	22,882

Area Equipped for Irrigation, by Country

Description

Irrigated agriculture is critically important for global food production, producing a disproportionate amount of food per unit area. This table shows the area of land "equipped for irrigation" by country, in thousand hectares, along with the year of the survey. The data are grouped by continental area and country. There is a tremendous diversity of experience with irrigation. Many countries report effectively no irrigated area; others, such as Egypt, are almost completely dependent on irrigation. As shown in the notes, the data for China and a few other countries are noted to be the area actually irrigated as opposed to the area equipped for irrigation.

Limitations

No information is provided here on actual irrigation (see previous table as well), which can vary tremendously from year to year with the availability of water, prices for food on global or local markets, and other factors. No information is provided here on the method of irrigation, which can also vary from precision drip systems to simple flood irrigation. The data in this table differ from the previous table and include areas of arable land capable of being irrigated but not necessarily in actual production at present. Surveys are not completed annually; thus some of these data are quite old as noted in the table itself. As a result, the data can not be easily or accurately compared.

SOURCE

All data from Aquastat of the Food and Agriculture Organization of the United Nations (FAO). http://www.fao.org/nr/water/aquastat/irrigationmap/index20.stm. (Updated as of June 2008.)

DATA TABLE 18   Area Equipped for Irrigation, by Country

Region and Country	Irrigated Area (ha)	Year(s) data is from or published	Notes regarding data
AFRICA			
Algeria	569,418	2001	
Angola	80,000	1975	It is believed that this data is still valid today.
Benin	12,258	2002	
Botswana	1,438	2002	
Burkina Faso	25,000	2001, 1998	
Burundi	21,430	2000	
Cameroon	25,654	2000	
Cape Verde	2,780	1997	
Central African Republic	135	1987	
Chad	30,273	2002	
Comoros	130	1987	
Congo	2,000	2003	
Congo, Democratic Republic	10,500	1995	
Côte d'Ivoire	72,750	1994	
Djibouti	1,012	1999	
Egypt	3,422,178	2002	
Equatorial Guinea	0		No irrigation because of climate conditions.
Eritrea	21,590	1993	
Ethiopia	289,530	2001	
Gabon	4,450	1987	
Gambia	2,149	1999	
Ghana	30,900	2000	
Guinea	94,914	2001	
Guinea-Bissau	22,558	1996	
Kenya	103,203	2003	
Lesotho	2,638	1999	
Liberia	2,100	1987	
Libya	470,000	2000	
Madagascar	1,086,291	2000	
Malawi	56,390	2002	
Mali	235,791	2000	
Mauritania	45,012	1994	
Mauritius	21,222	2002	
Morocco	1,484,160	2004	
Mozambique	118,120	2001–2003	
Namibia	7,573	2002	
Niger	73,663	2005	
Nigeria	293,117	2004	
Réunion	13,000	2005	
Rwanda	8,500	2000	
Sao Tome and Principe	9,700	1991	
Senegal	119,680	2002	
Seychelles	260	2003	
Sierra Leone	29,360	1992	

Region and Country	Irrigated Area (ha)	Year(s) data is from or published	Notes regarding data
Somalia	200,000	1984	It is believed that this data is still valid today.
South Africa	1,498,000	2000	
Sudan	1,863,000	2000	
Swaziland	49,843	2000	
Tanzania	184,330	2002	
Togo	7,300	1996	
Tunisia	394,063	2000	
Uganda	9,150	1998	
Zambia	155,912	2002	
Zimbabwe	173,513	1999	
ASIA			
Afghanistan	3,199,070	2003, 2004	
Armenia	286,027	1992	
Azerbaijan	1,453,318	2000, 2001	
Bahrain	4,060	2001	
Bangladesh	3,751,045	1995	
Bhutan	38,734	1989–1990, 1993–1994	
Brunei Darussalam	1,000	1995	
Cambodia	284,172	1990, 2001	
China	53,820,300	2000	Area actually irrigated.
Cyprus	55,813	2000, 1975	
Georgia	300,000	2001	
India	57,286,407	1999–2000	
Indonesia	4,459,000	1990	
Iran	6,913,800	1994	
Iraq	3,525,000	1990	
Israel	186,600	2000	
Japan	3,129,000	1993	
Jordan	76,912	2000	
Kazakhstan	1,855,200	1993	
Korea, Democratic People's Republic	1,460,000	1995	Recent data in indicate no change by 2002.
Korea Rep	880,365	2002	
Kuwait	6,968	1994	
Kyrgyzstan	1,075,040	1994	
Laos	295,535	2000	
Lebanon	117,113	1997	
Macao			No information available regarding irrigated area, assumed to be no irrigated area.
Malaysia	362,600	1994	
Maldives	0		"No irrigation of importance."
Mongolia	57,300	1994, 1995	

continues

DATA TABLE 18   *continued*

Region and Country	Irrigated Area (ha)	Year(s) data is from or published	Notes regarding data
ASIA (*continued*)			
Myanmar	1,841,320	1999–2000	
Nepal	1,168,349	2001–2002	
Oman	72,630	2001	
Pakistan	14,417,464	1990	
Palestinian Territory, incl. Israelian settlements	19,466	2001	
Philippines	1,550,000	1993	
Qatar	12,520	1993	
Saudi Arabia	1,730,767	1992, 2000	
Singapore			No information available regarding irrigated area.
Sri Lanka	570,000	1995	
Syrian Arab Republic	1,266,900	2001	
Taiwan, Province of China	525,528	1995	
Tajikistan	719,200	1994	Recent data indicate no change by 2002.
Thailand	4,985,708	2002	
Timor-Leste	1,400	1990	
Turkey	4,185,910	1994	
Turkmenistan	1,744,100	1994	
United Arab Emirates	280,341	2001	
Uzbekistan	4,223,000	1996	
Vietnam	3,000,000	1994	
Yemen	388,000	1996	
EUROPE			
Albania	340,000	2003, 2006	
Andorra	150	1990	Permanently irrigated area.
Austria	97,480	1995, 1997, 2000, 2003	
Belarus	115,000	2003	
Belgium	35,170	1997, 2000, 2003	
Bosnia and Herzegovina	4,630	2002	
Bulgaria	545,160	2003	
Croatia	5,790	2006	
Czech Republic	50,590	2003, 2005	
Denmark	467,000	1997, 2000 2003	
Estonia	1,363	2005	
Finland	103,800	2003	
France	2,906,081	2003	
Germany	491,620	2001	
Greece	1,544,530	1997, 2000, 2003	
Hungary	292,147	2001–2004	

Region and Country	Irrigated Area (ha)	Year(s) data is from or published	Notes regarding data
Iceland	0		No information available regarding irrigated area.
Ireland	1,100		
Italy	3,892,202	2000	
Latvia	1,150	2001	
Liechtenstein	0		No information available regarding irrigated area, assumed to be no irrigated area
Lithuania	4,416	2005	
Luxembourg	27	2002	Estimate, assuming that 75% of vegetable production is irrigated.
Macedonia	127,800	2006	
Moldova Republic	280,800	2002	
Montenegro	2,115	2005	Actual irrigated area, assumed to equal area equipped for irrigation.
Netherlands	476,315	2006	
Norway	134,396	1999	
Poland	134,050	2003, 2005	
Portugal	792,008	1999	
Romania	2,149,903	2004	
Russian Federation	4,454,100	2003	
Serbia	163,311	2001–2005, 2004	
Slovakia	225,310	2001	
Slovenia	15,643		
Spain	3,828,110	2003	
Sweden	188,470	2003	
Switzerland	40,000	2002	
Ukraine	2,395,500	1985	Recent data indicate little/no change.
United Kingdom	228,950	2003	
SOUTH AMERICA			
Argentina	1,767,784	1995, 2002	
Bolivia	128,240	1999	
Brazil	3,149,217	2001	
Chile	1,900,000	1996	
Colombia	900,000	1992, 1997	
Ecuador	863,370	1997	
French Guyana	2,000	1995	
Guyana	150,134	1991	
Paraguay	67,000	1998	
Peru	1,729,065	1994	
Suriname	51,180	1998	
Uruguay	217,593	2000	
Venezuela	570,219	1998	

continues

DATA TABLE 18   *continued*

Region and Country	Irrigated Area (ha)	Year(s) data is from or published	Notes regarding data
OCEANIA			
Australia	2,056,581	1996–1997	Area actually irrigated.
Fiji	3,000	1998	
New Zealand	577,882	2000	
Papua New Guinea			No irrigated area.
NORTH AND CENTRAL AMERICA, CARIBBEAN			
Antigua and Barbuda	130	1997	
Barbados	1,000	1989	
Belize	3,000	1997	
Canada	785,041	2000, 2001	
Costa Rica	103,084	1997	
Cuba	870,319	1996, 1997	
Dominica	0		Almost no irrigated area.
Dominican Republic	269,710	1994	
El Salvador	44,993	1997	
Grenada	219	1997	
Guadeloupe	2,000	1995	
Guatemala	129,803	1990	
Haiti	91,502	1991	
Honduras	73,210	1991	
Jamaica	25,214	1997	
Martinique	3,000	1995	
Mexico	6,435,800	2002–2005	
Nicaragua	61,365	1997	
Panama	34,626	1997	
Saint Kitts and Nevis	18	1997	
Saint Lucia	297	1997	
Saint Vincent and the Grenadines	0	1995	"No irrigation of importance."
Trinidad and Tobago	3,600	1997	
United States of America	55,311,236	2002	

Water Content of Things

Description

There is a growing interest in the resource implications of the goods and services that we all use, buy, and consume. How much energy, or how many greenhouse gases, or what amount of water is used to satisfy our demands for things? This table shows some estimates of the water implications, or "footprint," of a range of basic and manufactured goods, from a number of different sources. For a range of beverages, the data shown are the number of liters of freshwater required to produce a liter of beverage. For the other goods, the data are shown in liters of water per kilogram of product (or, since a liter of water weighs one kilogram, in kilograms of water per kilogram of product). There are very important uncertainties and limitations to these data, and we expect that improvements in measurement and reporting will continue over the next several years.

Limitations

These kinds of data are fraught with problems and uncertainties, and users should be extremely careful about using them for other than the most simple comparisons. When we can, we like to use ranges to try to bracket many of the uncertainties, but other sources rarely mention uncertainties or provide ranges of estimates. For example, the Water Footprint reports that 15,500 kg of water are required to produce beef, but work from the Pacific Institute reports a range of 15,000 to over 70,000 depending on diet, climate, the amount of product from each cow, and other variables. Similarly, the Water Footprint reports single estimates for the production of a range of vegetable and feed crops, but actual water requirements will vary dramatically with climate, soils, irrigation methods, and crop genetics.

Equally, if not more complicated, is evaluating the water required to produce manufactured items. For example, the water required to produce a liter of a soft drink may be as low as 2 to 4 liters per liter of product. But vast quantities of water are also consumed to produce the feedstocks, such as sugar or corn syrup, used in the same product. There are no consistent rules for where to draw the "supply chain" boundaries in such estimates, making it critical that users understand the assumptions that go into these values. This table, for example, lists 125 liters of water to make a kilogram of sheet paper, but it seems likely that this is the value for producing paper alone, and excludes the water required to grow the tree itself. Similarly, fewer than ten liters of water are required to process milk, but as many as 1,000 liters may be required if the water to produce the cow itself is included.

SOURCES

Gleick, P.H. Water in Crisis, Table H.17. New York: Oxford University Press.

Gleick, P.H. 2000. Water for Food: How Much Will Be Needed? In: Gleick, P.H. *The World's Water 2000–2001*. Washington DC: Island Press, pp. 63–91.

Pacific Institute, 2007. Bottled Water and Energy.
   http://www.pacinst.org/topics/integrity_of_science/case_studies/bottled_water_energy.html

Water Footprint.
   http://www.waterfootprint.org/

DATA TABLE 19 Water Content of Things

	Liters water	Comments/Notes/Sources
Beverages (per liter)		
Glass of beer	300	http://www.waterfootprint.org/; includes growing barley
Malt beverages (processing)	50	http://www.waterfootprint.org/; processing only
Glass of water	~1	http://www.waterfootprint.org/
Bottled Water	3 to 4	Pacific Institute estimate 2007; processing and water to make the plastic bottle
Milk	1,000	http://www.waterfootprint.org/; for the cow and processing
Milk (processing)	7	http://www.waterfootprint.org/; processing only
Cup of coffee	1,120	http://www.waterfootprint.org/
Cup of tea	120	http://www.waterfootprint.org/
Glass of wine	960	http://www.waterfootprint.org/; includes producing the grapes
Glass of apple juice	950	http://www.waterfootprint.org/; includes growing the apples
Glass of orange juice	850	http://www.waterfootprint.org/; includes growing the oranges
Assorted Produced Goods (per kilogram)		
Roasted coffee	21,000	to grow; http://www.waterfootprint.org/
Tea	9,200	to grow; http://www.waterfootprint.org/
Bread	1,300	http://www.waterfootprint.org/
Cheese	5,000	http://www.waterfootprint.org/
Cotton textile finished	11,000	http://www.waterfootprint.org/; assumes 45% crop use; 41% unproductive evaporation; 14% processing and wastewater
Sheet paper	125	http://www.waterfootprint.org/; Not including the water to grow tree
Potato chips	925	http://www.waterfootprint.org/
Hamburger	16,000	http://www.waterfootprint.org/
Leather shoes	16,600	http://www.waterfootprint.org/
Microchip	16,000	http://www.waterfootprint.org/
Assorted Crops (per kilogram)		To grow; depends on climate; depends on weight of finished crop versus total yield
Barley	1,300	http://www.waterfootprint.org/
Coconut	2,500	http://www.waterfootprint.org/
Corn	900	http://www.waterfootprint.org/
Sugar	1,500	http://www.waterfootprint.org/
Apple	700	http://www.waterfootprint.org/

continues

DATA TABLE 19   *continued*

	Liters water	Comments/Notes/Sources
Assorted Crops (per kilogram) *(continued)*		
Potato	500 to 1,500	Gleick 2000
Wheat	900 to 2,000	Gleick 2000
Alfalfa	900 to 2,000	Gleick 2000
Sorghum	1,100 to 1,800	Gleick 2000
Corn/Maize	1,000 to 1,800	Gleick 2000
Rice	1,900 to 5,000	Gleick 2000
Soybeans	1,100 to 2,000	Gleick 2000
Assorted Animals (per kilogram of meat)		Includes water for all feed
Sheep	6,100	http://www.waterfootprint.org/
Goat	4,000	http://www.waterfootprint.org/
Beef	15,000 to 70,000	Gleick 2000
Chicken	3,500 to 5,700	Gleick 2000
Eggs	3,300	http://www.waterfootprint.org/
Assorted Industrial Products (per kilogram)		Processing water; there is great variation depending on process
Steel	260	Gleick 1993
Primary Copper	440	Gleick 1993
Primary Aluminum	410	Gleick 1993
Phosphatic fertilizer	150	Gleick 1993
Nitrogenous fertilizer	120	Gleick 1993
Synthetic rubber	460	Gleick 1993
Inorganic pigments	410	Gleick 1993

Top Environmental Concerns of the American Public: Selected Years, 1997-2008

Description

This table presents a time series of the top environmental concerns of the American people as determined from consistent long-term polling. The data are expressed as the percentage of respondents who worried about a particular environmental problem "a great deal." For over a decade the Gallup polling organization has evaluated the perceptions, beliefs, and policy priorities of different public audiences on a wide range of issues. One of these is the environmental concerns of the average American. A series of consistent survey questions has been asked almost every year to elicit the environmental problems that Americans find most worrisome.

According to Gallup, the survey questions were presented as follows: "I'm going to read you a list of environmental problems. As I read each one, please tell me if you personally worry about this problem a great deal, a fair amount, only a little, or not at all. First, how much do you personally worry about — [problems presented in random order]?" The list of environmental problems read includes the following:

Pollution of drinking water
Pollution of rivers, lakes, and reservoirs
Contamination of soil and water by toxic waste
Maintenance of nation's supply of fresh water for household needs
Loss of natural habitat for wildlife
Air pollution
Damage to Earth's ozone layer
Loss of tropical rain forests
Extinction of plant and animal species
Urban sprawl and loss of open space
Greenhouse effect or global warming
Acid rain

Consistently, the most serious concerns have been expressed about water-related problems, including pollution of drinking water, pollution of rivers, lakes, and reservoirs, and maintenance of the nation's supply of freshwater for household needs.

Around half of all respondents worries "a great deal" about each of these three problems.

Limitations

All polls have limitations. According to Gallup, the results of these polls are typically based on telephone interviews with a sample of around 1,000 national adults, aged 18 and older. This sampling approach produces results with a 95% confidence level and a margin of sampling error of ±3 percentage points. They also note that in addition to sampling error, "question wording and practical difficulties in conducting surveys can introduce error or bias into the findings of public opinion polls."

SOURCES

Saad, L. Global warming on public's back burner, Gallup Poll News Service, April 20, 2004. http://www.gallup.com/poll/content/?ci=11398&pg=1; http://www.gallup.com/tag/Environment.aspx.

DATA TABLE 20   Top Environmental Concerns of the American Public: Selected Years, 1997–2008 (Percentage of People Who Worry About Issue "a great deal.")

Issue	1997	1999	2000	2001	2002	2003	2004	2006	2007	2008
Pollution of drinking water	NA	68	72	64	57	54	53	54	58	53
Pollution of rivers, lakes, and reservoirs	NA	61	66	58	53	51	48	52	53	50
Contamination of soil and water by toxic waste	NA	63	64	58	53	51	48	51	52	50
Maintenance of nation's supply of fresh water for household needs	NA	NA	42	35	50	49	47	49	51	48
Loss of natural habitat for wildlife										44
Air pollution	42	52	59	48	45	42	39	44	46	43
Damage to Earth's ozone layer	33	44	49	47	38	35	33	40	43	39
Loss of tropical rain forests	NA	49	51	44	38	39	35	40	43	40
Extinction of plant and animal species	NA	NA	45	43	35	34	36	34	39	37
Urban sprawl and loss of open space										33
Greenhouse effect or global warming	24	34	40	33	29	28	26	36	41	37
Acid rain	NA	29	34	28	25	24	20	24	25	23

NA = not available

Water Units, Data Conversions, and Constants

Water experts, managers, scientists, and educators work with a bewildering array of different units and data. These vary with the field of work: engineers may use different water units than hydrologists; urban water agencies may use different units than reservoir operators; academics may use different units than water managers. But they also vary with regions: water agencies in England may use different units than water agencies in France or Africa; hydrologists in the eastern United States often use different units than hydrologists in the western United States. And they vary over time: today's water agency in California may sell water by the acre-foot, but its predecessor a century ago may have sold miner's inches or some other now arcane measure.

These differences are of more than academic interest. Unless a common "language" is used, or a dictionary of translations is available, errors can be made or misunderstandings can ensue. In some disciplines, unit errors can be more than embarrassing; they can be expensive, or deadly. In September 1999, the $125 million Mars Climate Orbiter spacecraft was sent crashing into the face of Mars instead of into its proper safe orbit above the surface because one of the computer programs controlling a portion of the navigational analysis used English units incompatible with the metric units used in all the other systems. The failure to translate English units into metric units was described in the findings of the preliminary investigation as the principal cause of mission failure.

This table is a comprehensive list of water units, data conversions, and constants related to water volumes, flows, pressures, and much more. Most of these units and conversions were compiled by Kent Anderson and initially published in P. H. Gleick, 1993, *Water in Crisis: A Guide to the World's Fresh Water Resources*, Oxford University Press, New York.

Water Units, Data Conversions, and Constants

Prefix (Metric)	Abbreviation	Multiple	Prefix (Metric)	Abbreviation	Multiple
deka-	da	10	deci-	d	0.1
hecto-	h	100	centi-	c	0.01
kilo-	k	1000	milli-	m	0.001
mega-	M	10^6	micro-	μ	10^{-6}
giga-	G	10^9	nano-	n	10^{-9}
tera-	T	10^{12}	pico-	P	10^{-12}
peta-	P	10^{15}	femto-	f	10^{-15}
exa-	E	10^{18}	atto-	a	10^{-18}

LENGTH (L)

1 micron (μ)	$= 1 \times 10^{-3}$ mm	**10 hectometers**	= 1 kilometer
	$= 1 \times 10^{-6}$ m	**1 mil**	= 0.0254 mm
	$= 3.3937 \times 10^{-5}$ in		$= 1 \times 10^{-3}$ in
1 millimeter (mm)	= 0.1 cm	**1 inch (in)**	= 25.4 mm
	$= 1 \times 10^{-3}$ m		= 2.54 cm
	= 0.03937 in		= 0.08333 ft
1 centimeter (cm)	= 10 mm		= 0.0278 yd
	= 0.01 m	**1 foot (ft)**	= 30.48 cm
	$= 1 \times 10^{-5}$ km		= 0.3048 m
	= 0.3937 in		$= 3.048 \times 10^{-4}$ km
	= 0.03281 ft		= 12 in
	= 0.01094 yd		= 0.3333 yd
1 meter (m)	= 1000 mm		$= 1.89 \times 10^{-4}$ mi
	= 100 cm	**1 yard (yd)**	= 91.44 cm
	$= 1 \times 10^{-3}$ km		= 0.9144 m
	= 39.37 in		$= 9.144 \times 10^{-4}$ km
	= 3.281 ft		= 36 in
	= 1.094 yd		= 3 ft
	$= 6.21 \times 10^{-4}$ mi		$= 5.68 \times 10^{-4}$ mi
1 kilometer (km)	$= 1 \times 10^5$ cm	**1 mile (mi)**	= 1609.3 m
	= 1000 m		= 1.609 km
	= 3280.8 ft		= 5280 ft
	= 1093.6 yd		= 1760 yd
	= 0.621 mi	**1 fathom (nautical)**	= 6 ft
10 millimeters	= 1 centimeter	**1 league (nautical)**	= 5.556 km
10 centimeters	= 1 decimeter		= 3 nautical miles
10 decimeters (dm)	= 1 meter	**1 league (land)**	= 4.828 km
			= 5280 yd
10 meters	= 1 dekameter		= 3 mi
10 dekameters (dam)	= 1 hectometer	**1 international nautical mile**	= 1.852 km
			= 6076.1 ft
			= 1.151 mi

Water Units, Data Conversions, and Constants *(continued)*

AREA (L^2)

1 square centimeter	$= 1 \times 10^{-4} m^2$	**1 square foot (ft^2)**	$= 929.0 \text{ cm}^2$
(cm^2)	$= 0.1550 \text{ in}^2$		$= 0.0929 \text{ m}^2$
	$= 1.076 \times 10^{-3} \text{ ft}^2$		$= 144 \text{ in}^2$
	$= 1.196 \times 10^{-4} \text{ yd}^2$		$= 0.1111 \text{ yd}^2$
1 square meter	$= 1 \times 10^{-4} \text{ hectare}$		$= 2.296 \times 10^{-5} \text{ acre}$
(m^2)	$= 1 \times 10^{-6} \text{ km}^2$		$= 3.587 \times 10^{-8} \text{ mi}^2$
	$= 1 \text{ centare}$	**1 square yard (yd^2)**	$= 0.8361 \text{ m}^2$
	(French)		$= 8.361 \times 10^{-5}$
	$= 0.01 \text{ are}$		hectare
	$= 1550.0 \text{ in}^2$		$= 1296 \text{ in}^2$
	$= 10.76 \text{ ft}^2$		$= 9 \text{ ft}^2$
	$= 1.196 \text{ yd}^2$		$= 2.066 \times 10^{-4} \text{ acres}$
	$= 2.471 \times 10^{-4} \text{ acre}$		$= 3.228 \times 10^{-7} \text{ mi}^2$
1 are	$= 100 \text{ m}^2$	**1 acre**	$= 4046.9 \text{ m}^2$
1 hectare (ha)	$= 1 \times 10^4 m^2$		$= 0.40469 \text{ ha}$
	$= 100 \text{ are}$		$= 4.0469 \times 10^{-3} \text{ km}^2$
	$= 0.01 \text{ km}^2$		$= 43,560 \text{ ft}^2$
	$= 1.076 \times 10^5 \text{ ft}^2$		$= 4840 \text{ yd}^2$
	$= 1.196 \times 10^4 \text{ yd}^2$		$= 1.5625 \times 10^{-3} \text{ mi}^2$
	$= 2.471 \text{ acres}$	**1 square mile (mi^2)**	$= 2.590 \times 10^6 \text{ m}^2$
	$= 3.861 \times 10^{-3} \text{ mi}^2$		$= 259.0 \text{ hectares}$
1 square kilometer	$= 1 \times 10^6 m^2$		$= 2.590 \text{ km}^2$
(km^2)	$= 100 \text{ hectares}$		$= 2.788 \times 10^7 \text{ ft}^2$
	$= 1.076 \times 10^7 \text{ ft}^2$		$= 3.098 \times 10^6 \text{ yd}^2$
	$= 1.196 \times 10^6 \text{ yd}^2$		$= 640 \text{ acres}$
	$= 247.1 \text{ acres}$		$= 1 \text{ section (of land)}$
	$= 0.3861 \text{ mi}^2$	**1 feddan (Egyptian)**	$= 4200 \text{ m}^2$
1 square inch (in^2)	$= 6.452 \text{ cm}^2$		$= 0.42 \text{ ha}$
	$= 6.452 \times 10^{-4} \text{ m}^2$		$= 1.038 \text{ acres}$
	$= 6.944 \times 10^{-3} \text{ ft}^2$		
	$= 7.716 \times 10^{-4} \text{ yd}^2$		

(continues)

Water Units, Data Conversions, and Constants *(continued)*

VOLUME (L^3)

1 cubic centimeter	$= 1 \times 10^{-3}$ liter	**1 cubic foot (ft^3)**	$= 2.832 \times 10^4$ cm^3
(cm^3)	$= 1 \times 10^{-6}$ m^3		$= 28.32$ liters
	$= 0.06102$ in^3		$= 0.02832$ m^3
	$= 2.642 \times 10^{-4}$ gal		$= 1728$ in^3
	$= 3.531 \times 10^{-3}$ ft^3		$= 7.481$ gal
1 liter (1)	$= 1000$ cm^3		$= 0.03704$ yd^3
	$= 1 \times 10^{-3}$ m^3	**1 cubic yard (yd^3)**	$= 0.7646$ m^3
	$= 61.02$ in^3		$= 6.198 \times 10^{-4}$
	$= 0.2642$ gal		acre-ft
	$= 0.03531$ ft^3		$= 46656$ in^3
1 cubic meter (m^3)	$= 1 \times 10^6$ cm^3		$= 27$ ft^3
	$= 1000$ liter	**1 acre-foot**	$= 1233.48$ m^3
	$= 1 \times 10^{-9}$ km^3	**(acre-ft or AF)**	$= 3.259 \times 10^5$ gal
	$= 264.2$ gal		$= 43560$ ft^3
	$= 35.31$ ft^3	**1 Imperial gallon**	$= 4.546$ liters
	$= 6.29$ bbl		$= 277.4$ in^3
	$= 1.3078$ yd^3		$= 1.201$ gal
	$= 8.107 \times 10^{-4}$		$= 0.16055$ ft^3
	acre-ft	**1 cfs-day**	$= 1.98$ acre-feet
1 cubic decameter	$= 1000$ m^3		$= 0.0372$ in-mi^2
(dam^3)	$= 1 \times 10^6$ liter	**1 inch-mi^2**	$= 1.738 \times 10^7$ gal
	$= 1 \times 10^{-6}$ km^3		$= 2.323 \times 10^6$ ft^3
	$= 2.642 \times 10^5$ gal		$= 53.3$ acre-ft
	$= 3.531 \times 10^4$ ft^3		$= 26.9$ cfs-days
	$= 1.3078 \times 10^3$ yd^3	**1 barrel (of oil)**	$= 159$ liter
	$= 0.8107$ acre-ft	**(bbl)**	$= 0.159$ m^3
1 cubic hectometer	$= 1 \times 10^6$ m^3		$= 42$ gal
(ha^3)	$= 1 \times 10^3$ dam^3		$= 5.6$ ft^3
	$= 1 \times 10^9$ liter	**1 million gallons**	$= 3.069$ acre-ft
	$= 2.642 \times 10^8$ gal	**1 pint (pt)**	$= 0.473$ liter
	$= 3.531 \times 10^7$ ft^3		$= 28.875$ in^3
	$= 1.3078 \times 10^6$ yd^3		$= 0.5$ qt
	$= 810.7$ acre-ft		$= 16$ fluid ounces
1 cubic kilometer	$= 1 \times 10^{12}$ liter		$= 32$ tablespoons
(km^3)	$= 1 \times 10^9$ m^3		$= 96$ teaspoons
	$= 1 \times 10^6$ dam^3	**1 quart (qt)**	$= 0.946$ liter
	$= 1000$ ha^3		$= 57.75$ in^3
	$= 8.107 \times 10^5$		$= 2$ pt
	acre-ft		$= 0.25$ gal
	$= 0.24$ mi^3	**1 morgen-foot**	$= 2610.7$ m^3
1 cubic inch (in^3)	$= 16.39$ cm^3	**(S. Africa)**	
	$= 0.01639$ liter	**1 board-foot**	$= 2359.8$ cm^3
	$= 4.329 \times 10^{-3}$ gal		$= 144$ in^3
	$= 5.787 \times 10^{-4}$ ft^2		$= 0.0833$ ft^3
1 gallon (gal)	$= 3.785$ liters	**1 cord**	$= 128$ ft^3
	$= 3.785 \times 10^{-3}$ m^3		$= 0.453$ m^3
	$= 231$ in^3		
	$= 0.1337$ ft^3		
	$= 4.951 \times 10^{-3}$ yd^3		

Water Units, Data Conversions, and Constants *(continued)*

VOLUME/AREA (L^3/L^2)

1 inch of rain	= 5.610 gal/yd^2	**1 box of rain**	= 3,154.0 lesh
	= 2.715 × 10^4 gal/acre		

MASS (M)

1 gram (g or gm)	= 0.001 kg	**1 ounce (oz)**	= 28.35 g
	= 15.43 gr		= 437.5 gr
	= 0.03527 oz		= 0.0625 lb
	= 2.205 × 10^{-3} lb	**1 pound (lb)**	= 453.6 g
1 kilogram (kg)	= 1000 g		= 0.45359237 kg
	= 0.001 tonne		= 7000 gr
	= 35.27 oz		= 16 oz
	= 2.205 lb	**1 short ton (ton)**	= 907.2 kg
1 hectogram (hg)	= 100 gm		= 0.9072 tonne
	= 0.1 kg		= 2000 lb
1 metric ton (tonne or te or MT)	= 1000 kg	**1 long ton**	= 1016.0 kg
	= 2204.6 lb		= 1.016 tonne
	= 1. 102 ton	**1 long ton**	= 2240 lb
	= 0.9842 long ton		= 1.12 ton
1 dalton (atomic mass unit)	= 1.6604 × 10^{-24} g	**1 stone (British)**	= 6.35 kg
			= 14 lb
1 grain (gr)	= 2.286 × 10^{-3} oz		
	= 1.429 × 10^{-4} lb		

TIME (T)

1 second (s or sec)	= 0.01667 min	**1 day (d)**	= 24 hr
	= 2.7778 × 10^{-4} hr		= 86400 s
1 minute (min)	= 60 s	**1 year (yr or y)**	= 365 d
	= 0.01667 hr		− 8760 hr
1 hour (hr or h)	= 60 min		= 3.15 × 10^7 s
	= 3600 s		

DENSITY (M/L^3)

1 kilogram per cubic meter (kg/m^3)	= 10^{-3} g/cm^3	**1 metric ton per cubic meter (te/m^3)**	= 1.0 specific gravity
	= 0.062 lb/ft^3		= density of H$_2$O at 4°C
1 gram per cubic centimeter (g/cm^3)	= 1000 kg/m^3		= 8.35 lb/gal
	= 62.43 lb/ft^3	**1 pound per cubic foot (lb/ft^3)**	= 16.02 kg/m^3

(continues)

Water Units, Data Conversions, and Constants *(continued)*

VELOCITY (L/T)

1 meter per second (m/s)	= 3.6 km/hr = 2.237 mph = 3.28 ft/s	**1 foot per second (ft/s)**	= 0.68 mph = 0.3048 m/s
1 kilometer per hour (km/h or kph)	= 0.62 mph = 0.278 m/s	**velocity of light in vacuum (c)**	= 2.9979×10^8 m/s = 186,000 mi/s
1 mile per hour (mph or mi/h)	= 1.609 km/h = 0.45 m/s = 1.47 ft/s	**1 knot**	= 1.852 km/h = 1 nautical mile/hour = 1.151 mph = 1.688 ft/s

VELOCITY OF SOUND IN WATER AND SEAWATER
(assuming atmospheric pressure and sea water salinity of 35,000 ppm)

Temp, °C	Pure water, (meters/sec)	Sea water, (meters/sec)
0	1,400	1,445
10	1,445	1,485
20	1,480	1,520
30	1,505	1,545

FLOW RATE (L³/T)

1 liter per second (1/sec)	= 0.001 m³/sec = 86.4 m³/day = 15.9 gpm = 0.0228 mgd = 0.0353 cfs = 0.0700 AF/day	**1 cubic decameters per day (dam³/day)**	= 11.57 1/sec = 1.157×10^{-2} m³/sec = 1000 m³/day = 1.83×10^6 gpm = 0.264 mgd = 0.409 cfs = 0.811 AF/day
1 cubic meter per second (m³/sec)	= 1000 1/sec = 8.64×10^4 m³/day = 1.59×10^4 gpm = 22.8 mgd = 35.3 cfs = 70.0 AF/day	**1 gallon per minute (gpm)**	= 0.0631 1/sec = 6.31×10^{-5} m³/sec = 1.44×10^{-3} mgd = 2.23×10^{-3} cfs = 4.42×10^{-3} AF/day
1 cubic meter per day (m³/day)	= 0.01157 1/sec = 1.157×10^{-5} m³/sec = 0.183 gpm = 2.64×10^{-4} mgd = 4.09×10^{-4} cfs = 8.11×10^{-4} AF/day	**1 million gallons per day (mgd)**	= 43.8 1/sec = 0.0438 m³/sec = 3785 m³/day = 694 gpm = 1.55 cfs = 3.07 AF/day

Water Units, Data Conversions, and Constants *(continued)*

FLOW RATE (L^3/T) (continued)

1 cubic foot per second (cfs)	= 28.3 l/sec = 0.0283 m^3/sec = 2447 m^3/day = 449 gpm = 0.646 mgd = 1.98 AF/day	**1 miner's inch**	= 0.02 cfs (in Idaho, Kansas, Nebraska, New Mexico, North Dakota, South Dakota, and Utah) = 0.026 cfs (in Colorado)
1 acre-foot per day (AF/day)	= 14.3 l/sec = 0.0143 m^3/sec = 1233.48 m^3/day = 226 gpm = 0.326 mgd = 0.504 cfs	**1 weir** **1 quinaria (ancient Rome)**	= 0.028 cfs (in British Columbia) = 0.02 garcia = 0.47–0.48 l/sec
1 miner's inch	= 0.025 cfs (in Arizona, California, Montana, and Oregon: flow of water through 1 in^2 aperture under 6-inch head)		

ACCELERATION (L/T^2)

standard acceleration of gravity	= 9.8 m/s^2 = 32 ft/s^2

FORCE (ML/T^2 = Mass × Acceleration)

1 newton (N)	= $kg\text{-}m/s^2$ = 10^5 dynes = 0.1020 kg force = 0.2248 lb force	**1 dyne**	= $g{\cdot}cm/s^2$ = 10^{-5} N
		1 pound force	= lb mass × acceleration of gravity = 4.448 N

(continues)

Water Units, Data Conversions, and Constants *(continued)*

PRESSURE (M/L^2 = Force/Area)

| | | 1 kilogram per sq. centimeter (kg/cm^2) | = 14.22 lb/in^2 |

1 pascal (Pa) = N/m^2

1 bar = 1×10^5 Pa

= 1×10^6 dyne/cm^2

= 1019.7 g/cm^2

= 10.197 te/m^2

= 0.9869 atmosphere

= 14.50 lb/in^2

= 1000 millibars

1 kilogram per sq. centimeter (kg/cm^2) = 14.22 lb/in^2

1 inch of water at 62°F
= 0.0361 lb/in^2
= 5.196 lb/ft^3
= 0.0735 inch of mercury at 62°F

1 foot of water at 62°F
= 0.433 lb/in^2
= 62.36 lb/ft^2
= 0.833 inch of mercury at 62°F
= 2.950×10^{-2} atmosphere

1 atmosphere (atm)
= standard pressure
= 760 mm of mercury at 0°C
= 1013.25 millibars
= 1033 g/cm^2
= 1.033 kg/cm^2
= 14.7 lb/in^2
= 2116 lb/ft^2
= 33.95 feet of water at 62°F
= 29.92 inches of mercury at 32°F

1 pound per sq. inch (psi or lb/in^2)
= 2.309 feet of water at 62°F
= 2.036 inches of mercury at 32°F
= 0.06804 atmosphere
= 0.07031 kg/cm^2

1 inch of mercury at 32°F
= 0.4192 lb/in^2
= 1.133 feet of water at 32°F

TEMPERATURE

degrees Celsius or Centigrade (°C)
= (°F–32) × 5/9
= K–273.16

Kelvins (K)
= 273.16 + °C
= 273.16 + ((°F- 32) × 5/9)

degrees Fahrenheit (°F)
= 32 + (°C x 1.8)
= 32 + ((°K–273.16) × 1.8)

Water Units, Data Conversions, and Constants *(continued)*

ENERGY(ML^2/T^2 = Force $\times$ Distance)

1 joule (J)	$= 10^7$ ergs	**1 kilowatt-hour**	$= 3.6 \times 10^6$ J
	$=$ N·m	**(kWh)**	$= 3412$ Btu
	$=$ W·s		$= 859.1$ kcal
	$=$ kg·m^2/s^2	**l quad**	$= 10^{15}$ Btu
	$= 0.239$ calories		$= 1.055 \times 10^{18}$J
	$= 9.48 \times 10^{-4}$ Btu		$= 293 \times 10^9$ kWh
1 calorie (cal)	$= 4.184$ J		$= 0.001$ Q
	$= 3.97 \times 10^{-3}$ Btu		$= 33.45$ GWy
	(raises 1 g H$_2$O	**1 Q**	$= 1000$ quads
	l°C)		$\approx 10^{21}$ J
1 British thermal	$= 1055$ J	**1 foot-pound (ft-lb)**	$= 1.356$ J
unit (Btu)	$= 252$ cal (raises		$= 0.324$ cal
	1 lb H$_2$O l°F)	**1 therm**	$= 10^5$ Btu
	$= 2.93 \times 10^{-4}$ kWh	**1 electron-volt (eV)**	$= 1.602 \times 10^{-19}$ J
1 erg	$= 10^{-7}$ J	**1 kiloton of TNT**	$= 4.2 \times 10^{12}$ J
	$=$ g·cm^2/s^2	**1 10^6 te oil equiv.**	$= 7.33 \times 10^6$ bbl oil
	$=$ dyne·cm	**(Mtoe)**	$= 45 \times 10^{15}$ J
1 kilocalorie (kcal)	$= 1000$ cal		$= 0.0425$ quad
	$= 1$ Calorie (food)		

POWER (ML^2/T^3 = rate of flow of energy)

1 watt (W)	$=$ J/s	**1 horsepower**	$= 0.178$ kcal/s
	$= 3600$ J/hr	**(H.P. or hp)**	$= 6535$ kWh/yr
	$= 3.412$ Btu/hr		$= 33,000$ ft-lb/min
1 TW	$= 10^{12}$ W		$= 550$ ft-lb/sec
	$= 31.5 \times 10^{18}$ J		$= 8760$ H.P.-hr/yr
	$= 30$ quad/yr	**H.P. input**	$= 1.34 \times$ kW input
1 kilowatt (kW)	$= 1000$W		to motor
	$= 1.341$ horsepower		$=$ horsepower
	$= 0.239$ kcal/s		input to motor
	$= 3412$ Btu/hr	**Water H.P.**	$=$ H.P. required to
10^6 bbl (oil) /day	≈ 2 quads/yr		lift water at a
(Mb/d)	≈ 70 GW		definite rate to
1 quad/yr	$= 33.45$ GW		a given distance
	≈ 0.5 Mb/d		assuming 100%
1 horsepower	$= 745.7$W		efficiency
(H.P or hp)	$= 0.7457$ kW		$=$ gpm $\times$ total head
			(in feet)/3960

(continues)

Water Units, Data Conversions, and Constants *(continued)*

EXPRESSIONS OF HARDNESS[a]

1 grain per gallon	= 1 grain $CaCO_3$ per U.S. gallon	**1 French degree**	= 1 part $CaCO_3$ per 100,000 parts water
1 part per million	= 1 part $CaCO_3$ per 1,000,000 parts water	**1 German degree**	= 1 part CaO per 100,000 parts water
1 English, or Clark, degree	= 1 grain $CaCO_3$ per Imperial gallon		

CONVERSIONS OF HARDNESS

1 grain per U.S. gallon	= 17.1 ppm, as $CaCO_3$	**1 French degree**	= 10 ppm, as $CaCO_3$
1 English degree	= 14.3 ppm, as $CaCO_3$	**1 German degree**	= 17.9 ppm, as $CaCO_3$

WEIGHT OF WATER

1 cubic inch	= 0.0361 lb	**1 imperial gallon**	= 10.0 lb
1 cubic foot	= 62.4 lb	**1 cubic meter**	= 1 tonne
1 gallon	= 8.34 lb		

DENSITY OF WATER[a]

Temperature		Density
°C	°F	gm/cm³
0	32	0.99987
1.667	35	0.99996
4.000	39.2	1.00000
4.444	40	0.99999
10.000	50	0.99975
15.556	60	0.99907
21.111	70	0.99802
26.667	80	0.99669
32.222	90	0.99510
37.778	100	0.99318
48.889	120	0.98870
60.000	140	0.98338
71.111	160	0.97729
82.222	180	0.97056
93.333	200	0.96333
100.000	212	0.95865

Note: Density of Sea Water: approximately 1.025 gm/cm³ at 15°C.

[a]*Source:* van der Leeden, F., Troise, F. L., and Todd, D. K., 1990. *The Water Encyclopedia*, 2d edition. Lewis Publishers, Inc., Chelsea, Michigan.

Comprehensive Table of Contents

Volume 1
The World's Water 1998-1999: The Biennial Report on Freshwater Resources

DATA SECTION

Volume 2
The World's Water 2000-2001: The Biennial Report on Freshwater Resources

Foreword by Timothy E. Wirth   xiii

Acknowledgments   xv

Introduction   xvii

Volume 3
The World's Water 2002-2003: The Biennial Report on Freshwater Resources

DATA SECTION   237

Water Units, Data Conversions, and Constants   318

Index   329

Volume 4
The World's Water 2004–2005: The Biennial Report on Freshwater Resources

Foreword by Margaret Catley-Carlson xiii

Introduction xv

WATER UNITS, DATA CONVERSIONS, AND CONSTANTS   321

COMPREHENSIVE TABLE OF CONTENTS   331

COMPREHENSIVE INDEX   341

Volume 5
The World's Water 2006-2007: The Biennial Report on Freshwater Resources

Introduction   xv

ONE   Water and Terrorism   1
             by Peter H. Gleick

             Introduction   1
             The Worry   2
             Defining Terrorism   3
             History of Water-Related Conflict   5
             Vulnerability of Water and Water Systems   15
             Responding to the Threat of Water-Related Terrorism   22
             Water Security Policy in the United States   25
             Conclusion   25

TWO   Going with the Flow: Preserving and Restoring Instream
             Water Allocations   29
             by David Katz

             Environmental Flow: Concepts and Applications   30
             Legal Frameworks for Securing Environmental Flow   34
             The Science of Determining Environmental
                 Flow Allocations   38
             The Economics and Finance of Environmental
                 Flow Allocations   40
             Making It Work: Policy Implementation   43
             Conclusion   45

THREE   With a Grain of Salt: An Update on Seawater Desalination   51
             by Peter H. Gleick, Heather Cooley, Gary Wolff

             Introduction   51
             Background to Desalination   52
             History of Desalination   54
             Desalination Technologies   54
             Current Status of Desalination   55
             Advantages and Disadvantages of Desalination   66
             Environmental Effects of Desalination   76
             Desalination and Climate Change   80
             Public Transparency   81
             Summary   82
             Desalination Conclusions and Recommendations   83

DATA SECTION

WATER UNITS, DATA CONVERSIONS, AND CONSTANTS 319

COMPREHENSIVE TABLE OF CONTENTS 329

COMPREHENSIVE INDEX 343

Volume 6

The World's Water 2008–2009: The Biennial Report on Freshwater Resources

DATA SECTION

WATER UNITS, DATA CONVERSIONS, AND CONSTANTS   343

COMPREHENSIVE TABLE OF CONTENTS   353

COMPREHENSIVE INDEX   373

Comprehensive Index

KEY (book volume in boldface numerals)

1: The World's Water 1998–1999: The Biennial Report on Freshwater Resources

2: The World's Water 2000–2001: The Biennial Report on Freshwater Resources

3: The World's Water 2002–2003: The Biennial Report on Freshwater Resources

4: The World's Water 2004–2005: The Biennial Report on Freshwater Resources

5: The World's Water 2006–2007: The Biennial Report on Freshwater Resources

6: The World's Water 2008–2009: The Biennial Report on Freshwater Resources